Methods in Molecular Biology

Series Editor
John M. Walker
School of Life and Medical Sciences
University of Hertfordshire
Hatfield, Hertfordshire, AL10 9AB, UK

For further volumes:
http://www.springer.com/series/7651

Cartilage Tissue Engineering

Methods and Protocols

Edited by

Pauline M. Doran

Faculty of Science, Engineering and Technology, Swinburne University of Technology, Hawthorn, Melbourne, VIC, Australia

Editor
Pauline M. Doran
Faculty of Science, Engineering and Technology
Swinburne University of Technology
Hawthorn, Melbourne, VIC, Australia

ISSN 1064-3745 ISSN 1940-6029 (electronic)
Methods in Molecular Biology
ISBN 978-1-4939-2937-5 ISBN 978-1-4939-2938-2 (eBook)
DOI 10.1007/978-1-4939-2938-2

Library of Congress Control Number: 2015952786

Springer New York Heidelberg Dordrecht London
© Springer Science+Business Media New York 2015
This work is subject to copyright. All rights are reserved by the Publisher, whether the whole or part of the material is concerned, specifically the rights of translation, reprinting, reuse of illustrations, recitation, broadcasting, reproduction on microfilms or in any other physical way, and transmission or information storage and retrieval, electronic adaptation, computer software, or by similar or dissimilar methodology now known or hereafter developed.
The use of general descriptive names, registered names, trademarks, service marks, etc. in this publication does not imply, even in the absence of a specific statement, that such names are exempt from the relevant protective laws and regulations and therefore free for general use.
The publisher, the authors and the editors are safe to assume that the advice and information in this book are believed to be true and accurate at the date of publication. Neither the publisher nor the authors or the editors give a warranty, express or implied, with respect to the material contained herein or for any errors or omissions that may have been made.

Printed on acid-free paper

Humana Press is a brand of Springer
Springer Science+Business Media LLC New York is part of Springer Science+Business Media (www.springer.com)

Preface

Cartilage in articular joints is a relatively vulnerable tissue, being subject to common injuries and degenerative conditions such as arthritis. Motivated by the need to develop new treatment strategies, some of the earliest attempts at tissue engineering targeted cartilage as a feasible goal for in vitro synthesis. For more than 20 years, interdisciplinary teams of biologists, engineers, materials scientists, and clinicians have studied the culture and differentiation of cartilage cells and tissues. Many cornerstone technologies that distinguish tissue engineering from routine cell culture, such as three-dimensional culture systems and the use of scaffolds and bioreactors, were developed, tested, and widely adopted within the context of cartilage tissue engineering.

So far, the goal of producing laboratory-grown functional cartilage has eluded us but remains an active ambition. Irrespective of whether chondrocytes or stem cells are used as starting material, exerting adequate control over cellular differentiation is a major challenge. We do not yet know how to integrate engineered constructs with host cartilage in vivo and this continues to restrict clinical translation of cartilage engineering technology. Other important areas requiring further research include the response of chondrogenic cells to physical and mechanical stimuli, the heterogeneity of cell populations, and the complex molecular networks and regulatory cascades that direct cell lineage commitment and tissue development.

To answer all the outstanding questions in cartilage tissue engineering, further significant creative and intellectual input is required. This should come not only from established contributors but also, perhaps more importantly, from new and/or cross-disciplinary researchers in the area. How does a newcomer to cartilage tissue engineering become familiar with the techniques that underpin this field? I hope this question may be answered herein, as this book aims to describe clearly and in detail the key practical skills involved. Methods are outlined for isolation and expansion of chondrocytes and stem cells, differentiation, synthesis and application of three-dimensional scaffolds, design and operation of bioreactors, in vivo testing of engineered constructs, and molecular and functional analysis of cartilage cells and tissues. Frequently used techniques are covered, as well as more recent advances in inspirational areas such as "smart" biomaterial development, novel bioreactor design, –omics analysis, and genetic manipulation of matrix synthesis. The book does not attempt to be comprehensive; instead, it provides a snapshot of selected practical technologies that are either responsible for the progress already achieved in cartilage tissue engineering or indicative of the direction of future related research.

The chapters have been written by 45 authors and coauthors who have personal practical experience in cartilage tissue engineering. In the interests of informing the scientific community and expanding the engagement of researchers in this field, the contributors have provided careful and detailed protocols for experimental work covering a broad range of objectives for cartilage synthesis and regeneration. I thank all of the experts who have generously contributed their knowledge, insights, and valuable tips to prepare this volume. We hope that readers will find it a useful resource. I would also like to acknowledge the kind guidance and encouragement of Professor John Walker, Series Editor of *Methods in Molecular Biology*, throughout the duration of this project.

Melbourne, VIC, Australia ***Pauline M. Doran***

Contents

Contributors

MAURO ALINI • *Musculoskeletal Regeneration, AO Research Institute Davos, Davos Platz, Switzerland*

ANTHONY ATALA • *Wake Forest Institute for Regenerative Medicine, Wake Forest School of Medicine, Winston-Salem, NC, USA*

ALDO R. BOCCACCINI • *Department of Materials Science and Engineering, Institute of Biomaterials, University of Erlangen-Nuremberg, Erlangen, Germany*

PEGGY P.Y. CHAN • *Faculty of Science, Engineering and Technology, Swinburne University of Technology, Melbourne, VIC, Australia*

YU-HAN CHANG • *Department of Orthopaedic Surgery, Chang Gung Memorial Hospital, Linko, Taiwan*

CLARA R. CORREIA • *3B's Research Group—Biomaterials, Biodegradables and Biomimetics, University of Minho, Headquarters of the European Institute of Excellence on Tissue Engineering and Regenerative Medicine, Guimarães, Portugal; ICVS/3B's –PT Government Associate Laboratory, Guimarães, Braga, Portugal*

ANDREW DALY • *Trinity Centre for Bioengineering, Trinity Biomedical Sciences Institute, Trinity College Dublin, Dublin, Ireland; Department of Mechanical and Manufacturing Engineering, School of Engineering, Trinity College Dublin, Dublin, Ireland; Advanced Materials and Bioengineering Research Centre (AMBER), Trinity College Dublin, Dublin, Ireland*

PAULINE M. DORAN • *Faculty of Science, Engineering and Technology, Swinburne University of Technology, Melbourne, VIC, Australia*

HICHAM DRISSI • *Department of Orthopaedic Surgery, UConn Health, Farmington, CT, USA; Department of Genetics and Genome Sciences, UConn Health, Farmington, CT, USA; Stem Cell Institute, UConn Health, Farmington, CT, USA*

HUGO FERNANDES • *University of Maastricht – MERLN Institute for Technology Inspired Regenerative Medicine, Complex Tissue Regeneration Department, The Netherlands*

OLIVER F.W. GARDNER • *Musculoskeletal Regeneration, AO Research Institute Davos, Davos Platz, Switzerland; Cardiff University School of Biosciences, Cardiff, UK*

JASON D. GIBSON • *Department of Orthopaedic Surgery, UConn Health, Farmington, CT, USA*

ROSA M. GUZZO • *Department of Orthopaedic Surgery, UConn Health, Farmington, CT, USA; Stem Cell Institute, UConn Health, Farmington, CT, USA*

HYUN-WOOK KANG • *Wake Forest Institute for Regenerative Medicine, Wake Forest School of Medicine, Winston-Salem, NC, USA*

DANIEL JOHN KELLY • *Trinity Centre for Bioengineering, Trinity Biomedical Sciences Institute, Trinity College Dublin, Dublin, Ireland; Department of Mechanical and Manufacturing Engineering, School of Engineering, Trinity College Dublin, Dublin, Ireland; Advanced Materials and Bioengineering Research Center (AMBER), Trinity College Dublin, Dublin, Ireland*

TING TING LAU • *Division of Bioengineering, School of Chemical and Biomedical Engineering, Nanyang Technological University, Singapore*

WENYAN LEONG • *Division of Bioengineering, School of Chemical and Biomedical Engineering, Nanyang Technological University, Singapore*
NASTARAN MAHMOUDIFAR • *School of Biotechnology and Biomolecular Sciences, University of New South Wales, Sydney, NSW, Australia*
JOÃO F. MANO • *3B' s Research Group—Biomaterials, Biodegradables and Biomimetics, University of Minho, Headquarters of the European Institute of Excellence on Tissue Engineering and Regenerative Medicine, Guimarães, Portugal; ICVS/3B's –PT Government Associate Laboratory, Guimarães, Braga, Portugal*
KOICHI MASUDA • *Department of Orthopaedic Surgery, School of Medicine, University of California San Diego, La Jolla, CA, USA*
CLAUDIA N. MONTERO-MENEI • *Micro et Nanomédecines Biomimétiques (MINT), LUNAM Université, Angers, France; INSERM U1066, Angers, France*
MARIE MORILLE • *Institute Charles Gerhardt Montpellier, Equipe Matériaux Avancés pour la Catalyse et la Santé, Montpellier, France*
LORENZO MORONI • *University of Maastricht – MERLN Institute for Technology Inspired Regenerative Medicine, Complex Tissue Regeneration Department, The Netherlands*
HITOSHI NEMOTO • *Division of Otolaryngology-Head and Neck Surgery, Department of Surgery, School of Medicine, University of California San Diego, La Jolla, CA, USA; Department of Plastic and Reconstructive Surgery, Showa University Fujigaoka Hospital, Yokohama, Kanagawa, Japan*
PATCHARAKAMON NOOEAID • *Department of Materials Science and Engineering, Institute of Biomaterials, University of Erlangen-Nuremberg, Erlangen, Germany*
DINORATH OLVERA • *Trinity Centre for Bioengineering, Trinity Biomedical Sciences Institute, Trinity College Dublin, Dublin, Ireland; Department of Mechanical and Manufacturing Engineering, School of Engineering, Trinity College Dublin, Dublin, Ireland; Advanced Materials and Bioengineering Research Centre (AMBER), Trinity College Dublin, Dublin, Ireland*
JULIA THOM OXFORD • *Department of Biological Sciences, Biomolecular Research Center, Boise State University, Boise, ID, USA*
YVONNE PECK • *Division of Bioengineering, School of Chemical and Biomedical Engineering, Nanyang Technological University, Singapore*
ROMAN PEREZ • *Kennan Research Centre, Li Ka Shing Knowledge Institute, St. Michael's Hospital, Toronto, Canada; Institute of Tissue Regeneration Engineering (ITREN), Dankook University, Cheonan, Republic of Korea*
XINZHU PU • *Department of Biological Sciences, Biomolecular Research Center, Boise State University, Boise, ID, USA*
RUI L. REIS • *3B' s Research Group—Biomaterials, Biodegradables and Biomimetics, University of Minho, Headquarters of the European Institute of Excellence on Tissue Engineering and Regenerative Medicine, Guimarães, Portugal; ICVS/3B's –PT Government Associate Laboratory, Guimarães, Braga, Portugal*
GUNDULA SCHULZE-TANZIL • *Institute of Anatomy, Paracelsus Medical University, Salzburg and Nuremberg, Nuremberg, Germany; Department for Orthopedic, Trauma and Reconstructive Surgery, Charité-University of Medicine, Berlin, Nuremberg, Germany*
KIFAH SHAHIN • *Westmead Millennium Institute for Medical Research, University of Sydney, Westmead, NSW, Australia*

MARTIN J. STODDART • *Musculoskeletal Regeneration, AO Research Institute Davos, Davos Platz, Switzerland*
KAI SU • *Division of Bioengineering, School of Chemical and Biomedical Engineering, Nanyang Technological University, Singapore*
SHRADDHA THAKKAR • *Department of Biomedical Engineering, Eindhoven University of Technology, Eindhoven, The Netherlands*
MARIE-CLAIRE VENIER-JULIENNE • *Micro et Nanomédecines Biomimétiques (MINT), LUNAM Université, Angers, France; INSERM U1066, Angers, France*
STEPHEN D. WALDMAN • *Department of Mechanical and Materials Engineering, Queen's University, Kingston, Canada; Kennan Research Centre, Li Ka Shing Knowledge Institute, St. Michael's Hospital, Toronto, Canada; Department of Chemical Engineering, Ryerson University, Toronto, ON, Canada*
DONG-AN WANG • *Division of Bioengineering, School of Chemical & Biomedical Engineering, Nanyang Technological University, Singapore*
DEBORAH WATSON • *Division of Otolaryngology-Head and Neck Surgery, Department of Surgery, School of Medicine, University of California San Diego, La Jolla, CA, USA*
JOANNA F. WEBER • *Department of Mechanical and Materials Engineering, Queen's University, Kingston, Canada; Human Mobility Research Centre, Kingston General Hospital and Queen's University, Kingston, Canada; Kennan Research Centre, Li Ka Shing Knowledge Institute, St. Michael's Hospital, Toronto, Canada*
MIN-HSIEN WU • *Graduate Institute of Biochemical and Biomedical Engineering, Chang Gung University, Tao-Yuan, Taiwan*
REN-HE XU • *Department of Genetics and Genome Sciences, UConn Health, Farmington, CT, USA; Stem Cell Institute, UConn Health, Farmington, CT, USA*
JAMES J. YOO • *Wake Forest Institute for Regenerative Medicine, Wake Forest School of Medicine, Winston-Salem, NC, USA*
FENG ZHANG • *Division of Bioengineering, School of Chemical and Biomedical Engineering, Nanyang Technological University, Singapore*

Part I

Overview

Chapter 1

Cartilage Tissue Engineering: What Have We Learned in Practice?

Pauline M. Doran

Abstract

Many technologies that underpin tissue engineering as a research field were developed with the aim of producing functional human cartilage in vitro. Much of our practical experience with three-dimensional cultures, tissue bioreactors, scaffold materials, stem cells, and differentiation protocols was gained using cartilage as a model system. Despite these advances, however, generation of engineered cartilage matrix with the composition, structure, and mechanical properties of mature articular cartilage has not yet been achieved. Currently, the major obstacles to synthesis of clinically useful cartilage constructs are our inability to control differentiation to the extent needed, and the failure of engineered and host tissues to integrate after construct implantation. The aim of this chapter is to distil from the large available body of literature the seminal approaches and experimental techniques developed for cartilage tissue engineering and to identify those specific areas requiring further research effort.

Key words Bioreactor, Dedifferentiation, Hypertrophy, Scaffold, Stem cell, Three-dimensional culture, Tissue integration

1 Introduction

As one of the first tissues to be studied for tissue engineering, cartilage has played a crucial role in the development of methods and techniques now considered integral to the discipline. Because cartilage is generated by only one type of cell and contains neither blood vessels nor nerves, it was considered a suitable target for the earliest attempts at producing living, functional tissue constructs outside of the body. A compelling argument for engineering articular cartilage from both human health and commercial perspectives is that a large percentage of the population suffers chronic pain or disability at some stage in their life due to joint injury or degeneration [1–3]. This creates a significant potential market for tissue products suitable for implantation. Articular cartilage has the added advantage of being relatively thin (e.g., 0.9–3.0 mm in human lower joints: [4]), so that nutrients and oxygen can be

Pauline M. Doran (ed.), *Cartilage Tissue Engineering: Methods and Protocols*, Methods in Molecular Biology, vol. 1340, DOI 10.1007/978-1-4939-2938-2_1, © Springer Science+Business Media New York 2015

supplied readily from outside the construct in liquid-based in vitro culture systems.

To function successfully within joints, engineered cartilage must satisfy the most exacting biochemical and mechanical requirements. The extraordinary load-bearing, resilience, and low-friction properties of articular cartilage depend on its chemical composition and structure; tissues of inferior quality will not withstand the very high shear and compressive forces generated during normal joint movement. It is crucially important, therefore, that the properties of engineered cartilage match those of native articular cartilage. In adults, water accounts for approximately 70–80 % of the weight of cartilage tissue. The principal constituents of the dry matrix are collagen (50–75 % w/w) for tensile strength and proteoglycan (15–30 % w/w) for compressive stiffness, load distribution, and resilience [5, 6]. Several different collagen types are found; however, 90–95 % of the collagen in articular cartilage is type II. The principal proteoglycan is aggrecan, which consists of a core protein with many unbranched glycosaminoglycan (GAG) side chains. Chondrocytes, the cells responsible for synthesizing and maintaining cartilage, account for only about 1 % of the volume of mature tissue. In contrast, during fetal and early childhood growth, the concentration of chondrocytes in developing cartilage is 1–2 orders of magnitude greater than in adults, consistent with the need for active matrix synthesis and deposition. The composition of developing cartilage also differs significantly from that of adult tissue in terms of its collagen and GAG contents (Table 1).

Progress during the last 20–25 years of research into cartilage tissue engineering has been substantial. Innovative and pioneering work to develop new cell culture systems, scaffolds, bioreactors, differentiation techniques, and analytical methods has brought us closer to the goal of producing functional human cartilage in vitro. Using cartilage as a model system, researchers have created a substantial body of knowledge and developed an impressive skill-set of techniques for growing and regenerating tissues. Here, the most important practical lessons learned from cartilage tissue engineering are summarized. Critical areas where further research is needed to overcome the remaining barriers to clinical implementation are also highlighted.

2 Tissue Development Depends on a Three-Dimensional Culture Environment

Chondrocytes isolated from cartilage matrix tend to dedifferentiate when cultured in monolayer on flat, two-dimensional surfaces, resulting in downregulation of aggrecan and collagen type II synthesis and an increase in collagen type I production [7, 8]. Because expression of collagen type I leads to the development of mechanically inferior fibrocartilage, the consequences of cell attachment

Table 1
Measured composition of human adult and fetal cartilage. (Data from [61])

Cartilage tissue	Water content (weight %)	Cell concentration $\times 10^{-5}$ (per mg dry weight)	Glycosaminoglycan (GAG) concentration (% dry weight)	Total collagen concentration (% dry weight)	Collagen type II concentration (% dry weight)	Collagen type II as a % of total collagen
Adult	79.0 ± 0.3	0.4 ± 0.1	17.3 ± 0.51	54.0 ± 0.39	46.6 ± 6.3	86.3 ± 11
Fetal	91.2 ± 0.6	17.1 ± 0.8	46.9 ± 1.9	40.5 ± 1.8	24.7 ± 2.4	60.8 ± 3.5

and spreading on two-dimensional surfaces are highly undesirable for articular cartilage engineering. The ability of dedifferentiated chondrocytes to recover characteristics of the differentiated phenotype when returned to a favorable three-dimensional environment [7, 9–11] has underpinned an extensive research effort to produce scaffolds for cell attachment or entrapment that stimulate and support chondrogenesis and cartilage synthesis.

Many different scaffold materials have been studied for cartilage tissue engineering. These include porous foams and fibrous meshes made of biodegradable polymers such as poly(glycolic acid), poly(lactic acid), and poly(lactide-*co*-glycolide), and hydrogels based on polysaccharides such as alginate, agarose, hyaluronan, and chitosan or proteins such as collagen, gelatin, and fibrin (reviewed in [12, 13]). The focus of much scaffold development has been to mimic the native extracellular matrix (ECM) of cartilage where chondrocytes normally reside. To this end, complex scaffolds with physical and chemical gradients that imitate the zonal organization of articular cartilage, and scaffolds based on decellularized cartilage tissue itself, have been applied (reviewed in [14, 15]). High levels of sophistication have been achieved with the synthesis of advanced or "smart" scaffolds with features such as tunable material, surface, and pore properties and degradation rates. Multifunctional scaffolds can now be applied for delivery of trophic factors, regulatory molecules, or genetic components to control cellular differentiation, self-assembly of micro- and nanoscale surface patterning to enhance cell–scaffold interactions, and production of complex hierarchical structures for in situ optimization of scaffold conditions (reviewed in [16]).

Biochemical interactions and the regulatory responses of cells to surfaces have a major effect on cell attachment, orientation, shape, movement, distribution, proliferation, and differentiation (reviewed in [16]). The role of cell surface receptors to mediate the effects of the external environment, and thus determine whether cells grow, differentiate, switch between different lineages, or undergo apoptosis, is now well recognized. The observed sensitivity of chondrocytes and chondrogenic cells to cell–surface interactions highlights the importance of rational scaffold design and engineering to provide an appropriate material–biologic interface for tissue regeneration. Cartilage tissue engineering has been carried out using scaffolds with a wide variety of physicochemical and biological characteristics, all of which have the potential to influence cell behavior. A broad range of material and surface properties such as strength, stiffness, hydrophobicity, electrostatic charge, molecular functionality, and cell adhesiveness, and architectural properties such as porosity, pore size distribution, and micro- and nano-topography, has been examined. Similarly, a wide range of biodegradation mechanisms and kinetics, and ability to elicit inflammatory or immunogenic reactions, has been tested.

Despite these assiduous efforts, however, no single scaffold material or manner of fabrication has been identified as offering a clearly superior approach for cartilage synthesis. Development of new scaffolds that enhance the outcomes of cartilage tissue engineering continues to be an important research goal.

Extrinsic scaffolds are not essential for production of cartilaginous tissue, as scaffold-free three-dimensional culture systems can also be used. A potential disadvantage of using scaffolds is that they may induce changes in cell morphology that are unfavorable for chondrogenesis and cartilage synthesis [17]. To improve cellular adhesion to fibrous or gel scaffolds, chondrocytes tend to elongate and produce cytoplasmic extensions, thus destroying the spheroidal shape associated with the fully differentiated phenotype. To overcome this problem and more closely recapitulate the condensation phase of embryonic chondrogenesis, self-assembled or scaffold-free forms of three-dimensional cell culture have been developed (reviewed in [18]). Many cell types, including chondrocytes, are self-adherent and spontaneously form small aggregates under appropriate culture conditions; simple pellet or micromass cultures have been used extensively in tissue engineering studies. Close control is required over the size and cell density of low-porosity pellets to avoid necrosis within the aggregates [19]: in the absence of convective mass transfer networks such as vasculature, diffusional restrictions limit the practical size of cell aggregates. Although retention of the chondrocytic phenotype and ECM synthesis have been reported using scaffold-free systems [20, 21], other studies have shown reduced chondrogenic differentiation and cartilage development in pellet cultures compared with dynamic scaffold-based systems [22].

3 Dynamic Culture Is Better Than Static Culture

Static forms of cell culture provide a suboptimal environment for cartilage development, leading to the production of tissues of relatively poor quality in terms of their biochemical composition and mechanical properties. Several studies have established that fluid mixing enhances the development of cartilage tissues relative to static culture methods [23–25]. These findings reflect the critical role that liquid convection plays in cartilage maintenance and function in vivo, where joint movement during normal exercise drives the exchange and mixing of components between the synovial fluid bathing the joint and the interior of the cartilage tissue. The mechanisms by which fluid flow improves cartilage synthesis in vitro include physical effects, such as enhanced gas exchange and convective mass transfer to and from the cells, and direct biological effects as externally delivered hydrodynamic forces interact with mechanoreceptors on the cells to influence gene expression, cellular differentiation, and matrix deposition [26–28]. Fluid motion applied to developing tissue constructs needs to be regulated and applied judiciously in culture systems, as flow that is too vigorous or applied too early in the culture when there is little ECM present can lead to loss of cells and/or matrix components and poor construct quality [29–31].

Cell culture conditions incorporating some level of fluid flow can be achieved using simple well plate, Petri dish, or tissue flask systems if incubation is carried out on a shaking platform. However, because bioreactors can be designed to give high levels of control over fluid flow and mixing, mass transfer, gas exchange, and mechanical stimuli, they offer many advantages for engineering of cartilage tissue under reproducible conditions. A wide range of bioreactor configurations has been applied for cartilage production (reviewed in [5]), including spinner flask, rotating wall, perfusion, and wavy walled vessels [32]. One of the simplest designs is the perfusion system, in which recirculating medium is forced to flow through porous cell-seeded scaffolds inserted in the flow path. As long as the scaffold is fitted tightly against the walls of the bioreactor so that medium flows through the construct and not around the edges, direct contact is maintained between the moving fluid and the cells. Under these conditions, internal as well as external convective mass transport of nutrients and oxygen is achieved. Flow of medium through the scaffold in perfusion systems also generates hydrodynamic shear forces that provide mechanical stimulus to the cells; for a given scaffold, the magnitude of this stimulus is readily controlled by varying the medium recirculation flow rate.

In addition to the forces associated with fluids, such as hydrostatic pressure and hydrodynamic shear, cartilage cells and tissues respond to a variety of other mechanical stimuli. Specialized mechanobioreactors are required to exert direct compressive, tensile, mechanical shear and/or frictional forces on developing constructs (reviewed in [5]). The most commonly applied mechanical treatment in cartilage tissue engineering is uniaxial compression; however, because static compression has a detrimental effect on tissue development [33], it is important that dynamic or cyclical compression is applied. Dynamic compression enhances the synthesis of cartilage matrix in three-dimensional chondrocyte cultures [34, 35]; stimulatory effects on chondrogenesis have also been observed in scaffold-seeded stem cells [36–39]. The combination of dynamic compression with transient shear forces mimics the mechanical environment experienced by cartilage cells during the rolling and squeezing action of articular joints [40]. Mechanobioreactors that provide combined shear and compression stimuli have been shown to improve chondrocytic gene expression, cartilage synthesis, and/or construct mechanical properties compared with unstimulated controls [40–43].

Although it is clear that a dynamic culture environment, including fluid flow and mixing, mechanical stimulus, and adequate nutrient transport and gas exchange, produces higher quality engineered cartilage than static cultures, there is no consensus in the literature about the best culture system or bioreactor configuration required. Many different bioreactor types have been demonstrated to deliver culture conditions that support chondrogenesis and cartilage development.

4 Starting Cell Types: Many Contenders

Many different cell types have been investigated as source materials for engineering of human cartilage (Table 2). Comparative studies aimed at determining the relative merits of these cell types in terms of their proliferative capacity, chondrogenesis, and ability to synthesize cartilage components have also been carried out (Table 3). So far, however, no single cell type has been identified as a clearly superior starting point for production of engineered cartilage. To a large extent, this outcome reflects the wide diversity of isolation, storage, culture, and analytical procedures used to assess cellular performance, and the lack of standardized protocols for cell comparison. Given the major controlling influence of three-dimensional and dynamic culture conditions on chondrogenesis and cartilage deposition, it is questionable whether comparative studies carried out under distinctly suboptimal or inhibitory conditions, such as in static monolayer cultures, can provide useful results. A further complicating factor is that the relative performance of cell cultures in vitro may not be a reliable indicator of the performance achieved after in vivo transplantation [44–46].

4.1 Stem Cells

Embryonic stem cells, induced pluripotent stem cells, and mesenchymal stem cells are currently being investigated for cartilage tissue engineering.

- Embryonic stem cells are pluripotent cells obtained from the early mammalian embryo at the blastocyst stage, which occurs a few days after fertilization. When maintained in an undifferentiated state, embryonic stem cells can be propagated indefinitely while retaining the ability to differentiate along all primary differentiation lineages, viz. ectoderm, endoderm, and mesoderm, into any cell type.
- Induced pluripotent stem cells are obtained by manipulating adult somatic cells to produce selected transcription factors, including Oct3/4, Sox2, c-Myc, and Klf4, that play key roles in cell proliferation, pluripotency, and differentiation. Fibroblasts are often used as the starting cell type for induction of induced pluripotent stem cells; however, other types of

Table 2
Cell types used for human cartilage tissue engineering

Cell type	Example reference
Embryonic stem cells	Reviewed in [57]
Induced pluripotent stem cells	Reviewed in [57]
From dermal fibroblasts	[102]
From synovial cells	[103]
From chondrocytes	[104]
From fetal neural stem cells	[105]
Mesenchymal stem cells	Reviewed in [77, 106, 107]
From bone marrow	Reviewed in [108]
From adipose tissue	[22]
From umbilical cord blood	[109]
From peripheral blood	[110]
From amniotic fluid	[111]
From placenta	[112]
From umbilical cord matrix (Wharton's jelly)	[113]
From periosteum	[114]
From dental pulp	[115]
From synovium	[116]
From muscle	[117]
Fetal chondrocytes, from articular cartilage	[61]
Neonatal chondrocytes	[62]
Juvenile chondrocytes, from articular cartilage	[63]
Adult chondrocytes	
From articular cartilage	[64]
From nasal septal cartilage	[118]
From rib cartilage	[118]
From external ear cartilage	[119]

somatic cell have also been applied (Table 2). Like embryonic stem cells, induced pluripotent stem cells have an unlimited capacity for self-renewal as well as the ability to differentiate into all three germ layers.

Table 3
Examples of studies comparing the performance of different cell types for human cartilage tissue engineering

Cells compared	In vitro study	In vivo study	Reference
Embryonic stem cells and induced pluripotent stem cells from dermal fibroblasts	Yes	No	[120]
Embryonic stem cells and induced pluripotent stem cells from dermal fibroblasts	Yes	No	[102]
Embryonic stem cells and induced pluripotent stem cells from dermal fibroblasts	Yes	Yes	[121]
Induced pluripotent stem cells from chondrocytes and adult chondrocytes from articular cartilage	Yes	Yes	[104]
Mesenchymal stem cells from bone marrow and mesenchymal stem cells from adipose tissue	Yes	No	[122]
Mesenchymal stem cells from bone marrow, mesenchymal stem cells from adipose tissue, mesenchymal stem cells from periosteum, mesenchymal stem cells from synovium, and mesenchymal stem cells from skeletal muscle	Yes	No	[123]
Mesenchymal stem cells from bone marrow and mesenchymal stem cells from umbilical cord matrix (Wharton's jelly)	Yes	No	[113]
Mesenchymal stem cells from bone marrow, mesenchymal stem cells from adipose tissue, and mesenchymal stem cells from synovium	Yes	Yes	[88]
Mesenchymal stem cells from bone marrow, neonatal chondrocytes, and adult chondrocytes	Yes	No	[62]
Mesenchymal stem cells from bone marrow, mesenchymal stem cells from adipose tissue, adult chondrocytes from articular cartilage, adult chondrocytes from nasal cartilage, and adult chondrocytes from external ear cartilage	Yes	Yes	[46]
Mesenchymal stem cells from adipose tissue and fetal chondrocytes from articular cartilage	Yes	No	[69]
Adult chondrocytes from articular cartilage and adult chondrocytes from nasal cartilage	Yes	Yes	[124]
Juvenile chondrocytes from articular cartilage and adult chondrocytes from articular cartilage	Yes	Yes	[63]
Adult chondrocytes from external ear cartilage, adult chondrocytes from nasal septal cartilage, and adult chondrocytes from rib cartilage	Yes	No	[118]

- Mesenchymal stem cells are usually derived from adult tissues and have a high capacity for self-renewal. They are considered to be multipotent rather than pluripotent, with the ability to differentiate along standard mesenchymal lineages into chondrocytes, osteoblasts, adipocytes, and myocytes. However, differentiation into other cell types, including neural cells, cardiomyocytes, hepatocytes, pancreatic cells, and endothelial cells, has also been reported [47], suggesting that mesenchymal stem cells, or cells co-purifying with mesenchymal stem cells, may exhibit some tendency towards pluripotent characteristics [48].

Disadvantages associated with human embryonic stem cells include ethical concerns over the destruction of embryos to obtain the cells, the tendency of the cells to form tumors after implantation into patients [49, 50], and the potential for immune rejection of allogeneic grafts in vivo [51, 52]. Tumorigenicity is a major obstacle limiting the clinical application of embryonic stem cells, as formation of any type of tumor is unacceptable in medical practice. Because terminally differentiated cells are not tumorigenic, lineage commitment and completion of differentiation protocols prior to transplantation may, in theory, overcome this problem. However, because it is difficult to ensure that all residual pluripotent stem cells are excluded or eliminated, for example, by cell sorting or selective induction of apoptosis or necrosis, the risk of teratoma formation due to carryover of tumorigenic cells remains [49, 50]. Immune rejection occurs when the donor and recipient of cells or tissues are unrelated: the likelihood of rejection depends on allelic differences in transplant antigens expressed by the two individuals. Previously, because of their early stage of development, embryonic stem cells have been considered "immune-privileged" or potentially unrecognizable by the immune defences of recipient patients. There is increasing evidence, however, that even in their undifferentiated state, embryonic stem cells express enough antigens that are recognized by the immune system to render them susceptible to rejection mechanisms [51–53]. This is a serious impediment to clinical applications and strategies are being developed to address embryonic stem cell immunogenicity (reviewed in [51, 54]). For example, in the future, cell banks of immunotyped stem cells may be developed to provide a range of cell lines that are closely matched or sufficiently immunocompatible with the majority of a given population, so that only manageable, low-grade rejection responses occur [54].

The use of induced pluripotent stem cells overcomes the ethical issues relating to harvesting of human embryos; however, problems with tumor formation and immune rejection remain. The genetic and epigenetic characteristics of induced pluripotent stem cells in addition to their pluripotency make these cells more tumorigenic than embryonic stem cells [49]. Techniques to reduce this elevated risk of tumorigenesis are under investigation: these include new approaches to cellular reprogramming to eliminate the use of viral vectors, and new methods to either prevent permanent integration of transgenes or excise them from the host genome after reprogramming (reviewed in [55–57]). In principle, because induced pluripotent stem cells are generated using adult somatic cells, any problems with immune rejection could be overcome by using the patient's own cells as starting material. Currently, however, this is not a practical option because of the low success rate, inefficiency, and high cost of personalized cell line reprogramming [54]. Accordingly, methods for cell banking and inducing immune tolerance in transplant recipients that are currently being investigated for embryonic stem cells are also relevant for clinical application of induced pluripotent stem cells.

Mesenchymal stem cells offer several important advantages for cartilage tissue engineering. They do not require embryo harvesting, are not normally associated with tumor formation, and can be obtained readily from the individuals requiring treatment so that problems with immune rejection are avoided. Mesenchymal stem cells are available from a range of human tissues (Table 2); however, those derived from either bone marrow or adipose tissue are most commonly applied. As well as their capacity for chondrogenic differentiation, mesenchymal stem cells are recognized for their direct therapeutic value in vivo. After injection or infusion into animals, the cells secrete bioactive molecules that induce multiple panacrine effects with antiapoptotic, immunomodulatory, antiscarring, and chemoattractant functions (reviewed in [58–60]). The degree of immunomodulation achieved depends on the environment, particularly the matrix and surface surroundings and local inflammatory conditions [59, 60]. In tissue engineering applications, the immunosuppressive and enhanced reparative properties of mesenchymal stem cells may allow allogeneic constructs to be transplanted into patients without activating the full immune reaction responsible for tissue rejection. Before this can be implemented, however, further research is needed to understand the limits and opportunities associated with the therapeutic functions of mesenchymal stem cells in vivo.

4.2 Chondrocytes

Human chondrocytes from fetal [31, 61], neonatal [62], juvenile (<13 years old: [63]), and adult [64, 65] tissues have been used for cartilage tissue engineering. Of these, adult chondrocytes have most clinical relevance: younger allogeneic cells are unlikely to be used for treatment of patients without the need for immunosuppression therapy to prevent rejection. Nevertheless, human chondrocytes from cartilage at various stages of development represent a valuable tool in tissue engineering research. Chondrocytes have been shown in several studies to be better producers of cartilage matrix than chondro-induced mesenchymal stem cells [46, 66–71].

An important disadvantage associated with chondrocytes is that the number of cells available for isolation from native cartilage is generally very limited, so that ex vivo expansion is required.

Inevitably, dedifferentiation occurs during monolayer expansion prior to scaffold seeding [7, 8]. Subsequent three-dimensional culture has been reported to promote redifferentiation of expanded chondrocytes [7, 9–11]; however, all aspects of the mature chondrocytic phenotype may not be recovered after monolayer culture [72, 73].

5 Our Ability to Control Differentiation Is Currently Inadequate

Irrespective of the cell type employed, no tissue-engineered cartilage has yet been produced that replicates the biochemical and functional properties of native articular cartilage. The best constructs have been generated using chondrocytes; however, although these tissues contain the same or higher concentrations of GAG compared with adult cartilage, their collagen type II contents are substantially lower [74, 75]. Accordingly, collagen synthesis and accumulation currently represent the greatest biochemical limitations in cartilage tissue engineering.

The types of collagen found in cultured constructs are typically uncharacteristic of articular cartilage. In chondrocytes and mesenchymal stem cells, relatively high levels of collagen type I and relatively low levels of collagen type II are expressed [10, 76, 77]. This tendency towards production of fibrocartilage contributes to the poor biomechanical properties and low durability of engineered tissues. Overexpression of collagen type I is also often observed during chondrogenic differentiation of embryonic and induced pluripotent stem cells [78, 79]. Methods to suppress collagen type I synthesis using gene silencing and/or application of selected growth factors are being investigated [80, 81].

Chondrogenesis and the control of differentiation become a major challenge when stem cells are used for tissue engineering. Specific differentiation triggers are required in the culture environment to induce differentiation and maintain a chondrocytic phenotype. Members of the transforming growth factor-β (TGF-β) family of cytokines, especially TGF-β1 and TGF-β3, play a primary role in regulating cartilage development; however, many other growth factors such as those in the bone morphogenetic protein (BMP) and basic fibroblast growth factor (bFGF) groups also influence chondro-induction, either alone or in combination with TGF-β [82]. Indeed, a large number of biomolecules are known to modulate chondrogenesis [75], and specific genes must be transiently upregulated or downregulated in the correct order to achieve robust differentiation outcomes [83, 84]. With this in mind, attempts to control stem cell differentiation by applying just one or two growth factors have little prospect of long-term success.

When mesenchymal stem cells are used for chondrogenic differentiation, although key cartilage markers such as aggrecan and collagen type II may be expressed, the resulting cellular phenotype is more typical of hypertrophic cartilage than functional articular cartilage [77, 85]. Tissue mineralization and expression of hypertrophy markers such as collagen type X, matrix metalloproteinase 13, and alkaline phosphatase indicate that the differentiation pathways induced in vitro do not lead to a normal or stable chondrocytic state. Instead, differentiation proceeds towards endochondral ossification rather than terminating at the chondrocytic stage. Strategies investigated to control hypertrophy in chondro-induced stem cells include choice of scaffold material and cell source tissue, and application of regulatory factors, small molecule inhibitors, and hypoxic culture conditions [86–91]. Several studies have demonstrated that coculture of mesenchymal stem cells with differentiated chondrocytes suppresses the hypertrophic phenotype (reviewed in [92]). Exposure of the stem cells to conditioned medium from chondrocyte culture reduced hypertrophy, but stronger inhibition was achieved in cocultures providing direct contact between the two cell types [93]. Our understanding of the regulatory effects triggered by chondrocyte–stem cell interactions [93] is currently very sketchy.

6 Integration: A Remaining Challenge

Most researchers working on cartilage tissue engineering have not considered the problem of integration of the engineered tissue into the recipient's joint after implantation. Yet, strong and stable connections between the graft and native tissues are essential for functional integrity of the implant: poor integration compromises the mechanical strength and durability of the joint and could result in tissue degradation. Although vertical integration with subchondral bone can be achieved for full-thickness cartilage implants, lateral integration with the adjacent host cartilage and integration of smaller grafts away from the underlying bone typically fail [94]. This is a major obstacle to the success of tissue engineering strategies for cartilage repair.

Cell migration leading to cell and matrix accumulation at the interfacial zone is a key factor influencing tissue integration. Although chondrocytes migrating to the graft interface may originate from either the engineered or host tissue [95], chondrocyte movement is restricted in native cartilage due to location of the cells in lacunae. The dense networks of collagen fibrils and proteoglycans that make up cartilage matrix further impede cell migration and, because adult cartilage is avascular, transfer of progenitor cells from the blood stream or bone marrow to the graft site is also severely impaired. Enhancement of chondrocyte migration using gene- or protein-level induction of signaling pathways has been suggested as a strategy to improve integration outcomes [96].

Several other factors are known to contribute to poor cartilage integration, including low chondrocyte viability at the graft and host edges, differences in architecture and collagen cross-linking between the graft and host tissues, and the activity of potentially inhibitory compounds in native cartilage and synovial fluid [94]. Development of cartilage bioadhesive [97], application of lysyl-oxidase enzyme to promote cross-linking between engineered and native tissues [98], inhibition of chondrocyte death at the graft edge [99], and formation of cartilage constructs with collagen-based fibrous capsules [100] have been reported to improve integration outcomes in vitro. Application of a functionalized chondroitin sulfate bioadhesive to covalently link engineered and native cartilage has also been tested in vivo in animal joints [101]. Further research into improving the adhesion and integration of cartilage implants is needed for the translation of tissue engineering technologies into clinical practice.

7 Conclusions

Because chondrocytes dedifferentiate when cultured on two-dimensional surfaces, using three-dimensional culture systems to retain the biosynthetic capacity of the cells is well established in cartilage tissue engineering. This promoted a rapid expansion in biomaterials science for development of appropriate scaffolds to support cartilage production. Early work also showed that static culture of cells without mixing or motion of the culture medium generates constructs of much poorer quality than those produced in dynamic culture environments. As a result, bioreactors suitable for culture of three-dimensional cartilaginous tissues were designed and are now used widely in tissue engineering. Since cartilage engineering began as an active area of research, rapid developments in techniques for identifying, isolating, and culturing stem cells led to a wide range of cell types being used as starting material for cartilage production. Although our understanding of cellular differentiation has improved substantially over the period, knowledge in this area is as yet inadequate for practical purposes. Control over differentiation, and the development of new technologies for integrating engineered and native cartilage, are currently the most important challenges in the field. Research is ongoing into many aspects of cartilage tissue engineering that need further improvement.

References

1. Birk GT, DeLee JC (2001) Osteochondral injuries: clinical findings. Clin Sports Med 20:279–286
2. Friel NA, Chu CR (2013) The role of ACL injury in the development of posttraumatic knee osteoarthritis. Clin Sports Med 32: 1–12
3. Johnson VL, Hunter DJ (2014) The epidemiology of osteoarthritis. Best Pract Res Clin Rheumatol 28:5–15

4. Shepherd DET, Seedhom BB (1999) Thickness of human articular cartilage in joints of the lower limb. Ann Rheum Dis 58:27–34
5. Schulz RM, Bader A (2007) Cartilage tissue engineering and bioreactor systems for the cultivation and stimulation of chondrocytes. Eur Biophys J 36:539–568
6. Athanasiou K, Darling EM, Hu JC (2009) Articular cartilage tissue engineering. Morgan and Claypool, San Rafael, USA
7. Aulthouse AL, Beck M, Griffey E et al (1989) Expression of the human chondrocyte phenotype in vitro. In Vitro Cell Dev Biol 25: 659–668
8. Hardingham T, Tew S, Murdoch A (2002) Tissue engineering: chondrocytes and cartilage. Arthritis Res 4(Suppl 3):S63–S68
9. Benya PD, Shaffer JD (1982) Dedifferentiated chondrocytes reexpress the differentiated collagen phenotype when cultured in agarose gels. Cell 30:215–224
10. Murphy CL, Sambanis A (2001) Effect of oxygen tension and alginate encapsulation on restoration of the differentiated phenotype of passaged chondrocytes. Tissue Eng 7: 791–803
11. Schulze-Tanzil G, Mobasheri A, De Souza P et al (2004) Loss of chondrogenic potential in dedifferentiated chondrocytes correlates with deficient Shc–Erk interaction and apoptosis. Osteoarthritis Cartilage 12:448–458
12. Pan Z, Ding J (2012) Poly(lactide-*co*-glycolide) porous scaffolds for tissue engineering and regenerative medicine. Interface Focus 2:366–377
13. Zhao W, Jin X, Cong Y et al (2013) Degradable natural polymer hydrogels for articular cartilage tissue engineering. J Chem Technol Biotechnol 88:327–339
14. Klein TJ, Malda J, Sah RL et al (2009) Tissue engineering of articular cartilage with biomimetic zones. Tissue Eng Part B Rev 15: 143–157
15. Benders KEM, Van Weeren PR, Badylak SF et al (2013) Extracellular matrix scaffolds for cartilage and bone regeneration. Trends Biotechnol 31:169–176
16. Barrère F, Mahmood TA, De Groot K et al (2008) Advanced biomaterials for skeletal tissue regeneration: instructive and smart functions. Mat Sci Eng R 59:38–71
17. Nuernberger S, Cyran N, Albrecht C et al (2011) The influence of scaffold architecture on chondrocyte distribution and behavior in matrix-associated chondrocyte transplantation grafts. Biomaterials 32:1032–1040
18. Athanasiou KA, Eswaramoorthy R, Hadidi P et al (2013) Self-organization and the self-assembling process in tissue engineering. Annu Rev Biomed Eng 15:115–136
19. Ko EC, Fujihara Y, Ogasawara T et al (2011) Administration of the insulin into the scaffold atelocollagen for tissue-engineered cartilage. J Biomed Mater Res A 97:186–192
20. Ofek G, Revell CM, Hu JC et al (2008) Matrix development in self-assembly of articular cartilage. PLoS One 3(7):e2795
21. Mohanraj B, Farran AJ, Mauck RL et al (2014) Time-dependent functional maturation of scaffold-free cartilage tissue analogs. J Biomech 47:2137–2142
22. Mahmoudifar N, Doran PM (2010) Chondrogenic differentiation of human adipose-derived stem cells in polyglycolic acid mesh scaffolds under dynamic culture conditions. Biomaterials 31:3858–3867
23. Pazzano D, Mercier KA, Moran JM et al (2000) Comparison of chondrogenesis in static and perfused bioreactor culture. Biotechnol Prog 16:893–896
24. Gooch KJ, Kwon JH, Blunk T et al (2001) Effects of mixing intensity on tissue-engineered cartilage. Biotechnol Bioeng 72:402–407
25. Gemmiti CV, Guldberg RE (2006) Fluid flow increases type II collagen deposition and tensile mechanical properties in bioreactor-grown tissue-engineered cartilage. Tissue Eng 12:469–479
26. Mammoto A, Ingber DE (2009) Cytoskeletal control of growth and cell fate switching. Curr Opin Cell Biol 21:864–870
27. Arnsdorf EJ, Tummala P, Kwon RY et al (2009) Mechanically induced osteogenic differentiation—the role of RhoA, ROCKII and cytoskeletal dynamics. J Cell Sci 122:546–553
28. Yeatts AB, Choquette DT, Fisher JP (2013) Bioreactors to influence stem cell fate: augmentation of mesenchymal stem cell signaling pathways via dynamic culture systems. Biochim Biophys Acta 1830:2470–2480
29. Davisson T, Sah RL, Ratcliffe A (2002) Perfusion increases cell content and matrix synthesis in chondrocyte three-dimensional cultures. Tissue Eng 8:807–816
30. Freyria A-M, Yang Y, Chajra H et al (2005) Optimization of dynamic culture conditions: effects on biosynthetic activities of chondrocytes grown in collagen sponges. Tissue Eng 11:674–684
31. Shahin K, Doran PM (2011) Strategies for enhancing the accumulation and retention of extracellular matrix in tissue-engineered

cartilage cultured in bioreactors. PLoS One 6(8):e23119

32. Bueno EM, Bilgen B, Barabino GA (2005) Wavy-walled bioreactor supports increased cell proliferation and matrix deposition in engineered cartilage constructs. Tissue Eng 11:1699–1709
33. Grodzinsky AJ, Levenston ME, Jin M et al (2000) Cartilage tissue remodeling in response to mechanical forces. Annu Rev Biomed Eng 2:691–713
34. Kisiday JD, Jin M, DiMicco MA et al (2004) Effects of dynamic compressive loading on chondrocyte biosynthesis in self-assembling peptide scaffolds. J Biomech 37:595–604
35. Waldman SD, Spiteri CG, Grynpas MD et al (2004) Long-term intermittent compressive stimulation improves the composition and mechanical properties of tissue-engineered cartilage. Tissue Eng 10:1323–1331
36. Mouw JK, Connelly JT, Wilson CG et al (2007) Dynamic compression regulates the expression and synthesis of chondrocyte-specific matrix molecules in bone marrow stromal cells. Stem Cells 25:655–663
37. Terraciano V, Hwang N, Moroni L et al (2007) Differential response of adult and embryonic mesenchymal progenitor cells to mechanical compression in hydrogels. Stem Cells 25:2730–2738
38. Pelaez D, Huang C-YC, Cheung HS (2009) Cyclic compression maintains viability and induces chondrogenesis of human mesenchymal stem cells in fibrin gel scaffolds. Stem Cells Dev 18:93–102
39. Haugh MG, Meyer EG, Thorpe SD et al (2011) Temporal and spatial changes in cartilage-matrix-specific gene expression in mesenchymal stem cells in response to dynamic compression. Tissue Eng Part A 17: 3085–3093
40. Shahin K, Doran PM (2012) Tissue engineering of cartilage using a mechanobioreactor exerting simultaneous mechanical shear and compression to simulate the rolling action of articular joints. Biotechnol Bioeng 109: 1060–1073
41. Li Z, Yao S-J, Alini M et al (2010) Chondrogenesis of human bone marrow mesenchymal stem cells in fibrin–polyurethane composites is modulated by frequency and amplitude of dynamic compression and shear stress. Tissue Eng Part A 16:575–584
42. Grad S, Loparic M, Peter R et al (2012) Sliding motion modulates stiffness and friction coefficient at the surface of tissue engineered cartilage. Osteoarthritis Cartilage 20: 288–295
43. Huang AH, Baker BM, Ateshian GA et al (2012) Sliding contact loading enhances the tensile properties of mesenchymal stem cell-seeded hydrogels. Eur Cells Mater 24:29–45
44. Vinardell T, Sheehy EJ, Buckley CT et al (2012) A comparison of the functionality and in vivo phenotypic stability of cartilaginous tissues engineered from different stem cell sources. Tissue Eng Part A 18:1161–1170
45. Phillips MD, Kuznetsov SA, Cherman N et al (2014) Directed differentiation of human induced pluripotent stem cells toward bone and cartilage: in vitro versus in vivo assays. Stem Cells Transl Med 3:867–878
46. Pleumeekers MM, Nimeskern L, Koevoet WLM et al (2014) The in vitro and in vivo capacity of culture-expanded human cells from several sources encapsulated in alginate to form cartilage. Eur Cells Mater 27:264–280
47. Huang S-J, Fu R-H, Shyu W-C et al (2013) Adipose-derived stem cells: isolation, characterization, and differentiation potential. Cell Transplant 22:701–709
48. Jiang Y, Jahagirdar BN, Reinhardt RL et al (2002) Pluripotency of mesenchymal stem cells derived from adult marrow. Nature 418:41–49
49. Ben-David U, Benvenisty N (2011) The tumorigenicity of human embryonic and induced pluripotent stem cells. Nat Rev Cancer 11:268–277
50. Ghosh Z, Huang M, Hu S et al (2011) Dissecting the oncogenic and tumorigenic potential of differentiated human induced pluripotent stem cells and human embryonic stem cells. Cancer Res 71:5030–5039
51. Lui KO, Waldmann H, Fairchild PJ (2009) Embryonic stem cells: overcoming the immunological barriers to cell replacement therapy. Curr Stem Cell Res Ther 4:70–80
52. English K, Wood KJ (2011) Immunogenicity of embryonic stem cell-derived progenitors after transplantation. Curr Opin Organ Transplant 16:90–95
53. Swijnenburg R-J, Schrepfer S, Govaert JA et al (2008) Immunosuppressive therapy mitigates immunological rejection of human embryonic stem cell xenografts. Proc Natl Acad Sci USA 105:12991–12996
54. Taylor CJ, Bolton EM, Bradley JA (2011) Immunological considerations for embryonic and induced pluripotent stem cell banking. Philos Trans R Soc Lond B Biol Sci 366:2312–2322
55. Zhou Y, Zeng F (2013) Integration-free methods for generating induced pluripotent stem cells. Genomics Proteomics Bioinformatics 11:284–287

56. Diecke S, Jung SM, Lee J et al (2014) Recent technological updates and clinical applications of induced pluripotent stem cells. Korean J Intern Med 29:547–557
57. Park S, Im G-I (2014) Embryonic stem cells and induced pluripotent stem cells for skeletal regeneration. Tissue Eng Part B Rev 20: 381–391
58. Da Silva Meirelles L, Fontes AM, Covas DT et al (2009) Mechanisms involved in the therapeutic properties of mesenchymal stem cells. Cytokine Growth Factor Rev 20:419–427
59. Aronin CEP, Tuan RS (2010) Therapeutic potential of the immunomodulatory activities of adult mesenchymal stem cells. Birth Defects Res C Embryo Today 90:67–74
60. Griffin MD, Ritter T, Mahon BP (2010) Immunological aspects of allogeneic mesenchymal stem cell therapies. Hum Gene Ther 21:1641–1655
61. Mahmoudifar N, Doran PM (2005) Tissue engineering of human cartilage in bioreactors using single and composite cell-seeded scaffolds. Biotechnol Bioeng 91:338–355
62. Saha S, Kirkham J, Wood D et al (2010) Comparative study of the chondrogenic potential of human bone marrow stromal cells, neonatal chondrocytes and adult chondrocytes. Biochem Biophys Res Commun 401:333–338
63. Adkisson HD, Martin JA, Amendola RL et al (2010) The potential of human allogeneic juvenile chondrocytes for restoration of articular cartilage. Am J Sports Med 38:1324–1333
64. Jakob M, Démarteau O, Suetterlin R et al (2004) Chondrogenesis of expanded adult human articular chondrocytes is enhanced by specific prostaglandins. Rheumatology 43: 852–857
65. Schrobback K, Klein TJ, Schuetz M et al (2011) Adult human articular chondrocytes in a microcarrier-based culture system: expansion and redifferentiation. J Orthop Res 29:539–546
66. Mauck RL, Yuan X, Tuan RS (2006) Chondrogenic differentiation and functional maturation of bovine mesenchymal stem cells in long-term agarose culture. Osteoarthritis Cartilage 14:179–189
67. Connelly JT, Wilson CG, Levenston ME (2008) Characterization of proteoglycan production and processing by chondrocytes and BMSCs in tissue engineered constructs. Osteoarthritis Cartilage 16:1092–1100
68. Huang AH, Stein A, Mauck RL (2010) Evaluation of the complex transcriptional topography of mesenchymal stem cell chondrogenesis for cartilage tissue engineering. Tissue Eng Part A 16:2699–2708
69. Mahmoudifar N, Doran PM (2010) Extent of cell differentiation and capacity for cartilage synthesis in human adult adipose-derived stem cells: comparison with fetal chondrocytes. Biotechnol Bioeng 107:393–401
70. Meretoja VV, Dahlin RL, Wright S et al (2013) The effect of hypoxia on the chondrogenic differentiation of co-cultured articular chondrocytes and mesenchymal stem cells in scaffolds. Biomaterials 34:4266–4273
71. Saha S, Kirkham J, Wood D et al (2013) Informing future cartilage repair strategies: a comparative study of three different human cell types for cartilage tissue engineering. Cell Tissue Res 352:495–507
72. Darling EM, Athanasiou KA (2005) Rapid phenotypic changes in passaged articular chondrocyte subpopulations. J Orthop Res 23:425–432
73. Rackwitz L, Djouad F, Janjanin S et al (2014) Functional cartilage repair capacity of de-differentiated, chondrocyte- and mesenchymal stem cell-laden hydrogels in vitro. Osteoarthritis Cartilage 22:1148–1157
74. Kafienah W, Cheung FL, Sims T et al (2008) Lumican inhibits collagen deposition in tissue engineered cartilage. Matrix Biol 27:526–534
75. Mahmoudifar N, Doran PM (2012) Chondrogenesis and cartilage tissue engineering: the longer road to technology development. Trends Biotechnol 30:166–176
76. Steck E, Bertram H, Abel R et al (2005) Induction of intervertebral disc-like cells from adult mesenchymal stem cells. Stem Cells 23:403–411
77. Somoza RA, Welter JF, Correa D et al (2014) Chondrogenic differentiation of mesenchymal stem cells: challenges and unfulfilled expectations. Tissue Eng Part B Rev 20: 596–608
78. Hwang NS, Varghese S, Elisseeff J (2008) Derivation of chondrogenically-committed cells from human embryonic cells for cartilage tissue regeneration. PLoS One 3(6):e2498
79. Qu C, Puttonen KA, Lindeberg H et al (2013) Chondrogenic differentiation of human pluripotent stem cells in chondrocyte co-culture. Int J Biochem Cell Biol 45: 1802–1812
80. Yao Y, Zhang F, Zhou R et al (2010) Effects of combinational adenoviral vector-mediated TGFβ3 transgene and shRNA silencing type I collagen on articular chondrogenesis of synovium-derived mesenchymal stem cells. Biotechnol Bioeng 106:818–828

81. Perrier-Groult E, Pasdeloup M, Malbouyres M et al (2013) Control of collagen production in mouse chondrocytes by using a combination of bone morphogenetic protein-2 and small interfering RNA targeting *Col1a1* for hydrogel-based tissue-engineered cartilage. Tissue Eng Part C Methods 19:652–664
82. Freyria A-M, Mallein-Gerin F (2012) Chondrocytes or adult stem cells for cartilage repair: the indisputable role of growth factors. Injury 43:259–265
83. Sekiya I, Vuoristo JT, Larson BL et al (2002) In vitro cartilage formation by human adult stem cells from bone marrow stroma defines the sequence of cellular and molecular events during chondrogenesis. Proc Natl Acad Sci USA 99:4397–4402
84. Pelttari K, Winter A, Steck E et al (2006) Premature induction of hypertrophy during in vitro chondrogenesis of human mesenchymal stem cells correlates with calcification and vascular invasion after ectopic transplantation in SCID mice. Arthritis Rheum 54: 3254–3266
85. Pelttari K, Steck E, Richter W (2008) The use of mesenchymal stem cells for chondrogenesis. Injury 39(Suppl 1):S58–S65
86. Kim Y-J, Kim H-J, Im G-I (2008) PTHrP promotes chondrogenesis and suppresses hypertrophy from both bone marrow-derived and adipose tissue-derived MSCs. Biochem Biophys Res Commun 373:104–108
87. Varghese S, Hwang NS, Canver AC et al (2008) Chondroitin sulfate based niches for chondrogenic differentiation of mesenchymal stem cells. Matrix Biol 27:12–21
88. Dickhut A, Pelttari K, Janicki P et al (2009) Calcification or dedifferentiation: requirement to lock mesenchymal stem cells in a desired differentiation stage. J Cell Physiol 219:219–226
89. Weiss S, Hennig T, Bock R et al (2010) Impact of growth factors and PTHrP on early and late chondrogenic differentiation of human mesenchymal stem cells. J Cell Physiol 223:84–93
90. Lee H-H, Chang C-C, Shieh M-J et al (2013) Hypoxia enhances chondrogenesis and prevents terminal differentiation through PI3K/Akt/FoxO dependent anti-apoptotic effect. Sci Rep 3:2683. doi:10.1038/srep02683
91. Lee JM, Kim JD, Oh EJ et al (2014) PD98059-impregnated functional PLGA scaffold for direct tissue engineering promotes chondrogenesis and prevents hypertrophy from mesenchymal stem cells. Tissue Eng Part A 20:982–991
92. Hubka KM, Dahlin RL, Meretoja VV et al (2014) Enhancing chondrogenic phenotype for cartilage tissue engineering: monoculture and coculture of articular chondrocytes and mesenchymal stem cells. Tissue Eng Part B Rev 20:641–654
93. Fischer J, Dickhut A, Rickert M et al (2010) Human articular chondrocytes secrete parathyroid hormone-related protein and inhibit hypertrophy of mesenchymal stem cells in coculture during chondrogenesis. Arthritis Rheum 62:2696–2706
94. Khan IM, Gilbert SJ, Singhrao SK et al (2008) Cartilage integration: evaluation of the reasons for failure of integration during cartilage repair. Eur Cells Mater 16:26–39
95. Pabbruwe MB, Esfandiari E, Kafienah W et al (2009) Induction of cartilage integration by a chondrocyte/collagen-scaffold implant. Biomaterials 30:4277–4286
96. Lu Y, Xu Y, Yin Z et al (2013) Chondrocyte migration affects tissue-engineered cartilage integration by activating the signal transduction pathways involving Src, PLCγ1, and ERK1/2. Tissue Eng Part A 19:2506–2516
97. Allon AA, Ng KW, Hammoud S et al (2012) Augmenting the articular cartilage-implant interface: functionalizing with a collagen adhesion protein. J Biomed Mater Res A 100:2168–2175
98. Athens AA, Makris EA, Hu JC (2013) Induced collagen cross-links enhance cartilage integration. PLoS One 8(4):e60719
99. Gilbert SJ, Singhrao SK, Khan IM et al (2009) Enhanced tissue integration during cartilage repair in vitro can be achieved by inhibiting chondrocyte death at the wound edge. Tissue Eng Part A 15:1739–1749
100. Yang Y-H, Ard MB, Halper JT et al (2014) Type I collagen-based fibrous capsule enhances integration of tissue-engineered cartilage with native articular cartilage. Ann Biomed Eng 42:716–726
101. Wang D-A, Varghese S, Sharma B et al (2007) Multifunctional chondroitin sulphate for cartilage tissue–biomaterial integration. Nat Mater 6:385–392
102. Koyama N, Miura M, Nakao K et al (2013) Human induced pluripotent stem cells differentiated into chondrogenic lineage via generation of mesenchymal progenitor cells. Stem Cells Dev 22:102–113
103. Kim M-J, Son MJ, Son M-Y et al (2011) Generation of human induced pluripotent stem cells from osteoarthritis patient-derived synovial cells. Arthritis Rheum 63: 3010–3021

104. Wei Y, Zeng W, Wan R et al (2012) Chondrogenic differentiation of induced pluripotent stem cells from osteoarthritic chondrocytes in alginate matrix. Eur Cells Mater 23:1–12
105. Medvedev SP, Grigor'eva EV, Shevchenko AI et al (2011) Human induced pluripotent stem cells derived from fetal neural stem cells successfully undergo directed differentiation into cartilage. Stem Cells Dev 20:1099–1112
106. Gardner OFW, Archer CW, Alini M et al (2013) Chondrogenesis of mesenchymal stem cells for cartilage tissue engineering. Histol Histopathol 28:23–42
107. Ahmed TAE, Hincke MT (2014) Mesenchymal stem cell-based tissue engineering strategies for repair of articular cartilage. Histol Histopathol 29:669–689
108. Caplan AI (2007) Adult mesenchymal stem cells for tissue engineering versus regenerative medicine. J Cell Physiol 213:341–347
109. Pievani A, Scagliotti V, Russo FM et al (2014) Comparative analysis of multilineage properties of mesenchymal stromal cells derived from fetal sources shows an advantage of mesenchymal stromal cells isolated from cord blood in chondrogenic differentiation potential. Cytotherapy 16:893–905
110. Tondreau T, Meuleman N, Delforge A et al (2005) Mesenchymal stem cells derived from CD133-positive cells in mobilized peripheral blood and cord blood: proliferation, Oct4 expression, and plasticity. Stem Cells 23: 1105–1112
111. Kolambkar YM, Peister A, Soker S et al (2007) Chondrogenic differentiation of amniotic fluid-derived stem cells. J Mol Histol 38:405–413
112. Li F, Chen Y-Z, Miao Z-N et al (2012) Human placenta-derived mesenchymal stem cells with silk fibroin biomaterial in the repair of articular cartilage defects. Cell Reprogram 14:334–341
113. Baksh D, Yao R, Tuan RS (2007) Comparison of proliferative and multilineage differentiation potential of human mesenchymal stem cells derived from umbilical cord and bone marrow. Stem Cells 25:1384–1392
114. Jansen EJP, Emans PJ, Guldemond NA et al (2008) Human periosteum-derived cells from elderly patients as a source for cartilage tissue engineering? J Tissue Eng Regen Med 2: 331–339
115. Barachini S, Danti S, Pacini S et al (2014) Plasticity of human dental pulp stromal cells with bioengineering platforms: a versatile tool for regenerative medicine. Micron 67:155–168
116. Chang C-H, Chen C-C, Liao C-H et al (2014) Human acellular cartilage matrix powders as a biological scaffold for cartilage tissue engineering with synovium-derived mesenchymal stem cells. J Biomed Mater Res A 102:2248–2257
117. Andriamanalijaona R, Duval E, Raoudi M et al (2008) Differentiation potential of human muscle-derived cells towards chondrogenic phenotype in alginate beads culture. Osteoarthritis Cartilage 16:1509–1518
118. Tay AG, Farhadi J, Suetterlin R et al (2004) Cell yield, proliferation, and postexpansion differentiation capacity of human ear, nasal, and rib chondrocytes. Tissue Eng 10: 762–770
119. Park SS, Jin H-R, Chi DH et al (2004) Characteristics of tissue-engineered cartilage from human auricular chondrocytes. Biomaterials 25:2363–2369
120. Liu Y, Goldberg AJ, Dennis JE et al (2012) One-step derivation of mesenchymal stem cell (MSC)-like cells from human pluripotent stem cells on a fibrillar collagen coating. PLoS One 7(3):e33225
121. Cheng A, Kapacee Z, Peng J et al (2014) Cartilage repair using human embryonic stem cell-derived chondroprogenitors. Stem Cells Transl Med 3:1287–1294
122. Kim H-J, Im G-I (2009) Chondrogenic differentiation of adipose tissue-derived mesenchymal stem cells: greater doses of growth factor are necessary. J Orthop Res 27: 612–619
123. Sakaguchi Y, Sekiya I, Yagishita K et al (2005) Comparison of human stem cells derived from various mesenchymal tissues: superiority of synovium as a cell source. Arthritis Rheum 52:2521–2529
124. Kafienah W, Jakob M, Démarteau O et al (2002) Three-dimensional tissue engineering of hyaline cartilage: comparison of adult nasal and articular chondrocytes. Tissue Eng 8: 817–826

Part II

Cell Isolation, Expansion, and Differentiation

Chapter 2

Human Fetal and Adult Chondrocytes

Kifah Shahin, Nastaran Mahmoudifar, and Pauline M. Doran

Abstract

As the only cell type found in healthy adult cartilage, chondrocytes are the obvious and most direct starting point for cartilage tissue engineering. Human adult, juvenile, neonatal, and fetal chondrocytes have all been demonstrated to produce cartilage matrix components in vitro for production of engineered tissues. In this chapter, procedures are outlined for isolation of chondrocytes from human fetal and adult cartilage. Methods for expansion and cryopreservation of the cells and characterization of gene expression using quantitative polymerase chain reaction (Q-PCR) analysis are also described.

Key words Chondrocyte, Collagenase, Cryopreservation, Dedifferentiation, Human adult and fetal tissue, Q-PCR

1 Introduction

Human adult cartilage is an avascular and aneural connective tissue containing chondrocytes in lacunae embedded within a dense solid matrix. Chondrocytes are the cells that synthesize the biochemical and structural elements of cartilage. They have a characteristic rounded morphology and secrete an extracellular matrix rich in collagen type II and the proteoglycan, aggrecan. To generate cartilage in vitro, chondrocytic cells are needed for application in three-dimensional culture systems. Although human embryonic, induced pluripotent and mesenchymal stem cells have been investigated for chondrogenesis and cartilage production [1, 2], differentiated chondrocytes isolated from native cartilage tissue remain the most direct effectors of cartilage synthesis.

1.1 Availability and Properties of Human Chondrocytes for Tissue Engineering

Because cartilage has a very limited capacity for self-repair, obtaining healthy human adult chondrocytes for tissue engineering applications presents some difficulties, as there is a risk of inflicting an unacceptable level of irreparable injury when cartilage is harvested from the joints of living donors. Nevertheless, surgical procedures may be used to remove a biopsy of articular cartilage from

Pauline M. Doran (ed.), *Cartilage Tissue Engineering: Methods and Protocols*, Methods in Molecular Biology, vol. 1340, DOI 10.1007/978-1-4939-2938-2_2, © Springer Science+Business Media New York 2015

a low-weight-bearing location: 200–300-mg slivers of partial- or full-thickness articular cartilage are obtained routinely from patients to provide cells for autologous chondrocyte implantation [3]. Other options include harvesting articular cartilage from adult donors after death [4, 5] or limb amputation [6, 7], and collecting cartilage tissue removed during arthroplasty or joint surgery [7, 8]. Because most joint surgery is carried out in response to damage of articular cartilage by osteoarthritis or physical trauma, it is important that the cartilage tissues and chondrocytes recovered from arthroplasty or similar procedures are not detrimentally affected by disease or injury. When cultured in three-dimensional scaffolds, articular chondrocytes from patients with osteoarthritis were found to differentially express 184 genes compared with chondrocytes from normal donors [9]. In other studies, collagen synthesis by chondrocytes isolated from osteoarthritis-affected cartilage was significantly lower than for chondrocytes from unaffected tissue [10].

Alternatives to adult chondrocytes for human cartilage tissue engineering are fetal [11, 12], neonatal [13], and juvenile (<13 years old) [14] chondrocytes. These cells are suitable principally for research purposes: allogeneic chondrocytes and the engineered tissues derived from them may be expected to produce undesirable immune responses if used clinically, although there is evidence that fresh osteochondral allografts and juvenile allogeneic chondrocytes do not elicit graft rejection reactions [14]. Chondrocytes isolated from young (neonatal or juvenile) human cartilage expand in culture at a considerably faster rate than adult chondrocytes [13, 14]. Juvenile-derived chondrocytes also out-perform chondrocytes from adult donors in terms of their ability to synthesize cartilage matrix components, suggesting that chondrocytes from young tissues have a more strongly enhanced chondrogenic potential compared with adult cells [14]. Several studies have demonstrated that fetal, neonatal, and adult chondrocytes are more highly productive of cartilage matrix in three-dimensional cultures in vitro than chondro-induced mesenchymal stem cells [15–17].

1.2 Practical Aspects

Irrespective of whether fetal, neonatal, juvenile, or adult cartilage is used as a source of human chondrocytes, the number of cells that can be isolated from the available tissue is likely to be too low for direct tissue engineering applications, particularly when scaffold seeding is required. Accordingly, expansion of the cells must be carried out, typically using monolayer culture. It has been known for several decades that monolayer expansion is detrimental to chondrocyte differentiation [18–20]. Dedifferentiation to a fibroblast-like phenotype, including development of a flat spindle-like morphology, downregulation of cartilage-specific genes, and reduction in cartilage matrix synthesis, constitutes the usual response to monolayer culture. Representative gene expression data illustrating this effect are shown in Fig. 1. Dedifferentiation

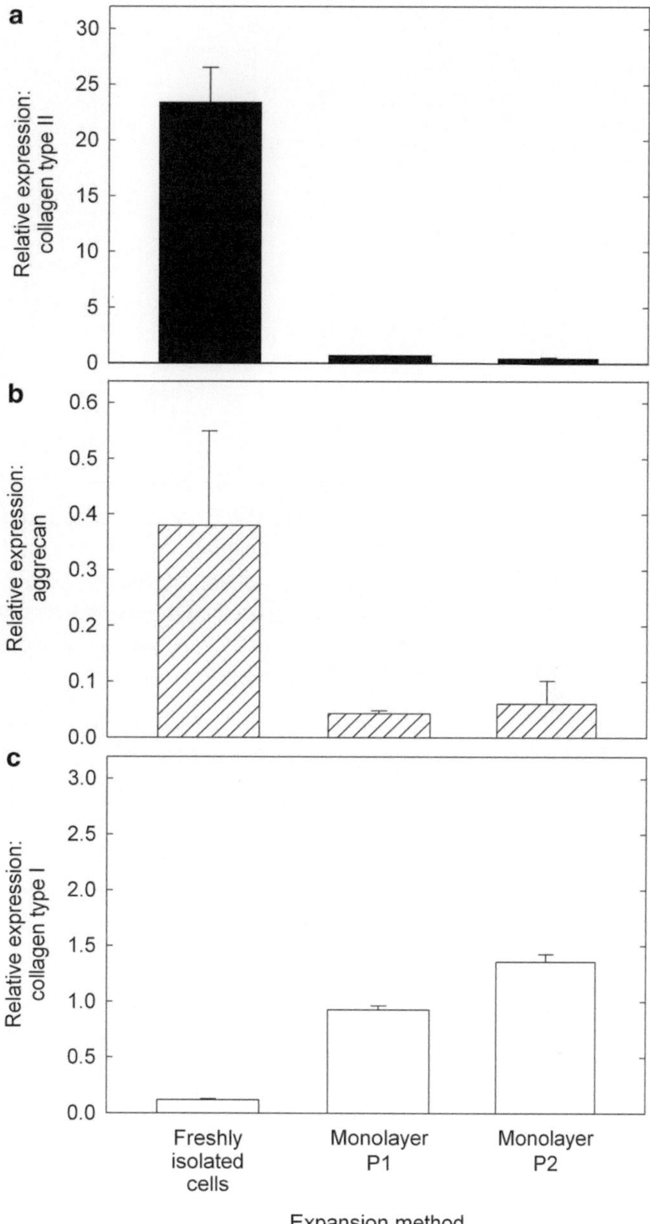

Fig. 1 Gene expression levels relative to β-actin as housekeeping gene for (**a**) collagen type II, (**b**) aggrecan, and (**c**) collagen type I. Gene expression levels were measured using PCR for freshly isolated fetal chondrocytes, fetal chondrocytes after one passage (P1: 12 days) of monolayer culture, and fetal chondrocytes after two passages (P2: 20 days) of monolayer culture. The error bars represent standard errors from four fetuses and three to five cultures. The typical response of chondrocytes to monolayer culture is downregulation of chondrocytic markers such as collagen type II and aggrecan, and upregulation of collagen type I

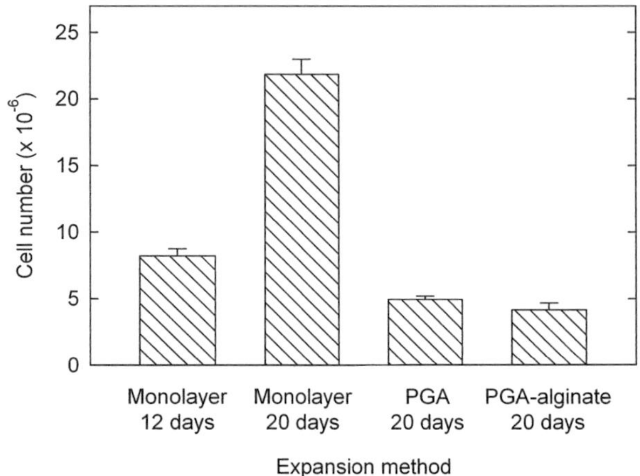

Fig. 2 Cell number after expansion of fetal chondrocytes in monolayer culture for one passage (12 days), monolayer culture for two passages (20 days), polyglycolic acid (PGA) scaffolds for 20 days, and PGA–alginate scaffolds [12] for 20 days. The initial number of freshly isolated cells in each culture was 1.5×10^6. The error bars represent standard errors from triplicate cultures. Expansion in monolayer culture for two passages generated 4.4–5.2 times the number of cells obtained using three-dimensional scaffold systems over the same period

is highly undesirable because dedifferentiated chondrocytes produce fibrocartilage, a less resilient type of extracellular matrix lacking the vital functional properties of articular cartilage that give joints their strength and durability. Practically, it means that expanded chondrocytes must be redifferentiated for tissue engineering purposes.

Several investigators have observed that dedifferentiated chondrocytes are capable of redifferentiating when cultured in three-dimensional hydrogels [18, 19, 21, 22]. However, results indicating that full chondrocytic function is irreversibly compromised after monolayer expansion have also been reported [23, 24]. Because cartilage synthesis capacity decreases as the number of population doublings increases [23, 24], in our work, we limit monolayer passaging of chondrocytes to Passage 2. Alternatives to monolayer expansion have been sought to allow better retention of the chondrocytic phenotype. Although cell differentiation is maintained to a greater extent in three-dimensional systems compared with surface culture, three-dimensional culture is not a feasible option for increasing cell numbers. As shown in Fig. 2, cell proliferation is greatly reduced when the cells are seeded in scaffolds compared with the level of expansion achieved under monolayer conditions.

Chondrocytes isolated from individual donors can differ substantially in their ability to synthesize cartilage matrix. As an example, levels of glycosaminoglycan (GAG) production by human chondrocytes from 52 different individuals varied by a factor of five- to six-fold when the cells were cultured as pellets in medium with growth factors [25]. Significant inter-individual variation was found between donors within the same age group as well as across different ages. The observed variations in chondrogenic potential may reflect genetic differences between individuals as well as lifestyle diversity and pharmaceutical history. When chondrocytes from individual patients are used for cartilage tissue engineering, there is a risk that the experimental findings will reflect the properties of the particular chondrocytes studied and will not be relevant more generally. To overcome this problem, chondrocytes isolated from as many individuals as possible are pooled for use in experiments. In this way, the results obtained reflect an average cellular phenotype, thus dampening the effects of individual variation.

1.3 Chondrocyte Isolation, Characterization, and Application

Here we describe methods for isolation, expansion, cryopreservation, and characterization of chondrocytes from human fetal and adult cartilage. The methods for cell isolation are based on enzymatic digestion of cartilage tissue using collagenase type II. General protocols for cell expansion are outlined using monolayer culture and serum-containing medium; cryopreservation in liquid nitrogen is carried out using dimethyl sulfoxide (DMSO) as cryoprotectant. Techniques for quantitative polymerase chain reaction (Q-PCR) analysis are provided for examining the expression of cartilage-specific genes in freshly isolated or expanded chondrocytes. The cells isolated and characterized using these methods are suitable for use in tissue engineering culture systems. As well as direct applications for production of cartilage matrix using porous scaffolds and bioreactors [11, 26], chondrocytes may also be used in coculture systems to improve the chondrogenic differentiation of mesenchymal stem cells [27] and for in vitro development of composite osteochondral constructs suitable for joint repair [28].

2 Materials

2.1 Clinical Samples

1. Human fetal knee or hip joint tissues obtained with informed parental consent (*see* **Note 1**).
2. Adult articular cartilage tissue pieces obtained with informed consent from patients undergoing knee or hip surgery (*see* **Note 1**).

2.2 Chondrocyte Isolation, Expansion, and Cryopreservation

1. DMEM base medium: Dulbecco's modified Eagle's medium (DMEM) containing 4.5 g/L of glucose and 584 mg/L of L-glutamine with 3.7 g of sodium hydrogen carbonate, 2.39 g of *N*-2-(hydroxyethyl)piperazine-*N'*-2-ethane sulfonic acid (HEPES), 0.046 g of L-proline, and 10 mL of 100× nonessential amino acid solution (Sigma-Aldrich) made up to 895 mL in Milli-Q water. Sterilize by filtration using 0.2-μm pressure-filtration units.

2. Antibiotic solution: 10,000 U/mL penicillin, 10 mg/mL streptomycin, and 25 μg/mL amphotericin B.
3. PBS (Dulbecco's phosphate buffered saline): 137 mM NaCl, 9.5 mM phosphate, and 2.7 mM KCl. Dissolve 8 g of NaCl, 0.2 g of KCl, 0.2 g of KH_2PO_4, and 1.14 g of Na_2HPO_4 in Milli-Q water to make up 1 L of solution. Sterilize by autoclaving and store at 4 °C.
4. Tissue dissection solution: 1 % (v/v) antibiotic solution in PBS.
5. Type II collagenase stock solution: for cartilage tissue digestion. Dissolve type II clostridial collagenase (Sigma-Aldrich) in DMEM base medium at a concentration equivalent to 3000 U/mL for fetal cartilage, or 10 mg/mL (approx. 4300 U/mL) for adult cartilage (*see* **Note 2**). Filter the solution using a 0.45-μm syringe filter to remove any undissolved materials, and then sterilize using a 0.2-μm single-use syringe filter. Aliquot and store at −20 °C for up to 1 year (*see* **Note 3**).
6. Digestion solution: 0.5 mL of fetal bovine serum (FBS), 0.1 mL of antibiotic solution, and 1 mL of Type II collagenase stock solution made up to 10 mL using DMEM base medium. Prepare the digestion solution fresh before use, making up 10 mL of digestion solution for every 300 mg of cartilage to be digested. Digestion solution contains FBS unless otherwise stated. The final concentration of collagenase in the digestion solution is equivalent to 300 U/mL for fetal cartilage, or 0.1 % w/v (approx. 430 U/mL) for adult cartilage (*see* **Note 2**).
7. Cell expansion culture medium (CECM): 895 mL of DMEM base medium with 100 mL of FBS and 5 mL of antibiotic solution added before use (*see* **Note 4**).
8. Viability stain: 0.4 % Trypan Blue in 0.81 % NaCl. Store at room temperature.
9. Trypsin–EDTA solution: Dissolve 0.04 % ethylenediaminetetraacetic acid disodium dihydrate (EDTA-$Na_2 \cdot 2H_2O$) in PBS, and add to it an equal volume of DMEM base medium containing 0.1 % trypsin. Sterilize using a 0.2-μm single-use syringe filter and store at −20 °C.
10. Chondrocyte freezing solution: a sterile solution of 20 % dimethyl sulfoxide (DMSO) in FBS (*see* **Note 5**).

2.3 Q-PCR

1. RNeasy® Mini kit, for total RNA extraction (Qiagen).
2. Superscript™ III First-Strand kit (Invitrogen), for reverse transcription of messenger RNA (mRNA) into complementary DNA (cDNA).
3. Primers for PCR amplification: Suitable primers for chondrocyte characterization are listed in Table 1 (*see* **Note 6**).

Table 1
Primers for Q-PCR analysis of chondrocytes

Gene of interest	Forward primer	Reverse primer
Collagen type I	CAGCCGCTTCACCTACAGC	TTTTGTATTCAATCACTGTCTTGCC
Collagen type II	GGCAATAGCAGGTTCACGTACA	CGATAACAGTCTTGCCCCACTT
Aggrecan	TCGAGGACAGCGAGGCC	TCGAGGGTGTAGCGTGTAGAGA
Versican	TGGAATGATGTTCCCTGCAA	AAGGTCTTGGCATTTTCTACAACAG
GAPDH	ATGGGGAAGGTGAAGGTCG	TAAAAGCAGCCCTGGTGACC
β-actin	TGAGCGCGGCTACAGCTT	TCCTTAATGTCACGCACGATTT

Prepare primers in autoclaved Milli-Q water at a concentration of 7.5 μM for collagen type I, aggrecan, versican, and GAPDH primers, 15 μM for the β-actin primers, and 22.5 μM for the collagen type II primers. Aliquot and store at −20 °C.

4. Platinum® SYBR® Green kit (Invitrogen). This kit includes a pre-mixed PCR reaction mixture (SuperMix-UDG) containing SYBR® Green I fluorescent dye, Platinum® Taq DNA polymerase, Mg^{2+}, uracil-DNA glycosylase (UDG), and deoxyribonucleotide triphosphates (dNTPs), with 2′-deoxyuridine 5′-triphosphate (dUTP) instead of 2′-deoxythymidine 5′-triphosphate (dTTP). The kit also includes a passive reference dye containing a glycine conjugate of 5-carboxy-X-rhodamine (ROX) (*see* **Note 7**).

3 Methods

3.1 Isolation, Expansion, and Cryopreservation of Human Fetal Chondrocytes

3.1.1 Cell Isolation

1. Transport the fetal joint tissue to the laboratory on ice in DMEM base medium containing 1 % (v/v) antibiotic solution (*see* **Note 8**).
2. Working in a biosafety cabinet, place the tissue sections, one by one, in disposable 90-mm Petri dishes. Wash with 2–4 mL of cold tissue dissection solution and dissect using sterile forceps and scalpels to expose the glossy white epiphyseal cartilage.
3. Cut out the soft cartilage tissue to separate it from the bone.
4. Carefully clean the cartilage and remove any attached fibrous tissue. Place the cartilage in a Petri dish containing cold tissue dissection solution while the digestion solution (Subheading 2.2) is being prepared.
5. Transfer the cartilage pieces to a new Petri dish and chop into fine pieces using scalpels in the presence of 1 mL of digestion solution (without FBS, *see* **Note 9**).

6. Transfer the minced tissue to 50-mL centrifuge tubes containing digestion solution (300 mg of tissue per 10 mL of solution per tube) using scalpels. Using a 5-mL pipette, rinse the Petri dish with 1–2 mL of digestion solution and aspirate to transfer the remaining tissue pieces (*see* **Note 10**).
7. Place the centrifuge tubes horizontally on their sides in a CO_2 incubator, making sure that all tissue pieces are in the solution. Incubate at 37 °C for 14–16 h.
8. Filter the digest through a sterile 150-μm nylon sieve to remove tissue debris and collect the released cells.
9. Wash the cells twice in DMEM base medium by centrifuging for 10 min at 450 × *g* (*see* **Note 11**). Resuspend in 20–40 mL of CECM for every 300 mg of cartilage.
10. To count the cells, mix 10 μL of cell suspension with an equal volume of viability stain and wait for 1 min. Load about 5–10 μL of the stained cell suspension into a hemocytometer and examine under a microscope to determine the cell number and viability (*see* **Note 12**).

3.1.2 Cell Expansion and Cryopreservation

1. Adjust the concentration of viable cells to 1.25×10^5 mL^{-1} of CECM. Culture the cells in vented-cap treated T-flasks at an area concentration of 2×10^4 cells per cm^2.
2. Place the T-flasks in a CO_2 incubator at 37 °C and leave undisturbed for 4 days to allow the chondrocytes to attach to the growth surface.
3. On Day 4, replace the medium with fresh CECM. Replace the medium every 3 days thereafter.
4. Examine the cultured cells for confluency using an inverted microscope. When the cells reach 90–100 % confluency (*see* **Note 13**), harvest the cells as described in **steps 5–9** below.
5. Remove the culture medium from the flask completely using a pipette, and then wash the cells three times with 5–10 mL of PBS (pre-warmed in a water bath to 37 °C).
6. Add 1 mL of trypsin–EDTA solution (pre-warmed to 37 °C) for every 20 cm^2 of flask surface area. Incubate for 7 min at 37 °C.
7. Collect the trypsin–EDTA from the flask using a pipette and add to a centrifuge tube containing at least an equal volume of cold CECM (*see* **Note 14**).
8. Return the flask to the incubator and incubate for another 10 min.
9. Add 5–10 mL of CECM to the flask and tap gently at the bottom of the flask to detach the remaining cells. Collect the detached cells using a 10-mL pipette and combine with the cells removed earlier.

10. Wash the cells twice in DMEM base medium by centrifuging for 10 min at 450 × *g*. Resuspend in DMEM base medium and count the cells as described in Subheading 3.1.1.
11. To expand the cells further, repeat **steps 1–10**, or cryopreserve the cells for later use.
12. To cryopreserve the chondrocytes, suspend them at a concentration of 10–20 × 10^6 mL^{-1} in DMEM base medium. Add an equal volume of chondrocyte freezing solution and mix quickly using a pipette. Aliquot into 2-mL cryovials, place in a freezing container, e.g., Nalgene™ Cryo 1 °C, and store at −70 °C overnight. Transfer the cells to liquid nitrogen for long-term storage (*see* **Note 15**).

3.2 Isolation and Expansion of Human Adult Chondrocytes

1. Place slices of adult articular cartilage obtained from knee or hip surgery into 50-mL plastic centrifuge tubes containing DMEM base medium supplemented with 1 % (v/v) antibiotic solution. Transfer the tissue to the laboratory on ice. Store the centrifuge tubes at 4 °C and process the cartilage samples on the same day or within 24 h (*see* **Note 8**).
2. Using sterile forceps, place the cartilage slices in a sterile 90-mm disposable Petri dish in a biosafety cabinet and rinse the slices twice with cold tissue dissection solution. Transfer the slices to a 50-mL centrifuge tube containing tissue dissection solution and place the tube on ice.
3. Transfer approximately 500 mg of cartilage to a fresh sterile 90-mm disposable Petri dish. Add 1 mL of digestion solution (without FBS, *see* **Note 9**) to the Petri dish and finely mince the cartilage into 1 mm^3 pieces using size 10 and 21 surgical blades.
4. Using a sterile spatula, transfer the minced cartilage to a 50-mL centrifuge tube containing 5 mL of digestion solution. Rinse the Petri dish with 4 mL of digestion solution and use a pre-wetted sterile 5-mL disposable pipette (*see* **Note 10**) to transfer all the remaining material to the centrifuge tube. The final volume of digestion solution in the centrifuge tube is 10 mL.
5. Place the centrifuge tube containing minced cartilage and digestion solution in a horizontal position on an orbital shaker operated at 40 rpm (*see* **Note 16**). Digest at 37 °C for 18–24 h until the cartilage matrix is digested and cells are free in suspension (*see* **Note 17**).
6. Repeat **steps 3–5** for all the cartilage slices.
7. Aspirate the cell suspension using a pre-wetted sterile 5-mL disposable pipette to break up the cell clumps. Filter the cell suspension through a sterile 150-μm nylon sieve to remove tissue debris.
8. Centrifuge the cell suspension at 450 × *g* for 10 min using a swing-bucket centrifuge (*see* **Note 11**) at room temperature or

4 °C. A creamy-white pellet is obtained in this way. Wash the pellet twice with DMEM base medium, resuspending the cells each time to remove the collagenase.

9. Count the isolated chondrocytes using a hemocytometer and viability stain as described in Subheading 3.1.1. Calculate the viable cell yield obtained from the adult cartilage slices (*see* **Note 18**).
10. Expand the cells by culturing in T-flasks as described in Subheading 3.1.2. The cells may also be cryopreserved as described in Subheading 3.1.2.

3.3 Determining the Chondrocytic Phenotype Using Q-PCR

1. Extract total RNA from cell pellets using an RNeasy® Mini kit according to the manufacturer's instructions. Use pellets of $3–5 \times 10^6$ chondrocytes, either freshly isolated or expanded in culture (*see* **Note 19**).
2. Measure the RNA concentration and RNA purity by reading the sample optical density at 260 and 280 nm using a microvolume spectrophotometer, e.g., NanoDrop® (NanoDrop Technologies) (*see* **Note 20**).
3. Synthesize cDNA from up to 5 μg of total RNA by reverse transcription using a Superscript™ III First-Strand kit according to the manufacturer's instructions (*see* **Note 21**). Include a negative (non-template) control containing water only and process it in the same way as for the test samples.
4. Dilute the resulting cDNA samples to a total volume of 200 μL with autoclaved Milli-Q water. Aliquot into 10-μL volumes and store at −20 °C.
5. Prepare a template of a 96-well PCR plate (*see* **Note 22**). Designate duplicate PCR wells for each test you plan to perform. Each cDNA sample should be amplified using all primer pairs to determine expression of the corresponding genes. Include a duplicate non-template control using the negative control from **step 3**.
6. For the first time only, determine the amplification efficiency of each target sequence: create a representative sample by mixing equal volumes of the cDNA test samples, and then serially dilute (1:2) with autoclaved Milli-Q water. Allocate duplicate reaction wells for each serially diluted sample for each primer pair.
7. Prepare a reagent master mixture for each primer pair. Each test reaction should contain 25 μL of Platinum® SYBR® Green qPCR SuperMix-UDG, 1 μL of ROX reference dye, 2 μL of forward primer, 2 μL of reverse primer and 10 μL of autoclaved Milli-Q water.
8. Load 40 μL of master mixture per PCR reaction.
9. Dilute aliquots of cDNA samples 1:100 with autoclaved Milli-Q water. Add 10 μL of each diluted sample to their

designated wells and seal using cap strips to prevent contamination and evaporation.

10. Mix the PCR plate for 2 min on a plate mixer, and then centrifuge briefly at 600 × *g*.
11. Program a Q-PCR thermal cycler to hold samples for 2 min at 50 °C followed by 10 min at 95 °C and then perform 45 amplification cycles, each consisting of 95 °C for 15 s and then 60 °C for 30 s.
12. Program the thermal cycler to generate melting curves at the end of the amplification cycles in three stages: 95 °C for 15 s followed by 60 °C for 20 s and then 95 °C for 15 s, with 20 min ramping time between stages 2 and 3.
13. Analyze the amplification and melting curves directly using the instrument computer or transfer the data to a separate computer and analyze using third party software.
14. Set a baseline fluorescence for your experiment (*see* **Note 23**).
15. Set a fluorescence threshold in such a way that all samples cross the threshold during their exponential amplification phase.
16. Use the software to calculate a threshold cycle number (*Ct*) for each sample reaction: *Ct* is the number of cycles required for the fluorescence intensity to reach the threshold value.
17. Melting curves should show one peak per reaction. Each peak corresponds to the melting temperature of the amplified product (*see* **Note 24**).
18. Plot efficiency curves as the measured increase in *Ct* after each 1:2 dilution (ΔCt) versus the logarithm (base 2) of the dilution factor. Calculate the amplification efficiency as the slope of the best linear fit of the efficiency curve (*see* **Note 25**).
19. To compare the expression levels of one gene between different samples, the expression level must first be normalized to the expression level of β-actin or GAPDH within each sample according to the following equation:

$$R_{\text{GOI}} = \frac{(1 + E_{\text{HKG}})^{Ct_{\text{HKG}}}}{(1 + E_{\text{GOI}})^{Ct_{\text{GOI}}}}$$

where R_{GOI} is the expression of the gene of interest (GOI) normalized to that of the housekeeping gene (HKG), E_{HKG} and E_{GOI} are the amplification efficiencies of the HKG and GOI target sequences, respectively, and Ct_{HKG} and Ct_{GOI} are the *Ct* values for the HKG and GOI, respectively. If the amplification efficiencies are 1 or close to 1, the formula may be simplified to (*see* **Note 26**):

$$\Delta Ct_{\text{GOI}} = Ct_{\text{HKG}} - Ct_{\text{GOI}}$$

4 Notes

1. Approval from an appropriate local institutional human research ethics committee must be obtained before human tissues can be collected for research purposes. Because of the potential variability between individual patients, tissues from as many donors as possible should be obtained to allow pooling of the isolated chondrocytes before application in tissue engineering experiments. The age of the donor or gestational age of the fetus should be recorded. Fetal tissues after 16–20 weeks of gestation provide sufficient joint material for chondrocyte isolation.
2. The activity of type II clostridial collagenase purchased from Sigma-Aldrich varies: we have observed batch variations from 350 to 450 U/mg solid. Accordingly, the mass concentration of collagenase in the stock solution should be adjusted to obtain 3000 U/mL for digestion of fetal cartilage or approximately 4300 U/mL for digestion of adult cartilage. The unit of collagenase enzyme activity used by Sigma is defined as the amount of enzyme that liberates peptides from collagen from bovine achilles tendon equivalent in ninhydrin color to 1.0 μmol of leucine in 5 h at 37 °C and pH 7.4 in the presence of calcium ions. The conversion factor from this Sigma unit to the Mandl unit is 1000 to 1.
3. Aliquot the stock solution into the volumes required before freezing, as repeated freeze–thaw cycles reduce enzyme activity. One milliliter of collagenase stock solution is required to digest 300 mg of fetal cartilage tissue or 500 mg of adult cartilage tissue.
4. Ascorbic acid is often added to chondrocyte culture media in various contexts. However, we have found that including ascorbic acid in the medium used to culture freshly isolated fetal chondrocytes results in complete cell death.
5. DMSO can be sterilized by filtration using a 0.2-μm single-use syringe filter.
6. GAPDH and β-actin are housekeeping genes. They are used as internal reference genes with the assumption that their expression levels are not affected by the experimental treatment. The use of up to five housekeeping genes is common and produces better validated results than when only a single housekeeping gene is applied [29].
7. ROX reference dye normalizes the fluorescent reporter signal. It normalizes non-PCR-related fluctuations in fluorescence from well-to-well that may occur due to artifacts such as pipetting errors or instrument limitations. Because ROX dye is formulated to work on Applied Biosystems (ABI) and Stratagene

real-time PCR instruments, it may not work in other Q-PCR systems.

8. The tissue samples should be processed as soon as possible after surgical excision. The cartilage may be processed within up to 48 h post operation; however, the cell viability drops if processing is carried out after 24 h post operation.
9. Addition of 1 mL of digestion solution without FBS helps in the mincing process by reducing the tendency of the fine cartilage pieces to stick to the Petri dish surface.
10. Pre-wet the pipette by aspirating DMEM base medium several times to prevent minced tissue sticking to the inside of the pipette.
11. Centrifugation using a swing-bucket centrifuge will result in a firm, compact cell pellet, which is easy to handle during subsequent washing steps. Use of a fixed-angle centrifuge causes smearing of the cell pellet at the lower end of the tube wall, which may lead to cell loss during repeated washing steps.
12. Dead cells are stained blue. The percentage viable cells can be calculated as:

$$\%\ \text{viable cells} = \left[1.00 - \left(\text{number of blue cells} / \text{total number of cells}\right)\right] \times 100$$

A typical cell yield from human fetal cartilage is 5×10^4 cells per mg of cartilage with >90 % viability.

13. Cells expand in number about fivefold during each passage. Cells in Passage 1 take about 12 days to become confluent. Passage 2 cells grow faster and reach confluency in about 8 days.
14. The FBS present in CECM is a trypsin inhibitor and protects cells from the damaging effects of prolonged exposure to trypsin.
15. The recovery of expanded cells following cryopreservation and thawing exceeds 90 %, while the recovery of freshly isolated chondrocytes is <40 %. Therefore, it is recommended to expand the cells before cryostorage unless the cells are specifically required in a non-expanded state.
16. Mixing during digestion of adult cartilage helps with mass transfer and disruption of the cartilage extracellular matrix (ECM). If a temperature-controlled orbital shaker is not available, digestion can be performed using a small orbital shaker placed inside a 37 °C incubator. Mixing is not required during fetal tissue digestion.
17. Collagenase disrupts collagen fibers and releases chondrocytes from the cartilage matrix. The solution in the centrifuge tubes becomes turbid due to released cells and the cartilage pieces become soft at the end of the digestion.

18. The viable cell yield can be estimated by dividing the number of viable cells obtained from one 50-mL centrifuge tube by the wet weight of cartilage tissue placed inside the tube in **step 3** (e.g., 0.5 g). A typical viable cell yield from human adult articular cartilage after 22 h of digestion is 2.46 ± 0.80 million cells per gram [8].
19. Chondrocytes lose their chondrocytic phenotype in monolayer culture and dedifferentiate gradually towards a fibroblastic phenotype. This transformation is characterized by reduced expression of collagen type II and aggrecan, concurrent with increased expression of collagen type I and versican. Differentiation indices are often reported as the relative gene expression levels of collagen II/collagen I and aggrecan/versican.
20. Absorbance at 260 nm (A260) reflects the RNA concentration, while A260/A280 indicates RNA purity. An A260/A280 value of 1.8–2.0 indicates that the RNA is pure.
21. Although not necessary, it is advisable to start your reverse transcription with similar amounts of total RNA in all samples so that you end up with comparable amounts of cDNA. You cannot determine your cDNA concentration on a spectrophotometer without purification because your cDNA sample will contain a mixture of DNA, RNA and nucleotides.
22. Depending on the thermal cycler used, alternative PCR reaction well models may be required.
23. Fluorescence levels may fluctuate due to changes in the reaction medium creating a background signal. The background signal is most evident during the initial cycles of the PCR. During these early cycles, the background signal in all wells is used to determine a baseline fluorescence across the entire reaction system. The fluorescence of each sample should be sufficiently above the background signal for accurate analysis.
24. Multiple peaks for one primer pair in the melting curve indicate the presence of more than one amplified DNA sequence. This may happen in contaminated samples or when nonspecific primers are used.
25. An efficiency of 1 means that each amplification cycle results in a twofold increase of DNA copy number. In other words, in one sample, a gene corresponding to a *Ct* value of 14 is expressed twice as much as a gene corresponding to a *Ct* value of 15.
26. A ΔCt difference of one gene between two samples or between two genes in one sample is often referred to as $\Delta\Delta Ct$. A $\Delta\Delta Ct$ value of 1 indicates twice the level of expression.

References

1. Oldershaw RA (2012) Cell sources for the regeneration of articular cartilage: the past, the horizon and the future. Int J Exp Pathol 93:389–400
2. Park S, Im G-I (2014) Embryonic stem cells and induced pluripotent stem cells for skeletal regeneration. Tissue Eng Part B Rev 20:381–391
3. Brittberg M (2008) Autologous chondrocyte implantation—technique and long-term follow-up. Injury 39(S1):S40–S49
4. Schulze M, Kuettner KE, Cole AA (2000) Human adult articular chondrocytes cultured in alginate beads maintain their phenotype in preparation for transplantation. Orthopäde 29:100–106
5. Jakob M, Démarteau O, Suetterlin R et al (2004) Chondrogenesis of expanded adult human articular chondrocytes is enhanced by specific prostaglandins. Rheumatology 43:852–857
6. Jakob M, Démarteau O, Schäfer D et al (2001) Specific growth factors during the expansion and redifferentiation of adult human articular chondrocytes enhance chondrogenesis and cartilaginous tissue formation in vitro. J Cell Biochem 81:368–377
7. Schrobback K, Klein TJ, Schuetz M et al (2011) Adult human articular chondrocytes in a microcarrier-based culture system: expansion and redifferentiation. J Orthop Res 29:539–546
8. Jakob M, Démarteau O, Schäfer D et al (2003) Enzymatic digestion of adult human articular cartilage yields a small fraction of the total available cells. Connect Tissue Res 44:173–180
9. Dehne T, Karlsson C, Ringe J et al (2009) Chondrogenic differentiation potential of osteoarthritic chondrocytes and their possible use in matrix-associated autologous chondrocyte transplantation. Arthritis Res Ther 11:R133. doi:10.1186/ar2800
10. Tallheden T, Bengtsson C, Brantsing C et al (2005) Proliferation and differentiation potential of chondrocytes from osteoarthritic patients. Arthritis Res Ther 7:R560–R568. doi:10.1186/ar1709
11. Mahmoudifar N, Doran PM (2005) Tissue engineering of human cartilage in bioreactors using single and composite cell-seeded scaffolds. Biotechnol Bioeng 91:338–355
12. Shahin K, Doran PM (2011) Improved seeding of chondrocytes into polyglycolic acid scaffolds using semi-static and alginate loading methods. Biotechnol Prog 27:191–200
13. Saha S, Kirkham J, Wood D et al (2010) Comparative study of the chondrogenic potential of human bone marrow stromal cells, neonatal chondrocytes and adult chondrocytes. Biochem Biophys Res Commun 401:333–338
14. Adkisson HD, Martin JA, Amendola RL et al (2010) The potential of human allogeneic juvenile chondrocytes for restoration of articular cartilage. Am J Sports Med 38:1324–1333
15. Mahmoudifar N, Doran PM (2010) Extent of cell differentiation and capacity for cartilage synthesis in human adult adipose-derived stem cells: comparison with fetal chondrocytes. Biotechnol Bioeng 107:393–401
16. Saha S, Kirkham J, Wood D et al (2013) Informing future cartilage repair strategies: a comparative study of three different human cell types for cartilage tissue engineering. Cell Tissue Res 352:495–507
17. Pleumeekers MM, Nimeskern L, Koevoet WLM et al (2014) The *in vitro* and *in vivo* capacity of culture-expanded human cells from several sources encapsulated in alginate to form cartilage. Eur Cell Mater 27:264–280
18. Benya PD, Shaffer JD (1982) Dedifferentiated chondrocytes reexpress the differentiated collagen phenotype when cultured in agarose gels. Cell 30:215–224
19. Aulthouse AL, Beck M, Griffey E et al (1989) Expression of the human chondrocyte phenotype in vitro. In Vitro Cell Dev Biol 25:659–668
20. Hardingham T, Tew S, Murdoch A (2002) Tissue engineering: chondrocytes and cartilage. Arthritis Res 4(S3):S63–S68
21. Murphy CL, Sambanis A (2001) Effect of oxygen tension and alginate encapsulation on restoration of the differentiated phenotype of passaged chondrocytes. Tissue Eng 7:791–803
22. Schulze-Tanzil G, Mobasheri A, De Souza P et al (2004) Loss of chondrogenic potential in dedifferentiated chondrocytes correlates with deficient Shc–Erk interaction and apoptosis. Osteoarthritis Cartilage 12:448–458
23. Darling EM, Athanasiou KA (2005) Rapid phenotypic changes in passaged articular chondrocyte subpopulations. J Orthop Res 23:425–432
24. Rackwitz L, Djouad F, Janjanin S et al (2014) Functional cartilage repair capacity of de-differentiated, chondrocyte- and mesenchymal stem cell-laden hydrogels in vitro. Osteoarthritis Cartilage 22:1148–1157
25. Barbero A, Grogan S, Schäfer D et al (2004) Age related changes in human articular chondrocyte yield, proliferation and post-expansion chondrogenic capacity. Osteoarthritis Cartilage 12:476–484
26. Shahin K, Doran PM (2011) Strategies for enhancing the accumulation and retention of

extracellular matrix in tissue-engineered cartilage cultured in bioreactors. PLoS One 6(8):e23119. doi:10.1371/journal.pone.0023119

27. Hubka KM, Dahlin RL, Meretoja VV et al (2014) Enhancing chondrogenic phenotype for cartilage tissue engineering: monoculture and coculture of articular chondrocytes and mesenchymal stem cells. Tissue Eng Part B Rev 20:641–654
28. Mahmoudifar N, Doran PM (2005) Tissue engineering of human cartilage and osteochondral composites using recirculation bioreactors. Biomaterials 26:7012–7024
29. Vandesompele J, De Preter K, Pattyn F et al (2002) Accurate normalization of real-time quantitative RT-PCR data by geometric averaging of multiple internal control genes. Genome Biol 3(7), research0034.1–0034.11

Chapter 3

Mesenchymal Stem Cells Derived from Human Bone Marrow

Oliver F.W. Gardner, Mauro Alini, and Martin J. Stoddart

Abstract

Mesenchymal stem cells are found in a number of tissues and have the potential to differentiate into a range of mesenchymal lineages. This ready availability and multipotent character means that mesenchymal stem cells have become a focus for the field of tissue engineering, particularly for the repair of bone and cartilage.

This chapter describes the isolation of mesenchymal stem cells from human bone marrow tissue, as well as expansion of the cells and characterisation of their multipotency.

Key words Mesenchymal stem cell, Osteogenesis, Chondrogenesis, Adipogenesis

1 Introduction

Mesenchymal stem cells (MSCs) can be isolated from a number of tissues and are characterized by their ability to differentiate into a number of different mesenchymal cell lineages [1, 2]. The accepted definition of an MSC is that prescribed by the International Society for Cellular Therapy: an MSC must be plastic adherent, 95 % or more of the cells of a colony must express CD105, CD73, and CD90, less than 2 % of the cells may express CD45, CD434, CD14, CD11b, CD79α, CD19, or HLA class II, and an MSC must be capable of forming cells of at least the osteogenic, chondrogenic, and adipogenic lineages [3].

The first description of MSCs came in the form of colony forming fibroblasts (CFU-F) described by Friedenstein et al. in the early 1970s [4]. Friedenstein and colleagues produced a series of papers characterizing a small population of clonal fibroblastic bone marrow cells which could be isolated from bone marrow using their ability to adhere to culture vessels and showed their capacity for osteogenic differentiation [4–6]. Further work on these cells showed that they had the ability to differentiate down different cell lineages and in 1999 Pittenger et al. showed that there was a

Pauline M. Doran (ed.), *Cartilage Tissue Engineering: Methods and Protocols*, Methods in Molecular Biology, vol. 1340, DOI 10.1007/978-1-4939-2938-2_3, © Springer Science+Business Media New York 2015

homogeneous population of cells that could be isolated using adhesion that were capable of differentiation down the osteogenic, chondrogenic, and adipogenic lineages.

The isolation of MSC populations is based on their adherence and subsequent proliferation when placed into culture flasks. Most isolation techniques use either a density separation step or a red blood cell lysis buffer to separate the mononuclear fraction of cells from the whole bone marrow [7]. The mononuclear fraction can then be collected and the number of cells counted before plating onto tissue culture plastic. A number of the cell types within the mononuclear fraction will initially adhere, but only fibroblast-like MSCs will subsequently proliferate. As a result, the removal of media during medium changes removes any remaining non-MSC cells.

The medium used to expand MSCs after isolation contains fibroblast growth factor 2 (FGF2 or bFGF). The reason for this is that expansion in the presence of FGF2 leads to an increased rate of cell proliferation, an increased life span of the cells, and an improved retention of multi-lineage differentiation over the course of expansion [8, 9]. The effect of FGF2 on MSCs in culture may be due to the selection of a pluripotent subpopulation of cells which are at a more progenitor-like stage of development, rather than those that may already have a predisposition towards a particular lineage [8]. Despite the wide use of FGF2 in MSC expansion there is not an optimum concentration that has been empirically determined: the protocol in this chapter will use FGF2 at a concentration of 5 ng/ml.

Serum (the liquid that remains after collected blood is allowed to clot) is commonly used in cell culture as a source of growth factors, hormones and nutrients that allow for the culture of cells in vitro. As serum is collected from humans or animals the levels of these constituents can vary greatly from batch to batch [10]. These changes can have dramatic effects on the proliferation and subsequent differentiation of MSCs as demonstrated by the varying degree of cartilage matrix deposition by MSCs expanded in media containing different batches of serum (*see* Fig. 1). This means that each batch of serum should be tested before being used (Subheading 3.8). This involves characterization of the isolation, expansion, and differentiation of MSCs in the serum to show that cells can be isolated, expanded, and differentiated down the bone, cartilage, and adipogenic lineages as would be expected.

2 Materials

2.1 MSC Isolation

1. Alpha-MEM + 10 % FBS: Alpha-MEM, 10 % MSC-qualified fetal bovine serum (FBS) (*see* **Note 1**), 25 ml/l HEPES, and 1 % penicillin–streptomycin.

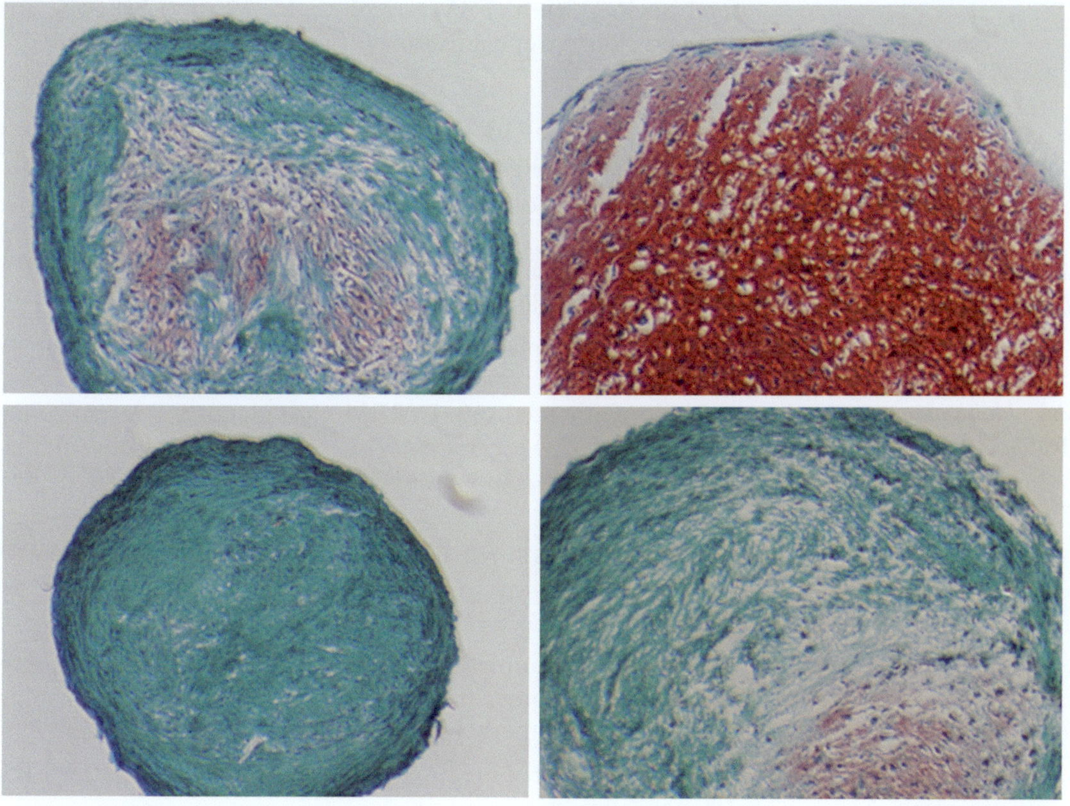

Fig. 1 Human bone marrow derived MSCs were isolated by way of attachment to tissue culture plastic and monolayer expanded. Fresh bone marrow from one donor was exposed to four different sera during both the isolation and expansion. The harvested cells then underwent chondrogenesis in defined serum-free conditions (Subheading 3.4.1). The chondrogenic response, as defined by glycosaminoglycan accumulation (*orange*), varied greatly between the sera

2. Phosphate buffered saline, calcium and magnesium free (PBS).
3. Ficoll.
4. Red blood cell lysis buffer: 0.15 M ammonium chloride, 10 mM potassium bicarbonate, and 0.1 mM EDTA in dH_2O.
5. Methylene blue solution: 0.3 % (w/v) dye, alkaline according to Löffler (Sigma-Aldrich).
6. FGF2 solution: 50 μg/ml FGF2 in 150 mM NaCl, 0.1 % w/v bovine serum albumin, and 5 mM TRIS in ddH_2O (10,000× stock).
7. Trypsin–EDTA solution: 0.05 % trypsin–EDTA in phosphate buffered saline.
8. Trypan blue solution: 0.4 % trypan blue in phosphate buffer saline.

2.2 MSC Expansion

1. Alpha-MEM + 10 % FBS: Alpha-MEM, 10 % MSC-qualified fetal bovine serum (FBS) (*see* **Note 1**), 25 ml/l HEPES, and 1 % penicillin–streptomycin.
2. 1× PBS.
3. FGF2 solution: 50 μg/ml in 150 mM NaCl, 0.1 % w/v bovine serum albumin, and 5 mM TRIS in ddH_2O (10,000× stock).
4. Trypsin–EDTA solution: 0.05 % in phosphate buffered saline.

2.3 MSC Cryopreservation

1. Cryopreservation buffer: 92 % MSC-qualified FBS and 8 % dimethyl sulfoxide (DMSO).
2. Cryotubes, 1.0 ml.

2.4 Multipotency Testing

1. Chondrogenic medium: DMEM high glucose, 1 % penicillin–streptomycin, 50 μg/ml ascorbic acid, 1×10^{-7} M dexamethasone, 1 % ITS+, 1 % nonessential amino acids, 10 ng/ml TGF-β1.
2. Chondrogenic control medium: DMEM high glucose, 1 % penicillin–streptomycin, 50 μg/ml ascorbic acid, 1×10^{-7} M dexamethasone, 1 % ITS+, 1 % nonessential amino acids.
3. Adipogenic medium: DMEM high glucose, 10 % FBS, 1 % penicillin–streptomycin, 1 μM dexamethasone, 10 μM insulin, 0.5 mM isobutylmethylxanthine, 0.2 mM indomethacin.
4. Adipogenic control medium: DMEM high glucose, 10 % FBS, 1 % penicillin–streptomycin.
5. Osteogenic medium: DMEM low glucose, 10 % FBS, 1 % penicillin–streptomycin, 50 μg/ml ascorbic acid, 10 nM dexamethasone, 5 mM betaglycerol-2-phosphate.
6. Osteogenic control medium: DMEM low glucose, 10 % FBS, 1 % penicillin–streptomycin.
7. PBS.
8. Methanol solution: 70 % methanol in ddH_2O at 4 °C.

2.5 Histological Evaluation of Chondrogenesis, Adipogenesis, and Osteogenesis

1. Methanol solution: 70 % methanol in ddH_2O.
2. PBS containing 5 % sucrose.
3. Cryocompound.
4. Safranin O solution: 0.1 % Safranin O in ddH_2O.
5. Weigert's hematoxylin.
6. Acetic acid solution: 1 % acetic acid in ddH_2O.
7. Fast Green solution: 0.02 % Fast Green in 0.1 % acetic acid.
8. Ethanol solution: 96 % in ddH_2O.
9. 100 % ethanol.
10. Xylene.

11. Eukitt mounting medium.
12. PBS.
13. Formalin solution: 4 % formalin in 0.1 M Phosphate buffer, pH 7.4.
14. Oil Red O solution: After diluting 1 % Oil Red O in isopropanol, mix three parts of the isopropanol solution with two parts of ddH_2O. Leave for 1 h then filter on paper.
15. Alizarin Red stain solution: Dilute 1 % w/v Alizarin Red in 0.5 N ammonium hydroxide, pH 4.1, in distilled water. Filter on paper.

3 Methods

3.1 MSC Isolation

If the marrow has been stored in a refrigerator it should be allowed to reach room temperature. Alpha-MEM + 10 % FBS should be heated up to 37 °C in a water bath.

Note should be taken of any donor information provided with the sample, e.g. date of birth, sex or the anatomical location of harvest (*see* **Note 2**).

3.1.1 MSC Harvest from Bone Marrow: Method 1

Before isolation begins, Ficoll should be allowed to warm up to room temperature.

1. Measure the volume of marrow with a pipette.
2. Transfer into a 50-ml Falcon tube (*see* **Note 3**).
3. Add a volume of PBS equal to twice the volume of the marrow sample. Pipette up and down repeatedly to remove any lumps.
4. Pass through a 70-μm cell strainer (*see* **Note 4**).
5. Wash the 50-ml Falcon tube with a volume of PBS equal to the original marrow volume and then pass this through the cell strainer as well. This results in a mixture of one part marrow to three parts PBS.
6. Add 2.6 ml of Ficoll per milliliter of original bone marrow to a 50-ml Falcon tube (*see* **Note 5**).
7. Use a 10-ml pipette to very slowly run the marrow–PBS mixture down the side of the tube to apply it on top of the Ficoll. Take care not to mix the layers.
8. Centrifuge at 800 × *g* for 20 min with the acceleration and brake on the centrifuge set to the lowest setting.
9. Remove the bulk of the top layer, leaving easy access to the interphase.
10. Use a pipette to collect the interphase.
11. After collection, transfer the harvested interphase to a 50-ml Falcon tube.

12. Add 5 ml of alpha-MEM + 10 % FBS per milliliter of collected interphase.
13. Centrifuge at 400 × *g* for 15 min under normal acceleration and braking.
14. Remove the supernatant and resuspend the cell pellet in 5 ml of alpha-MEM + 10 % FBS.
15. Repeat this wash step two more times with the same volume of alpha-MEM + 10 % FBS.
16. Resuspend the pellet in a volume of alpha-MEM + 10 % FBS according to pellet size: 5 ml for a small pellet, 10 ml for a normal pellet, and 20 ml for a large pellet (*see* **Note 6**).

3.1.2 MSC Harvest from Bone Marrow: Method 2

An alternative method of isolating the mononuclear fraction of the whole marrow is the use of a red blood cell lysis buffer. This approach uses ammonium chloride to lyse the erythrocyte population without damaging mononuclear cells [11]. This results in an increase in the number of harvested cells, as cells such as granulocytes, which would be removed by density centrifugation, are not removed by red cell lysis [11]. The behavior of mononuclear cells is comparable whether isolated using density centrifugation or red blood cell lysis in terms of the formation of colony forming units and the differentiation of MSCs after isolation [11].

The following method can be found in Horn et al. [11]:

1. Add marrow to red blood cell lysis buffer at a ratio of 2.8 ml of lysis buffer per ml of marrow.
2. Incubate for 10 min at room temperature on a shaker.
3. Centrifuge cells at 400 × *g* for 10 min under normal acceleration and braking.
4. Resuspend the cell pellet in 10 ml of pre-warmed PBS.
5. Centrifuge cells at 400 × *g* under normal acceleration and braking.
6. Resuspend the cells in 10 ml of alpha-MEM + 10 % FBS and count before seeding.

3.1.3 Counting Mononuclear Cells: Method 1

After resuspension, the number of isolated mononuclear cells needs to be determined. This can be done via manual counting using methylene blue:

1. Take 50 μl of the cell suspension.
2. Add 50 μl of methylene blue solution and incubate for 10 min at room temperature.
3. Count the stained cells using a hemocytometer. Only count the regular, evenly stained cells. The cells are now considered to be at passage 0.

3.1.4 Counting Mononuclear Cells: Method 2

Alternatively, the number of isolated mononuclear cells may be determined using a Sceptre cell counter (Merck Millipore). To count with the Sceptre system:

1. Dilute the cell suspension 1:20 in PBS (50 μl of cell suspension: 950 μl of PBS).
2. Take a sample using the Sceptre cell counter and a 40-μl tip.
3. Once the measurement has been made, set the lower gate to 8 μm and then note the cell count. Setting the gate in this way excludes any non-mononuclear cells. The cells are now considered to be at passage 0.

3.2 Adhesion and Expansion

After counting, the mononuclear fraction should be plated into culture flasks for adhesion and expansion. These cells are considered to be passage 0 (P0) as they have not yet been exposed to an enzymic harvest step.

1. Seed cells at a density of approximately 50,000 cells/cm^2 in alpha-MEM + 10 % FBS and 0.1 ng/ml FGF2. Use 1.6 ml of medium per cm^2.
2. Following seeding, leave the flask unmoved for 4 days to allow cells to attach.
3. After this initial period, change the medium three times a week until the cells are 80 % confluent (approximately 10–14 days). Care should be taken when handling the flasks and removing/adding liquids to ensure attachment of cells.
4. Once the cells are ready, wash with PBS (1.6 ml/cm^2).
5. Trypsinize cells using 0.05 % trypsin–EDTA (0.5 ml/cm^2). Monitor cells and stop the enzymic reaction when most cells are rounded and floating.
6. Inactivate trypsin with an equal volume of alpha-MEM + 10 % FBS.
7. Centrifuge the cell suspension at 300 × *g* for 10 min.
8. Resuspend the cell pellet in 10 ml of alpha-MEM + 10 % FBS.
9. Count the cells using either a hemocytometer and trypan blue or a Sceptre cell counter using a 60-μm sensor (*see* Subheadings 3.1.3 and 3.1.4).

Cells can either be reseeded or cryopreserved for future use. Cells should be reseeded at a density of 3000 cells/cm^2: they are now considered passage 1. The cells should be monitored carefully and the medium changed three times a week. Once the cells reach 80 % confluency they should be trypsinized and either reseeded or cryopreserved.

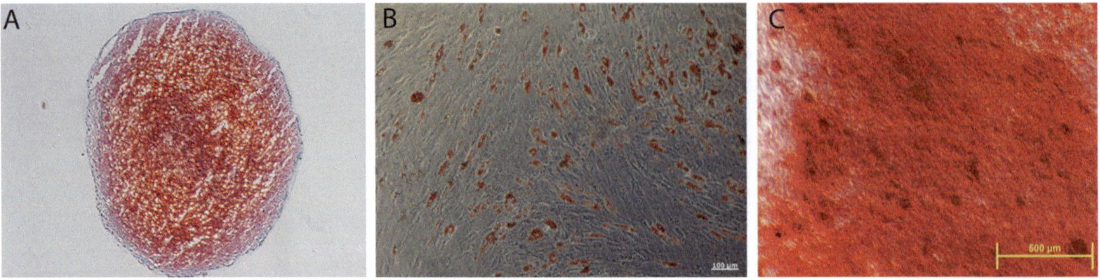

Fig. 2 Monolayer-expanded human MSCs were differentiated into the chondrogenic, adipogenic, and osteogenic phenotypes to assess multipotency. Chondrogenic differentiation (**a**) was assessed using Safranin O/Fast Green staining, adipogenic differentiation (**b**) was assessed using Oil Red O staining, and osteogenic differentiation (**c**) was assessed using Alizarin Red S staining

3.3 Cryopreservation of Expanded Cells

1. Centrifuge the cells for cryopreservation at 300 × *g* for 10 min.
2. Resuspend the cell pellet in cryopreservation buffer at a concentration of two million cells/ml.
3. Fill cryotubes with 1 ml of cell suspension and place in a controlled freezing container that cools at a rate of 1 °C/min (such as Mr Frosty) at −80 °C for 48–72 h.
4. Transfer to liquid nitrogen storage.

3.4 Multipotency Testing

The multipotent capacity of harvested MSCs can be tested using the following differentiation assays for chondrogenesis, adipogenesis, and osteogenesis (*see* Fig. 2).

3.4.1 Chondrogenic

For chondrogenic differentiation in pellet cultures:

1. Trypsinize and count cells for use.
2. Place 750,000 cells in each of two separate 50-ml Falcon tubes.
3. Centrifuge cells at 400 × *g* for 10 min under normal acceleration and braking.
4. Resuspend the cell pellet with 3 ml of medium to produce 250,000 cells/ml using chondrogenic medium for one tube and chondrogenic control medium for the second tube.
5. Place 1 ml of cell suspension into each of three 1.5-ml Eppendorf tubes for both the control and chondrogenic groups.
6. Spin in a microcentrifuge at 400 × *g* for 10 min.
7. Leave for 3 days to round up, then gently tip the tubes to remove the cell pellets from the sides of the tubes.
8. Aspirate medium and replace with 1 ml of fresh medium. Change media 3× a week (remove and replace all 1 ml). Care should be taken not to touch the pellets.

9. Terminate cultures on Day 21. Aspirate medium and add 1 ml of PBS.
10. Remove PBS and add 1 ml of 70 % methanol solution at 4 °C.
11. Pellets should be left for a minimum of 24 h at 4 °C to fix before sectioning and histological analysis (Subheading 3.5.1).

3.4.2 Adipogenic

For adipogenic differentiation:

1. Plate cells at a density of 50,000 cells/cm^2 in a 24-well plate (100,000 cells/well).
2. Add 200 μl of adipogenic medium to stimulated cells, or 200 μl of adipogenic control medium to unstimulated controls.
3. Refeed cells three times per week.
4. Terminate cultures on Day 21 for fixing and histological analysis (Subheading 3.6).

3.4.3 Osteogenic

For osteogenic differentiation:

1. Plate cells at a density of 20,000 cells/cm^2 in a 24-well plate (40,000 cells/well).
2. Per well, add 500 μl of osteogenic medium to stimulated cells, or 500 μl of osteogenic control medium to unstimulated controls.
3. Refeed cells three times per week.
4. Terminate cultures on Day 21 for fixing and histological analysis (Subheading 3.7).

3.5 Evaluation of Chondrogenic Differentiation

Safranin O is used to detect the presence of sulfated proteoglycan, a classical marker of chondrogenesis, in cryosectioned pellets. Fast green provides a counterstain.

3.5.1 Cryosectioning of Pellets

1. Remove pellets from 70 % methanol solution and place in PBS containing 5 % sucrose at least 12 h before sectioning.
2. In order to section, place pellets into a mold filled with cryocompound and then freeze.
3. After freezing, the block containing the pellet should be frozen to a chuck before being mounted on a cryostat.
4. Section the mounted pellets all the way through in 12-μm sections. Collect the sections on microscope slides.
5. Once cut, select the middle sections for staining.

3.5.2 Safranin O/Fast Green Staining

Let the slides reach room temperature before staining begins, otherwise the sections may peel off the slides. Whilst the slides are coming up to temperature, prepare the stain solutions and place them into staining baths. Once the slides are up to temperature and the solutions are prepared, the following stain can be performed.

1. Place in distilled water for 10 min to remove the cryocompound. Change the water two or three times.
2. Submerge in Weigert's hematoxylin for 12 min.
3. Place in lukewarm tap water for 10 min to "blue" the hematoxylin. Change the water two or three times.
4. Dip into a bath of distilled water to remove remaining tap water.
5. Submerge in Fast Green solution for 5 min.
6. Place in 1 % acetic acid solution for 30 s.
7. Submerge in Safranin O solution for 5 min.
8. Place in 96 % ethanol solution for 1 min. Repeat this step.
9. Place in 100 % ethanol for 2 min. Repeat this step.
10. Place in xylene for 2 min and then transfer to xylene for coverslipping (*see* **Note 7**).

3.5.3 Coverslipping

1. Remove the slide for coverslipping from the xylene bath using flat-ended forceps. Let the excess xylene run off and place the slide on a paper towel.
2. Use a crucible to pour Eukitt mounting medium onto the slide.
3. Take an appropriately sized cover slip and dip the edge in xylene.
4. Place the xylene-covered edge on the glass slide and allow it to contact the mounting medium.
5. Align the edge of the coverslip with the near edge of the glass slide and lower it gently towards the far edge of the slide, avoiding the formation of any bubbles (*see* **Note 8**).
6. Allow to dry for 24 h.

3.6 Evaluation of Adipogenic Differentiation

Oil Red O staining is used to detect lipid-filled vacuoles as a marker for adipogenesis [12]:

1. Remove culture media.
2. Wash cells twice with 1 ml of PBS.
3. Add 4 % formalin solution and incubate at room temperature for 10 min.
4. Remove formalin and wash two times with PBS.
5. Add Oil Red O solution and incubate for 30 min at room temperature.
6. Wash wells with PBS until the negative wells are clear.

3.7 Evaluation of Osteogenic Differentiation

Alizarin Red staining is used to detect calcium deposits as a marker of osteogenesis [12]:

1. Aspirate media and wash twice with 1 ml of PBS.
2. Add 1 ml of 4 % neutral buffered formalin solution and fix at room temperature for 10 min.
3. Aspirate formalin and wash two times with 1 ml of PBS.
4. Apply 1 ml of Alizarin Red solution and leave for 30 min at 37 °C.
5. Wash repeatedly with distilled water until stain is removed from the negative control wells.

3.8 Serum Testing

It is very important to test the effect of each batch of serum on the isolation, growth and multilineage potential of MSCs. In order to do this, every step from adhesion during isolation to differentiation must be carried out with each batch of serum.

1. To test adhesion, seed the resuspended mononuclear fraction of cells (Subheading 3.1.1 or 3.1.2) into culture flasks containing an equal volume of alpha-MEM + 10 % FBS, with each flask containing FBS from each individual batch being tested.
2. Repeatedly passage the cells to determine the kinetics of cell growth in culture. Count the number of cells at each passage and determine the number of population doublings over the course of the culture:

$$\text{Number of doublings} = 3.32\left[\log(\text{cells counted at passage}) - \log(\text{cells seeded into flask})\right]$$

3. As only one flask needs to be reseeded at each passage for the determination of cell kinetics, use the remaining cells collected at passage 1 to test the multilineage potential of each cell line (Subheadings 3.4–3.7). For osteogenic differentiation this will require the seeding of three wells for culture in differentiation media and three wells for culture in control media (Subheading 3.4.3), this is also the same for adipogenic differentiation (Subheading 3.4.2). Confirmation of chondrogenic differentiation capacity will require three pellets for culture in control media and three pellets for culture in chondrogenic media containing 10 ng/ml TGF-β1 (Subheading 3.4.1).

The results of the differentiation assays should show that the serum batch will allow for differentiation of cells down all three tested lineages and will not favor one in particular.

The growth kinetics data will show the behavior of the cells over the course of extended culture and will show if a serum leads to premature slowing of cell proliferation. It is useful if possible to perform a characterization of a “current” previously tested batch of serum against a potential new batch to provide an idea of how the cells should behave.

4 Notes

1. All serum to be used for monolayer expansion should be tested to ensure the maintenance of multipotency.
2. Depending on the time of arrival, marrow can be kept overnight in the fridge.
3. Marrow taken from one patient but collected in different tubes can be pooled at this point.
4. Pushing the pipette tip against the membrane makes passing the sample through the strainer easier.
5. If the volume is small, a 15-ml tube can be used to increase the size of the interphase.
6. The size of pellet is a relatively arbitrary quantification that becomes more apparent with experience.
7. All work with xylene should be performed in a fume hood using nitrile gloves.
8. If bubbles do form they may be moved away from sections by applying gentle pressure with a pair of forceps or a coverslip.

References

1. Barry FP, Murphy JM (2004) Mesenchymal stem cells: clinical applications and biological characterization. Int J Biochem Cell Biol 36: 568–584
2. Pittenger MF, Mackay AM, Beck SC et al (1999) Multilineage potential of adult human mesenchymal stem cells. Science 284:143–147
3. Dominici M, Le BK, Mueller I et al (2006) Minimal criteria for defining multipotent mesenchymal stromal cells. The International Society for Cellular Therapy position statement. Cytotherapy 8:315–317
4. Friedenstein AJ, Gorskaja JF, Kulagina NN (1976) Fibroblast precursors in normal and irradiated mouse hematopoietic organs. Exp Hematol 4:267–274
5. Friedenstein AJ, Chailakhjan RK, Lalykina KS (1970) The development of fibroblast colonies in monolayer cultures of guinea-pig bone marrow and spleen cells. Cell Tissue Kinet 3: 393–403
6. Friedenstein AJ, Chailakhyan RK, Latsinik NV et al (1974) Stromal cells responsible for transferring the microenvironment of the hemopoietic tissues. Cloning in vitro and retransplantation in vivo. Transplantation 17:331–340
7. Haynesworth SE, Goshima J, Goldberg VM et al (1992) Characterization of cells with osteogenic potential from human marrow. Bone 13:81–88
8. Bianchi G, Banfi A, Mastrogiacomo M et al (2003) Ex vivo enrichment of mesenchymal cell progenitors by fibroblast growth factor 2. Exp Cell Res 287:98–105
9. Tsutsumi S, Shimazu A, Miyazaki K et al (2001) Retention of multilineage differentiation potential of mesenchymal cells during proliferation in response to FGF. Biochem Biophys Res Commun 288:413–419
10. Honn KV, Singley JA, Chavin W (1975) Fetal bovine serum: a multivariate standard. Proc Soc Exp Biol Med 149:344–347
11. Horn P, Bork S, Diehlmann A et al (2008) Isolation of human mesenchymal stromal cells is more efficient by red blood cell lysis. Cytotherapy 10:676–685
12. Gregory CA, Prockop DJ (2007) Fundamentals of culture and characterisation of mesenchymal stem/progenitor cells (MSCs) from bone marrow stroma. In: Freshney RI, Stacey GN, Auerbach JM (eds) Culture of human stem cells. Wiley, Hoboken, NJ, pp 207–232

Chapter 4

Mesenchymal Stem Cells Derived from Human Adipose Tissue

Nastaran Mahmoudifar and Pauline M. Doran

Abstract

Human adult mesenchymal stem cells are present in fat tissue, which can be obtained using surgical procedures such as liposuction. The multilineage capacity of mesenchymal stem cells makes them very valuable for cell-based medical therapies. In this chapter, we describe how to isolate mesenchymal stem cells from human adult fat tissue, propagate the cells in culture, and cryopreserve the cells for tissue engineering applications. Flow cytometry methods are also described for identification and characterization of adipose-derived stem cells and for cell sorting.

Key words Mesenchymal stem cells, Human adipose tissue, Stem cell culture, Cell isolation, Cell expansion, Flow cytometry

1 Introduction

Stem cells are valuable tools in tissue engineering and other cell-based medical therapies because they have the capacity to differentiate into various lineages. Tissue-derived mesenchymal stem cells have many practical advantages for cartilage tissue engineering compared with other stem cell types. Use of mesenchymal stem cells avoids ethical concerns over embryo harvesting and the safety issues relating to tumor formation in recipient patients by embryonic and induced pluripotent stem cells [1, 2]. There is increasing evidence of an additional benefit of immune-privilege, in that allogeneic mesenchymal stem cells fail to activate host immune responses that are typically responsible for rejection of implanted cells and organs [3, 4]. This feature reflects the role of mesenchymal stem cells in early tissue repair and remodeling when inflammation control is required; it also overcomes the immune-rejection problems associated with embryonic stem cells [5]. Further research is required to understand the mechanisms behind the

Pauline M. Doran (ed.), *Cartilage Tissue Engineering: Methods and Protocols*, Methods in Molecular Biology, vol. 1340, DOI 10.1007/978-1-4939-2938-2_4, © Springer Science+Business Media New York 2015

immunomodulatory and immunosuppressive functions of mesenchymal stem cells; nevertheless, these properties could open the way for clinical application of "off-the-shelf" replacement organs and tissues produced using cells that are not patient-specific.

Human adult mesenchymal stem cells are present in abundance in adipose tissue, which can be obtained as a waste material from liposuction surgery. The availability of stem cells in fat is much greater than in bone marrow. Mesenchymal stem cells comprise about 2 % of nucleated cells in lipoaspirate compared with only 0.001–0.004 % in bone marrow, the number of stem cells per milliliter of lipoaspirate is about eightfold higher than in bone marrow, and the volume of lipoaspirate typically obtained under local anesthesia is at least fivefold greater than is possible for bone marrow [6]. The relative ease with which large numbers of adipose-derived stem cells can be obtained for tissue engineering applications is an important advantage. The expression of cell surface markers used to identify cells with multilineage potential, such as CD90, CD105, and CD166, in adipose-derived stem cells is similar to that in stem cells obtained from bone marrow [6, 7].

Adipose-derived stem cells have been shown to differentiate along classical mesenchymal lineages towards cartilage, bone, muscle, and fat [8]. We have previously demonstrated chondrogenic [9] and osteogenic [10] differentiation of these cells in three-dimensional culture systems. In vitro differentiation into other cell types, including neuronal cells, cardiomyocytes, hepatocytes, pancreatic cells, and endothelial cells, has also been reported (reviewed in ref. 7), suggesting that adipose-derived stem cells have some degree of multilineage plasticity across different germ layers. Specific signaling molecules and culture conditions are required to achieve stem cell differentiation, extracellular matrix synthesis, and maintenance of the differentiated cell phenotype. For chondrogenesis and cartilage production, growth factors in the transforming growth factor β superfamily (TGF-βs) play an especially important role [11].

In this chapter, we describe methods to isolate stem cells from human adipose tissue, culture the isolated cells in monolayer to increase cell numbers, and freeze and store the cells in liquid nitrogen for future use. Flow cytometry techniques are also described for identification and purification of the stem cells. The cells isolated and characterized using these methods are suitable for subsequent application of growth factors to induce chondrogenesis, and for seeding and culture in three-dimensional scaffold systems.

2 Materials

2.1 Cell Isolation and Expansion

1. Human adipose tissue in the form of waste aspirate from liposuction surgery (*see* **Note 1**).

2. Antibiotic solution: 10,000 U/mL of penicillin, 10 mg/mL of streptomycin, and 25 μg/mL of amphotericin B. Divide into 10-mL portions, freeze and store at −20 °C.
3. DMEM base medium: DMEM (Dulbecco's modified Eagle's medium) containing 4500 mg/L of glucose and 584 mg/L of glutamine with 3.5 g of sodium hydrogen carbonate and 2.38 g of HEPES (*N*-(2-hydroxyethyl)piperazine-*N'*-2-ethane sulfonic acid) made up to 890 mL in Milli-Q water, pH 7.2. Filter sterilize using a 0.2-μm filter and store at 4 °C. This solution is used as a basis for preparing tissue collection medium and cell culture medium.
4. Tissue collection medium: DMEM base medium with 10 mL of antibiotic solution added before use.
5. PBS (Dulbecco's phosphate buffered saline): 0.2 g of KCl, 0.2 g of KH_2PO_4, 8 g of NaCl, and 2.16 g of $Na_2HPO_4{\cdot}7H_2O$ per liter of solution in Milli-Q water, pH 7.2. Sterilize at 121 °C (15 psi) for 20 min, cool at room temperature, and store at 4 °C.
6. Collagenase solution: 13.5 mg/mL of collagenase type IA in PBS (*see* **Note 2**). Dissolve the collagenase in PBS at room temperature. Filter the solution using a 0.45-μm syringe-driven filter unit to remove undissolved materials, and then filter sterilize using a 0.2-μm syringe-driven filter unit.
7. Digestion solution: Prepare 300 mL of digestion solution fresh before use for every 100 mL of washed and dissected adipose tissue. First, dissolve 6 g of BSA (bovine serum albumin) in 267 mL of PBS and filter sterilize using a 0.2-μm filter. Add 30 mL of collagenase solution and 3 mL of antibiotic solution (*see* **Note 3**).
8. FBS (fetal bovine serum): Divide into 100-mL portions, freeze and store at −20 °C.
9. Cell culture medium: DMEM base medium with 100 mL of FBS and 10 mL of antibiotic solution added before use.
10. Lysis buffer: 8.24 g of NH_4Cl, 1 g of $KHCO_3$, and 38 mg of $Na_4EDTA{\cdot}2H_2O$ made up to 1 L in Milli-Q water, pH 7.4. Filter sterilize using a 0.2-μm filter and store at 4 °C.
11. Trypsin-EDTA solution: 0.25 % w/v trypsin and 0.02 % w/v EDTA. Dissolve 40 mg of $Na_4EDTA{\cdot}2H_2O$ in 100 mL of PBS (*see* **Note 4**). Dissolve 0.5 g of trypsin (Sigma: 1000–2000 U/mg solid) in 100 mL of DMEM base medium. Combine the two solutions together and filter sterilize using a 0.2-μm filter. Divide into 10-mL portions, freeze and store at −20 °C.
12. Trypan Blue solution: 0.81 g of NaCl and 0.4 g of Trypan Blue in 100 mL of Milli-Q water. Store at room temperature.

2.2 Cell Cryopreservation

1. Freezing solution 1: 20 % v/v DMEM base medium and 80 % v/v FBS. Under sterile conditions, add 2 mL of DMEM base medium to 8 mL of FBS. Prepare fresh before use and store on ice until needed.
2. Freezing solution 2: 20 % v/v DMSO (dimethylsulfoxide) and 80 % v/v FBS. Under sterile conditions, add 2 mL of DMSO to 8 mL of FBS (*see* **Note 5**). Prepare fresh before use and store on ice until needed.

2.3 Flow Cytometry

1. Paraformaldehyde solution: 4 % w/v paraformaldehyde in PBS. Weigh 4 g of paraformaldehyde in a 250-mL conical flask (*see* **Note 6**). Add 100 mL of PBS to the flask, cover the top with parafilm, and place the flask on a magnetic stirrer under a fume hood. Stir the contents while heating (low heat) until the paraformaldehyde is dissolved and the solution becomes clear (*see* **Note 7**). Switch off the heater and continue to stir until cool. Pour the solution into a screw-cap bottle and store at 4 °C.
2. Flow cytometry buffer (FCB): 2 mL of FBS and 0.2 g of Tween-20 made up to 100 mL in PBS. Use FCB for diluting antibodies and as a washing buffer.
3. Blocking buffer: 0.2 mL of FBS and 0.2 g of BSA made up to 10 mL in PBS. Use blocking buffer to block nonspecific sites.
4. Primary antibody solutions: Dilute primary antibodies just before use.
   (a) Anti-vimentin antibody solution: 1:100 mouse monoclonal anti-human antibody against vimentin (clone V9, Zymed Laboratories) in FCB. Add 10 μL of antibody to 990 μL of FCB in a 1.5-mL microcentrifuge tube and aspirate gently to mix.
   (b) Anti-CD90 antibody solution: 1:100 mouse monoclonal anti-human antibody against CD90 (Thy-1) (clone F15-42-1, Biosource International) in FCB. Add 10 μL of antibody to 990 μL of FCB in a 1.5-mL microcentrifuge tube and aspirate gently to mix.
   (c) Anti-SMA (smooth muscle actin) antibody solution: 1:50 mouse monoclonal anti-human antibody against SMA (clone IA4, Zymed Laboratories) in FCB. Add 20 μL of antibody to 980 μL of FCB in a 1.5-mL microcentrifuge tube and aspirate gently to mix.
   (d) Anti-Factor VIII antibody solution: Mouse monoclonal anti-human antibody against Factor VIII (clone Z002, Zymed Laboratories) as diluted by the manufacturer.
5. Secondary antibody solution: Rabbit anti-mouse FITC-conjugated IgG (H+L) (Zymed Laboratories).

3 Methods

3.1 Isolation of Mesenchymal Stem Cells from Human Adipose Tissue

1. Transfer the human adipose tissue to the laboratory in screw-cap bottles on ice (*see* **Note 1**). Process the adipose samples immediately if possible. Otherwise, store the bottles at 4 °C and process the samples on the same day (*see* **Note 8**).
2. In a biosafety cabinet, pour off the medium from each bottle leaving the settled adipose tissue at the bottom, and add a volume of PBS similar to the volume of the retained tissue. Resuspend the tissue and then pour off the wash solution. Wash at least four times using PBS until the solution becomes clear of blood.
3. Using a sterile spatula, place 10–20 mL of washed adipose tissue in a sterile 100-mm disposable Petri dish. Dissect the yellow adipose tissue free from the white fibrous tissue and blood vessels using disposable sterile scalpels. Mince the adipose tissue finely using sterile surgical scissors.
4. Measure and make a note of the total volume of the dissected adipose tissue. Place 20-mL aliquots (*see* **Note 9**) of the dissected tissue into 250-mL conical flasks (*see* **Note 10**), each containing 60 mL of digestion solution. Place the conical flasks in a water bath shaker at 37 °C and shake at 20 rpm for 60–90 min to digest the tissue (*see* **Note 11**).
5. Pour the contents of the conical flasks into 50-mL centrifuge tubes and centrifuge the cell suspension for 5 min at 400 × *g* (*see* **Note 12**). Decant the liquid fat layer and floating adipocytes. Wash the pellet in each tube by resuspending it in 10 mL of cell culture medium. Centrifuge the tubes again.
6. Resuspend the pellet in each centrifuge tube in 20 mL of lysis buffer. Incubate the tubes at room temperature for 10 min to lyse any red blood cells.
7. Centrifuge the tubes to remove the lysed cells and resuspend the pellets in DMEM base medium. Filter the cell suspension in all centrifuge tubes through a sterile 150-μm nylon filter to remove any residual tissue debris. Collect the cells in the filtrate.
8. Centrifuge the cell suspension (filtrate). Combine and wash the pellets (*see* **Note 13**) at least twice using DMEM base medium to remove the lysis buffer.
9. Resuspend the pellet in 5–10 mL of DMEM base medium. Combine 200 μL of cell suspension with 200 μL of Trypan Blue solution in a 1.5-mL microcentrifuge tube. Leave for 1–2 min, introduce into a hemocytometer, and perform a cell count using an inverted phase-contrast light microscope (*see* **Note 14**). Calculate the viable cell yield per volume of tissue by dividing the total number of viable cells by the volume of the dissected adipose tissue measured in **step 4** (*see* **Note 15**).

3.2 Cell Culture and Expansion

1. Culture the cells in 225-cm^2 T-flasks at a density of 20,000 cells per cm^2 using 35 mL of cell culture medium in each flask. Place the flasks in a CO_2 incubator at 37 °C and 95 % humidity. Leave the flasks undisturbed for 3 days to allow the cells to adhere to the bottom surface.
2. Remove the medium after 3 days, rinse the cell monolayers using PBS to eliminate nonadherent cells, and add fresh medium. Change the medium every 3 days until the cells are confluent (about 2 weeks).
3. Rinse the monolayers twice with PBS to remove FBS.
4. Add 8–10 mL of trypsin-EDTA solution to each flask and incubate at 37 °C for 8 min. Remove most of the trypsin solution containing the cells using a pre-wetted pipette, leaving a residual 1 mL of solution in each flask. Transfer the removed trypsin solution to a 50-mL centrifuge tube containing 10 mL of cell culture medium to neutralize the trypsin. Place the centrifuge tubes on ice.
5. Incubate the remaining 1 mL of trypsin solution in the flasks at 37 °C for 12 min. Place each flask on an inverted phase-contrast light microscope and observe the cells (*see* **Note 16**). Tap the bottom of each flask to assist with cell detachment: incubate for longer if necessary for the cells to detach. Add 10 mL of cell culture medium to each flask to neutralize the trypsin, collect the cells using a pre-wetted pipette, and combine with the cells in the 50-mL centrifuge tube on ice. Rinse each flask using 10 mL of DMEM base medium and combine with the cells in the 50-mL centrifuge tube.
6. Centrifuge the cell suspension for 5 min at 400 × *g*. These cells are designated passage zero (P0) cells (*see* **Note 17**). They can be expanded to increase cell numbers, or frozen and stored in liquid nitrogen for later use.
7. To expand the cells, culture P0 cells in cell culture medium in 225-cm^2 T-flasks in a CO_2 incubator at 37 °C and 95 % humidity for two passages, dividing the cells from each flask into three flasks (split ratio of 1:3) for each passage. Change the medium every 3 days until the cells become confluent (*see* **Note 18**).

3.3 Cryopreservation of the Cells

1. Resuspend the pellet harvested from one 225-cm^2 T-flask in 0.9 mL of freezing solution 1 and place on ice.
2. Place 0.9 mL of freezing solution 2 in a 2-mL cryogenic vial. Add the cell suspension (in solution 1) to solution 2, using a pre-wetted 1-mL pipette. Close the vial, place in a freezing container, freeze overnight at −80 °C, and then store in liquid nitrogen until needed.

3. To culture frozen cells, thaw the cells quickly by placing the vials in a water bath at 37 °C. Pour the contents of each vial into a centrifuge tube containing cell culture medium prewarmed to 37 °C. Divide the cell suspension between three 225-cm^2 T-flasks (split ratio 1:3). Change the medium completely the next day to remove DMSO, and then every 3 days until the cells become confluent.

3.4 Flow Cytometry

Flow cytometry is performed to assess the purity of the adipose-derived stem cell population at P0. Vimentin, which is an intermediate filament protein expressed in mesenchymal cells, and CD90, which is a cell surface marker for mesenchymal cells and fibroblasts, are used to identify stem cells. Antibody against smooth muscle actin (SMA) is used to identify smooth muscle cells and pericytes; antibody against Factor VIII is used to identify endothelial cells. Other positive and negative markers suitable for identifying mesenchymal stem cells may also be used.

1. Harvest P0 cells (Subheading 3.2) from three to four 225-cm^2 T-flasks to obtain approximately 1×10^7 cells (*see* **Note 17**). Perform a cell count as described in Subheading 3.1 and place 1×10^7 cells (*see* **Note 19**) in a 50-mL centrifuge tube. Centrifuge the cell suspension for 5 min at $400 \times g$ (*see* **Note 12**).
2. Wash the cells once using 10 mL of PBS. Loosen the pellet by gently tapping the tube several times and then resuspend the pellet using a 10-mL pipette. Centrifuge for 5 min at $400 \times g$.
3. Resuspend the cells in 10 mL of PBS and add 10 mL of ice-cold paraformaldehyde solution to give a final paraformaldehyde concentration of 2 % w/v. Gently mix the cell suspension using a 10-mL pipette and place the tube on ice for 15 min to fix the cells. Centrifuge for 5 min at $400 \times g$.
4. Wash the cells twice as described in **step 2** using 10 mL of FCB for each wash (*see* **Note 20**).
5. Resuspend the cells in 10 mL of blocking buffer and place the tube on ice for 15 min to block nonspecific sites.
6. Prepare primary antibody solutions as described in Subheading 2.3.
7. Centrifuge the tube from **step 5** and resuspend the pellet in 10 mL of FCB. Place 1-mL aliquots of cell suspension containing 10^6 cells into 10-mL centrifuge tubes. Centrifuge the tubes and add 1 mL of appropriate antibody solution to the cell pellet in each tube. Loosen the pellet and resuspend the cells as described in **step 2**. Incubate the tubes on ice for 40 min, agitating the tubes manually every 10 min to mix the cells with antibody. Include a sample of cells not stained with primary and secondary antibodies as a negative control. Include a sample of cells stained with secondary antibody only to assess nonspecific binding.

8. Centrifuge the tubes and wash the cells three times as described in **step 2**, using 1 mL of FCB for each wash. Aspirate the supernatant carefully using a Pasteur pipette. Avoid touching the pellet with the tip of the pipette.
9. Add 990 μL of FCB to each tube and resuspend the cells. Add 10 μL of secondary antibody solution to each tube to give a 1:100 dilution and incubate on ice for 30–40 min (*see* **Note 21**). Agitate the tubes manually every 10 min to mix the cells with antibody.
10. Centrifuge the tubes and wash the cells three times as described in **step 2**, using 1 mL of FCB for each wash. Aspirate the supernatant carefully using a Pasteur pipette. Avoid touching the pellet with the tip of the pipette.
11. Resuspend the cells in 1 mL of FCB and filter using a 38-μm cell strainer (*see* **Note 22**) into sterile 5-mL flow cytometry tubes.
12. Perform flow cytometry, e.g., using a Becton Dickinson flow cytometer (*see* **Note 23**).

As well as identifying and characterizing adipose-derived stem cells, flow cytometry can also be used for cell sorting. For example, CD90-positive cells may be sorted or isolated from the rest of the population to obtain a more pure starting cell line for further expansion. In this case, all solutions should be prepared sterile and all steps should be performed under aseptic conditions. Also, the cells should not be fixed or permeabilized as these treatments are harmful (*see* **Note 24**). The sorted cells are collected in cell culture medium on ice and cultured in T-flasks as described in Subheading 3.2.

4 Notes

1. Human lipoaspirate may be obtained from medical clinics specializing in subcutaneous adipose tissue removal. Approval from the local institutional research ethics committee is required before the tissue can be collected with informed consent. The lipoaspirate is harvested by clinical staff into a sterile container and refrigerated immediately after removal from the patient. Approximately 250 mL of lipoaspirate is placed into a 500-mL screw-cap bottle containing 200 mL of tissue collection medium. The tissue should be collected within 1–2 h of the surgery for transport to the laboratory in a sealed container on ice. After arrival in the laboratory, the material can be stored for a limited time at 4 °C.
2. The collagenase type IA we used is from Sigma (from *Clostridium histolyticum*, ≥125 U/mg of solid). However, the activity of this enzyme varies from batch to batch. We have purchased different batches of collagenase type IA with activities of 367, 495, 514, and 680 U/mg solid. Make

up the collagenase solution at 13.5 mg/mL if the enzyme activity is 367 or approximately 400 U/mg solid, otherwise adjust the concentration. For example, if the collagenase activity is 495 U/mg solid, make up the solution at a concentration of 10 mg/mL. In our work, the collagenase solution was made fresh each time it was required; however, the solution could be aliquoted and stored at −20 °C.

3. The final concentrations of BSA and collagenase in the digestion mixture, which is comprised of 300 mL of digestion solution plus 100 mL of adipose tissue, are 1.5 % w/v and 0.1 % w/v (approx. 400 U/mL), respectively.
4. Warm the PBS solution to help dissolve the EDTA, then cool it down before combining with the trypsin solution.
5. DMSO can be purchased sterile. However, if the bottle has been opened so that sterility may have been compromised, sterilize 10 mL of DMSO in an autoclave.
6. Paraformaldehyde is a skin irritant and suspected carcinogen and safety precautions must be observed while preparing this solution.
7. Avoid boiling and overflow of the paraformaldehyde solution.
8. It is possible to store the adipose tissue at 4 °C overnight and process it the next day, but the viable cell yield will decrease. The adipose tissue will float to the top when stored overnight and will not be submerged in nutrient medium, which could be one of the reasons for achieving a lower viable cell yield.
9. Use a 50-mL graduated centrifuge tube to make an approximate measurement of the volume of dissected tissue.
10. Coat the flasks with Coatasil glass treatment solution to prevent cell attachment to the glass. In a fume hood, pour about 5 mL of Coatasil into a conical flask. Coat the inside of the flask by moving the Coatasil around for 1–2 min. Pour the leftover Coatasil back into the bottle for reuse. Allow the Coatasil to evaporate, then wash and sterilize the flasks as usual.
11. The end of the digestion period is marked by a change in the appearance of the digestion mixture. Minced adipose tissue disappears, the fat becomes solubilised, and the digestion solution becomes turbid with cells.
12. Use of a swing-bucket centrifuge is recommended for all procedures involving cell centrifugation, as this produces a compact cell pellet that is easy to handle during subsequent washing steps. Use of fixed-angle rotors causes smearing of the cells on the tube wall, thus increasing the risk of cell loss during repeated washing steps.
13. The pellets are creamy white at this stage.
14. The basis of the Trypan Blue staining test is that viable cells exclude the dye whereas dead cells are permeable and take up the stain. The number of viable cells in the sample is the

difference between the total number of cells and the number of stained cells counted under the microscope.

15. In our work, the yield of isolated stem cells was $(5.7 \pm 1.6) \times 10^5$ per mL of washed and dissected adipose tissue, which is comparable to approximately 4.0×10^5 per mL of lipoaspirate tissue reported in the literature as an average from several studies [12].
16. At this stage, the cells are mostly round in shape and detached. The remaining cells detach easily if the bottom of the flask is tapped gently.
17. In our work, the average number of P0 cells from one 225-cm^2 T-flask was $(2.9 \pm 0.3) \times 10^6$.
18. In our work, the average number of P2 cells from one 225-cm^2 T-flask was $(6.1 \pm 0.1) \times 10^6$. We cultured P2 cells on polyglycolic acid (PGA) mesh scaffolds under dynamic culture conditions to study their differentiation into chondrogenic [9] and osteogenic [10] lineages. Quantitative real-time PCR was used to assess differentiation of the stem cells cultured on PGA.
19. If it is difficult to obtain 1×10^7 cells, this method can be scaled down for fewer cells by reducing the volumes used in Subheading 3.4, **step** 7. For instance, the pellet can be resuspended in 2 mL of FCB instead of 10 mL, and 200-μL aliquots of cell suspension containing 10^6 cells per mL can be placed in the 10-mL centrifuge tubes. Centrifuge the tubes and add 200 μL of appropriate antibody solution to the cell pellet in each tube and continue as described in **step** 7. Also in **step 9**, scale down the volume of secondary antibody by resuspending the cells in 198 μL of FCB and adding 2 μL of secondary antibody solution per tube.
20. The presence of Tween-20 in FCB aids permeabilization of the cells, which is needed when antigens, e.g., vimentin, are intracellular.
21. FITC-conjugated antibody is light-sensitive, so cover the ice bucket during incubation and perform all of the following steps under dim light.
22. Passing the cell suspension through cell strainers will break up the cell clumps that would otherwise block the flow cytometer nozzle. We used Becton Dickinson disposable sterile 38-μm cell strainers.
23. Using this method, we found that 96 % of P0 cells were positive for CD90, 85 % were positive for vimentin, only 4 % reacted positively with antibody against SMA, and staining with antibody against Factor VIII was negative [9]. Figure 1 shows typical flow cytometry histograms for cells stained using antibodies against vimentin, CD90, SMA, and Factor VIII.
24. For cell sorting applications, omit **steps 3** and **4** in Subheading 3.4, and remove Tween-20 from the FCB.

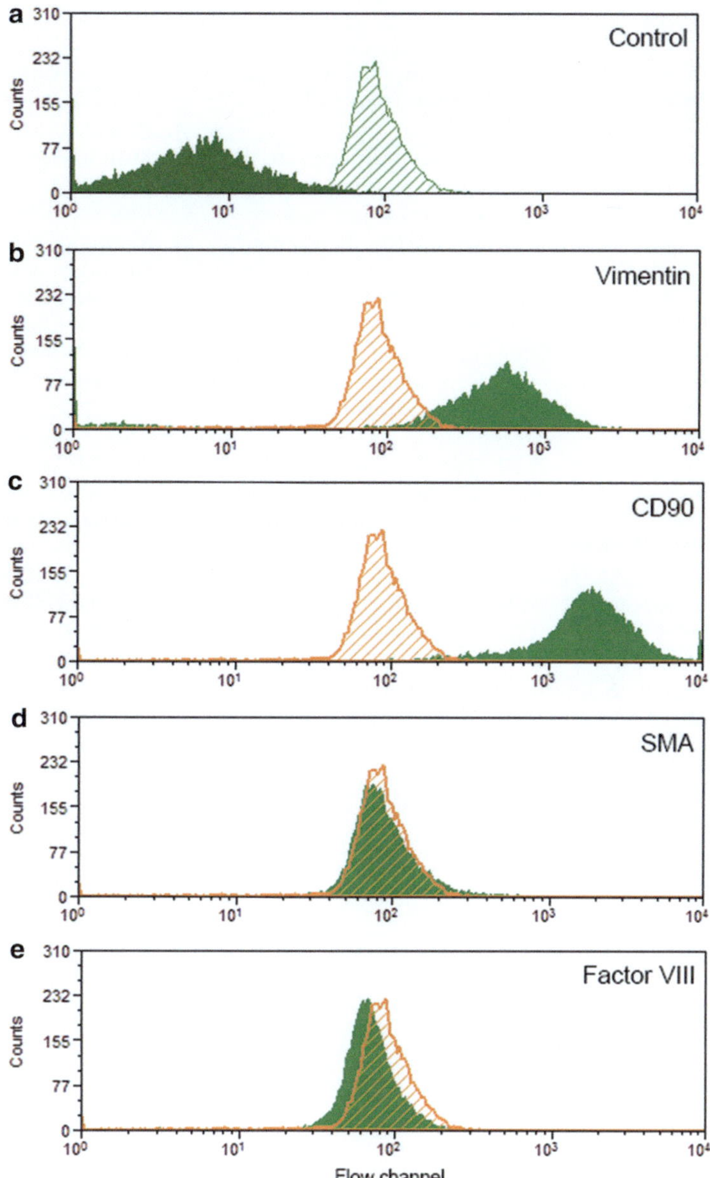

Fig. 1 Typical flow cytometry histograms for cells stained using antibodies against vimentin, CD90, SMA, and Factor VIII. The *solid shaded area* in (**a**) represents unstained cells (negative control). In (**b–e**), the *solid shaded areas* represent cells stained with primary and secondary antibodies. The *hatched areas* in (**a–e**) represent cells stained with secondary antibody only (nonspecific binding control). For analysis of the data, the number of cells stained with secondary antibody only is subtracted from the number of cells stained with primary and secondary antibodies at each flow channel to eliminate the effects of nonspecific binding and thus determine the number of cells positive for a specific primary antibody. The results indicate strong positive staining for vimentin and CD90 within the cell population, and weak or no staining for SMA and Factor VIII

References

1. Blum B, Benvenisty N (2008) The tumorigenicity of human embryonic stem cells. Adv Cancer Res 100:133–158
2. Hanley J, Rastegarlari G, Nathwani AC (2010) An introduction to induced pluripotent stem cells. Br J Haematol 151:16–24
3. Aronin CEP, Tuan RS (2010) Therapeutic potential of the immunomodulatory activities of adult mesenchymal stem cells. Birth Defects Res C Embryo Today 90:67–74
4. Griffin MD, Ritter T, Mahon BP (2010) Immunological aspects of allogeneic mesenchymal stem cell therapies. Hum Gene Ther 21:1641–1655
5. English K, Wood KJ (2011) Immunogenicity of embryonic stem cell-derived progenitors after transplantation. Curr Opin Organ Transplant 16:90–95
6. Strem BM, Hicok KC, Zhu M et al (2005) Multipotential differentiation of adipose tissue-derived stem cells. Keio J Med 54: 132–141
7. Huang S-J, Fu R-H, Shyu W-C et al (2013) Adipose-derived stem cells: isolation, characterization, and differentiation potential. Cell Transplant 22:701–709
8. Zuk PA, Zhu M, Mizuno H et al (2001) Multilineage cells from human adipose tissue: implications for cell-based therapies. Tissue Eng 7:211–228
9. Mahmoudifar N, Doran PM (2010) Chondrogenic differentiation of human adipose-derived stem cells in polyglycolic acid mesh scaffolds under dynamic culture conditions. Biomaterials 31:3858–3867
10. Mahmoudifar N, Doran PM (2013) Osteogenic differentiation and osteochondral tissue engineering using human adipose-derived stem cells. Biotechnol Prog 29: 176–185
11. Freyria A-M, Mallein-Gerin F (2012) Chondrocytes or adult stem cells for cartilage repair: the indisputable role of growth factors. Injury 43:259–265
12. Guilak F, Lott KE, Awad HA et al (2006) Clonal analysis of the differentiation potential of human adipose-derived adult stem cells. J Cell Physiol 206:229–237

Chapter 5

Derivation and Chondrogenic Commitment of Human Embryonic Stem Cell-Derived Mesenchymal Progenitors

Hicham Drissi, Jason D. Gibson, Rosa M. Guzzo, and Ren-He Xu

Abstract

The induction of human embryonic stem cells to a mesenchymal-like progenitor population constitutes a developmentally relevant approach for efficient directed differentiation of human embryonic stem (hES) cells to the chondrogenic lineage. The initial enrichment of a hemangioblast intermediate has been shown to yield a replenishable population of highly purified progenitor cells that exhibit the typical mesenchymal stem cell (MSC) surface markers as well as the capacity for multilineage differentiation to bone, fat, and cartilage. Herein, we provide detailed methodologies for the derivation and characterization of potent mesenchymal-like progenitors from hES cells and describe in vitro assays for bone morphogenetic protein (BMP)-2-mediated differentiation to the chondrogenic lineage.

Key words Human embryonic stem cells, Mesenchymal stem cells, Chondrogenic differentiation, Hemangioblast, Pellet culture, BMP-2

1 Introduction

Current cell-based strategies to promote articular cartilage defect repair for the prevention of posttraumatic osteoarthritis have focused primarily on the use of autologous somatic cells, such as articular chondrocytes and bone marrow-derived progenitor cells [1–3]. While extensive studies have demonstrated the formation of cartilage extracellular matrix from articular chondrocytes and bone marrow-derived mesenchymal stem cells, these adult cell sources are increasingly viewed as inadequate for restoring articular cartilage defects due to their limited supply and expansion, loss of chondrogenicity, and phenotypic instability [4–7]. Thus, due to their unlimited self-renewal capacity and their ability to differentiate into target cells, such as chondrocytes, human pluripotent stem cells have been proposed as an alternative cell source for the development of cell-based cartilage repair.

Directed differentiation of pluripotent stem cells to chondrocytes, at the expense of other lineages and without the carry-over

Pauline M. Doran (ed.), *Cartilage Tissue Engineering: Methods and Protocols*, Methods in Molecular Biology, vol. 1340, DOI 10.1007/978-1-4939-2938-2_5, © Springer Science+Business Media New York 2015

of pluripotent cells, poses a significant challenge. Induction of mesenchymal-like progenitors from human embryonic stem (hES) cells has been employed as a developmentally relevant approach to promote efficient differentiation to the chondrogenic lineage. Recent reports have described a variety of methods for deriving a renewable source of mesenchymal-like stem cells from differentiating hES cells, including coculture with OP9 cells, cell sorting, and manual selection of cellular outgrowths within adherent cultures [8–13]. However, many of the labor-intensive approaches bear limited efficacy due to low efficiency derivation and the formation of highly heterogeneous populations from differentiating hES cells. We, and others, have recently developed efficient, straightforward methodologies for the large-scale production of functional mesenchymal progenitors from human pluripotent stem cells, including (1) a direct plating method without the requirement of embryoid body formation or homogenous cell selection; and (2) a hemangioblast-enriching method [14–16]. Using these well established methods of derivation [14–16], the mesenchymal-like progenitor population obtained from hES cells displayed the typical morphological and immunophenotypic features of bone marrow (BM)-derived mesenchymal stem cells, as well as the capacity for multilineage differentiation to bone, fat, and cartilage.

Herein we describe methodologies for the derivation, characterization, and expansion of mesenchymal-like progenitors from human ES cells via the initial enrichment of a hemangioblast intermediate. For efficient chondrogenic induction and differentiation, we describe a pellet culture system in a chemically defined medium supplemented with human recombinant BMP-2. Chondrogenic differentiation is evaluated by histological and gene expression analyses.

2 Materials

2.1 Cell Culture

1. Human embryonic stem cell (hESC) line cultured under feeder-dependent conditions.
2. Dulbecco's Modified Eagle Medium (DMEM)/F12 medium: DMEM/F12 medium with L-glutamine, HEPES, and Phenol Red.
3. Human basic fibroblast growth factor (bFGF) stock solution: 10 μg of lyophilized bFGF powder dissolved in 1 mL of PBS containing 0.1 % BSA.
4. hESC medium: DMEM/F12 medium supplemented with 20 % knockout serum replacement, 1 mM L-glutamine, 0.1 mM non-essential amino acids, 0.55 mM β-mercaptoethanol, and 4 ng/mL bFGF.
5. hES cell cloning and recovery supplement: 1000× (Stemgent).

6. DMEM high glucose medium: DMEM medium with 4.5 g/L of D-glucose.
7. Penicillin-streptomycin solution: 10,000 U/mL penicillin and 10,000 μg/mL streptomycin.
8. Mouse embryonic fibroblasts (MEFs).
9. MEF (mouse embryonic fibroblast) medium: DMEM high glucose medium supplemented with 10 % defined FBS (*see* **Note 1**), 1 % non-essential amino acids, and 1 % penicillin-streptomycin solution.
10. mTeSR1 medium (Stem Cell Technologies).
11. Dispase solution: 15 mg of Dispase II (Life Technologies) powder in 30 mL of DMEM/F12 medium. Filter with a 0.22-μm polyethersulfone (PES) membrane. Store at 4 °C for up to 1 month.
12. Matrigel® Growth Factor Reduced (GFR) Basement Membrane Matrix (Corning)-coated 6-well plates.
13. Rho-associated kinase (ROCK) inhibitor solution: 10 μM Y27632 in DMSO.
14. Stemline II medium (Sigma-Aldrich).
15. Phosphate buffered saline (PBS).
16. Dulbecco's Ca^{2+}- and Mg^{2+}-free PBS (D-PBS).
17. 0.1 % bovine serum albumin (BSA) solution: 0.1 % BSA Fraction V in D-PBS. Filter with a 0.22-μm PES membrane. Store at 4 °C.
18. Bone morphogenetic protein (BMP)-4 stock solution: 50 μg/mL BMP-4 in sterile 4 mM HCl containing 0.1 % BSA.
19. Vascular endothelial growth factor (VEGF) stock solution: 50 μg/mL VEGF in 0.1 % BSA solution.
20. Stemline II medium with BMP-4 and VEGF: 50 ng/mL BMP-4 and 50 ng/mL VEGF in Stemline II medium.
21. Stemline II medium with BMP-4, VEGF, and bFGF: 50 ng/mL BMP-4, 50 ng/mL VEGF, and 45 ng/mL bFGF in Stemline II medium.
22. 0.05 % trypsin–EDTA: with phenol red.
23. 0.25 % trypsin–EDTA: with phenol red.
24. Thrombopoietin (TPO) stock solution: 50 μg/mL TPO in 0.1 % BSA solution.
25. Flt3-ligand stock solution: 50 μg/mL Flt3-ligand in 0.1 % BSA solution.
26. Blast cell growth medium (BGM): Methocult™ H4536 serum-free methylcellulose colony forming cell (CFC) medium (Stem Cell Technologies), supplemented with 50 ng/mL

VEGF, 50 ng/mL TPO, 50 ng/mL Flt3-ligand, 30 ng/mL bFGF, 1 % EX-CYTE Growth Enhancement Media Supplement (Millipore), and 1 % penicillin-streptomycin solution. Mix well, prepare 3-mL aliquots, and store at −20 °C.

27. HB-MSC growth medium: α-Minimum Essential Medium (α-MEM) supplemented with 20 % defined FBS.
28. L-Ascorbic acid-2-phosphate (AA2P) stock solution: 50 mg/mL AA2P in sterile water, 100×.
29. Dexamethasone stock solution: 10^{-3} M dexamethasone in sterile water.
30. L-Proline stock solution: 4 mM L-proline in sterile PBS.
31. Sodium pyruvate stock solution: 100 mM sodium pyruvate in sterile water.
32. DMEM low glucose medium: DMEM medium with 1.0 g/L of D-glucose.
33. Osteogenic medium: DMEM low glucose medium, 10 % defined FBS, 50 μg/mL AA2P, 10 mM β-glycerophosphate, 1 mM sodium pyruvate, 10^{-7} M dexamethasone, and 1 % penicillin-streptomycin solution.
34. Adipogenic medium: DMEM high glucose medium, 10 % FBS, 1 mM sodium pyruvate, 10^{-6} M dexamethasone, 10 μg/mL recombinant human insulin, 0.5 mM 3-isobutyl-1-methylxanthine (IBMX), 200 μM indomethacin, and 1 % penicillin-streptomycin solution.
35. Trypan blue solution: 0.4 % trypan blue in aqueous solution.
36. Chondrogenic medium: DMEM high glucose medium, 1 % insulin-transferrin-selenium^{+1} 100× (ITS^{+1}), 40 μg/mL L-PROLINE, 1 mM sodium pyruvate, 1 % MEM non-essential amino acids 100×, 2 mM Glutamax, 50 μg/mL AA2P, 10^{-7} M dexamethasone, and 1 % penicillin-streptomycin solution.
37. BMP-2 stock solution: 100 μg/mL of human recombinant BMP-2 in sterile 4 mM HCl containing 0.1 % BSA.
38. Chondrogenic medium with BMP-2: 100 ng/mL of BMP-2 in chondrogenic medium.

2.2 Flow Cytometry

1. 0.25 % trypsin–EDTA: with phenol red.
2. FACS staining buffer (*see* **Note 2**).
3. Dulbecco's Ca^{2+}- and Mg^{2+}-free PBS (D-PBS).
4. 40-μm nylon cell strainer (Falcon).
5. Round-bottom 5-mL polystyrene test tubes with cell-strainer cap (BD Falcon).
6. Propidium iodide (PI) solution: 10 μg/mL PI in PBS stored at 4 °C in the dark.

2.3 Histochemical Analyses

1. Formalin solution: 10 % formalin in water.
2. PBS.
3. Alkaline phosphatase staining solution: Alkaline phosphatase red membrane substrate solution (Sigma).
4. Isopropanol solution: 60 % isopropanol in water.
5. Oil Red O solution: 0.5 % Oil Red O in isopropanol solution.
6. Acetic acid solution: 3 % acetic acid in water.
7. Alcian Blue solution: 1 % Alcian Blue in 3 % acetic acid solution, pH 2.5.
8. Ethanol solution series: 95, 70, and 50 % ethanol in water.
9. Ethanol/xylene solution: 1:1 100 % ethanol:xylene.
10. CytoSeal™ Mounting Medium (Richard-Allen Scientific).
11. Nuclear Fast Red Kernechtrot solution: 0.1 % Nuclear Fast Red.
12. Safranin O solution: 0.1 % Safranin O in water.

2.4 Gene Expression Analysis

1. Linear acrylamide solution: Ambion®, 5 mg/mL.
2. Ultrapure water: RNase-, DNase-, and pyrogen-free distilled water.
3. Ethanol/ultrapure water solution: 75 % molecular-grade ethanol in ultrapure water.
4. DNAse I, amplification grade (Invitrogen).
5. iScript™ cDNA Synthesis Kit (BioRad).
6. SYBR™ Green Master Mix (Life Technologies).

3 Methods

3.1 Maintenance and Expansion of Human Embryonic Stem Cells Using Feeder Layers

Human embryonic stem cell (hESC) lines are initially maintained for expansion in hESC medium on mouse embryonic fibroblasts (MEFs) and cultured in a water-jacketed 5 % CO_2 incubator at 37 °C. All medium is prewarmed to 37 °C, and all enzyme solutions are brought to room temperature prior to use.

1. Seed mouse embryonic fibroblasts (MEFs) at a density of 2×10^4 cells/cm^2 onto 0.1 % gelatin-coated 6-well tissue culture-treated plates containing 2 mL/well of MEF medium, and culture in a water-jacketed 5 % CO_2 incubator at 37 °C. One day prior to seeding hESCs, remove the MEF medium, wash each well three times with 1× PBS, and replace with 2 mL/well of hESC medium, incubating overnight.
2. Seed and culture hESCs in 0.1 % gelatin-coated 6-well tissue culture-treated plates containing adherent irradiated mouse embryonic fibroblasts (2×10^4 cells/cm^2) and 2 mL/well of hESC medium.

3. Replace the hESC medium every day and routinely monitor the cultures (*see* **Note 3**).
4. Grow the hESC colonies to confluency, with colony sizes ranging from 400 to 800 μm.
5. Wash the hESCs twice with D-PBS and enzymatically passage using Accumax™ via incubation at 37 °C for 5–10 min (*see* **Note 4**).
6. Pellet the harvested hESCs and resuspend in hESC medium containing 1× hES cell cloning and recovery supplement.
7. Seed the hESCs onto freshly plated irradiated MEFs (2×10^4 cells/cm^2) in 0.1 % gelatin-coated 6-well tissue culture-treated plates containing 2 mL/well hESC medium. Replace the hESC seeding medium with fresh hESC medium within 24 h.

3.2 Feeder-Free Human Embryonic Stem Cell Culture

hESC lines cultured under feeder-dependent conditions are transitioned to feeder-free conditions and cultured in a water-jacketed 5 % CO_2 incubator at 37 °C. Medium is replaced every other day and colonies are mechanically passaged with a pipette using Dispase solution (*see* **Note 5**).

1. Replace standard hESC culture medium with prewarmed mTeSR1 medium 3 days prior to passaging. Add 2 mL/well in the 6-well plate of hESC and exchange the medium daily.
2. Once the cultures reach 75 % confluence, aspirate the medium and incubate the cells with prewarmed Dispase solution at 1 mL/well for 5–20 min at 37 °C or until the edges of the cell colonies start to curl up.
3. Using a cooled pipette, prepare a Matrigel-coated plate(s) by pipetting 0.5 mL of Matrigel® Growth Factor Reduced (GFR) Basement Membrane Matrix onto each well of a 6-well plate on ice, and then incubate the plate(s) at 37 °C for 30 min.
4. Aspirate the Dispase solution from **step 2** and rinse the cells three times with 1 mL/well of warmed DMEM/F12 medium.
5. Using 2 mL of mTeSR1 medium and a serological pipette, gently scrape the cells with the pipette while slowly expelling the 2 mL of mTeSR1 medium from the pipette into the well.
6. Aspirate the excess Matrigel solution from the 6-well plate(s) in **step 3**, and pipette 2 mL/well of prewarmed mTeSR1 medium into the wells of the Matrigel-coated 6-well plate(s).
7. Transfer equal volumes of the cell suspension from **step 4** into each well of the Matrigel-coated 6-well plate(s) and incubate overnight to allow cell attachment. To promote cell survival, ROCK inhibitor solution may be added for up to 24 h of culture following passaging. Cells should be passaged so that roughly 3×10^5 cells are seeded into each fresh well of a Matrigel-coated 6-well plate.

3.3 Hemangioblast-Enriching Method for Generation of Mesenchymal Progenitors

1. To induce embryoid body (EB) formation, digest the hESC colonies with Dispase solution for 5–10 min, wash the cells with Stemline II medium, harvest the colonies by gently scraping with a 5-mL pipette, and centrifuge the cell clumps at 200 × *g* (1000 rpm) for 5 min.
2. Resuspend the clumps in 10 mL of Stemline II medium with BMP-4 and VEGF, and seed the hESCs in 60-mm ultra-low attachment (suspension) plates for 2 days to form EB aggregates, incubating in a water-jacketed 5 % CO_2 incubator at 37 °C.
3. Remove half of the medium after 48 h, and add 5 mL of Stemline II medium containing BMP-4, VEGF, and bFGF (for a final bFGF concentration of 22.5 ng/mL). Incubate at 37 °C for 2 days.
4. Collect EB aggregates in a 15-mL conical tube, and allow aggregates to settle.
5. Aspirate the medium and wash aggregates in 5 mL of PBS.
6. Dissociate the aggregates with 0.05 % trypsin–EDTA for 2–5 min. Pipette to disrupt the clumps, and inactivate the trypsin with 10 % FBS. Prepare a single-cell suspension by passing through a 16-G needle three to five times, and then through a 40-μm cell strainer.
7. Centrifuge (200 × *g*) for 5 min, and resuspend in Stemline II medium at 5×10^6 cells/mL.
8. Mix the single-cell suspension with 3 mL of BGM medium (5×10^4 cells/mL), and plate 1 mL of the BGM-cell suspension into each well of a 6-well plate.
9. Incubate for 4–6 days at 37 °C to allow for mesenchymal-hemangioblast progenitor formation [17, 18].
10. Monitor blast colony formation. Blast colonies are typically visible after 4 days. By day 6, the colonies typically become loose and large.
11. Harvest the loosened culture of mesenchymal-hemangioblast progenitors and collect into a 15-mL conical tube. Rinse the colonies 3× with PBS, digest the colonies with 0.05 % trypsin–EDTA, and centrifuge at 200 × *g* for 5 min.
12. Resuspend the hESC-derived mesenchymal-hemangioblast progenitors in HB-MSC growth medium and plate the cells (5×10^4 cells/well) onto Matrigel-coated 6-well plates.
13. Remove suspended hemangioblasts and then culture adherent mesenchymal progenitors with the HB-MSC growth medium within 24 h.
14. Grow hESC-derived MSC in HB-MSC growth medium on Matrigel-coated plates, exchanging the medium every 2–3 days.

15. When the cells reach 80–90 % confluency, passage using 0.25 % trypsin–EDTA. Seed cells onto Matrigel-coated 6-well plates until passage 4. hESC-MSC cells can be passaged subsequently for expansion using noncoated 6-well plates [14, 15] (*see* **Note 6**).

3.4 Flow Cytometry Analysis of Mesenchymal Stem Cell-Like Populations

1. Harvest the hESC-MSCs with 0.25 % trypsin–EDTA, wash 2× with D-PBS, and resuspend in ice-cold FACS staining buffer (*see* **Note 2**).
2. Singularize the cells by passing them through a 40-μm cell strainer (*see* **Note 7**). Aliquot ~1×10^6 cells/tube for each antibody stain.
3. Centrifuge (1000 rpm) for 5 min at 4 °C. Resuspend the cell pellet in 100 μL of ice-cold FACS staining buffer.
4. Incubate for 30 min at 4 °C in the dark (for fluorescent labels) with primary conjugated antibodies for cell surface antigens specific for mesenchymal, hematopoietic, and epithelial cell progenitors (Table 1).

Table 1
Flow cytometry antibodies

Antibody	Isotype
Cell surface protein-specific antibodies	
PE Conjugated Mouse Anti-human CD73	Mouse IgG1, κ
PE Conjugated Mouse Anti-Human CD166	Mouse IgG1, κ
PE Conjugated Mouse Anti-Human CD90	Mouse IgG1, κ
PE Conjugated Mouse Anti-Human CD29	Mouse IgG2a, κ
PE Conjugated Mouse Anti-Human HLA-DR	Mouse IgG2a, κ
FITC Conjugated Mouse anti-Human CD105	Mouse IgG1, κ
FITC Conjugated Mouse Anti-Human HLA-ABC	Mouse IgG1, κ
FITC Conjugated Mouse Anti-Human CD31	Mouse IgG1 κ
FITC Conjugated Mouse Anti-Human CD45	Mouse IgG1, κ
FITC Conjugated Mouse Anti-Human CD44	Mouse IgG2b, κ
Isotype-matched monoclonal antibodies	
PE Mouse IgG1, κ Isotype Control	Mouse IgG1, κ
PE Mouse IgG2a, κ Isotype Control	Mouse IgG2a, κ
FITC Mouse IgG1, κ Isotype Control	Mouse IgG1, κ
FITC Mouse IgG2b, κ Isotype Control	Mouse IgG2b, κ

5. Incubate additional cell aliquots with isotype-matched monoclonal antibodies as controls for nonspecific fluorescence (Table 1). Include an unstained negative control for gating the size of each cell population being examined.
6. Add 500 μL of ice-cold FACS staining buffer and centrifuge at 1000 rpm for 5 min at 4 °C. Wash samples 3×, then resuspend in 300–500 μL of FACS staining buffer. Strain into round-bottom test tubes with cell-strainer caps for flow analysis.
7. Keep samples on ice (*see* **Note 8**) and collect at least 10,000 events on a flow cytometer instrument (e.g., FACS Calibur, BD Biosciences) using appropriate software for data collection (e.g., FACS Diva, BD Biosciences) and for performing the corresponding gating analysis (e.g., FlowJo, Tree Star, Inc.) (*see* **Note 9**).

3.5 Multipotential Assessment of Mesenchymal Stem Cells Derived from hESCs

1. To induce osteogenesis, seed hESC-derived MSCs at 120,000 cells/cm^2 in tissue culture-treated plates and culture cells in osteogenic medium in a water-jacketed 5 % CO_2 incubator at 37 °C. Exchange the medium every 2–3 days. After 21 days, fix the cultures in formalin solution for 15 min, wash 2× in PBS, and stain for 10 min in alkaline phosphatase staining solution for detection of alkaline phosphatase activity, an early enzymatic marker of osteogenesis. Insoluble, diffuse red dye deposits viewed by microscopy of the cells indicate sites of alkaline phosphatase activity.
2. To induce adipogenic differentiation, seed hESC-MSCs at 120,000 cells/cm^2 and culture in adipogenic medium in a water-jacketed 5 % CO_2 incubator at 37 °C. Exchange the adipogenic medium every 2–3 days. After 21 days in culture, fix the cultures in formalin solution for 15 min, wash the cells 2× in PBS, rinse the cells with 60 % isopropanol solution, and stain with Oil Red O solution for 10 min for detection of lipid-filled vacuoles in the cells. Punctate red staining viewed by microscopy of the cells is indicative of adipocyte differentiation.
3. To induce chondrogenic differentiation, culture hESC-derived MSC progenitors as high-density pellet cultures in chondrogenic medium in a water-jacketed 5 % CO_2 incubator at 37 °C as described in the following Subheading 3.6.

3.6 Chondrogenic Commitment of MSC-Like Progenitors

1. To induce chondrogenic differentiation of hESC-derived MSCs, harvest cells using 0.25 % trypsin–EDTA. Spin the cells at 1000 rpm and resuspend in HB-MSC growth medium.
2. Count viable cells via microscopy on a hemocytometer, or other cell counter equipment, using trypan blue solution in a 1:2 dilution of the cell suspension from **step 1** to exclude lysed cells stained with trypan blue. Aliquot 2.5×10^5 viable

hESC-derived MSC cells in 0.5 mL of HB-MSC growth medium into 15-mL conical tubes for pellet cultures. Centrifuge the cells at 1000 rpm and incubate in 15-mL conical tubes overnight at 37 °C and 5 % CO_2 with caps loosened for gas exchange [16] (*see* **Note 10**).

3. Within 24–48 h of pellet formation, examine the cell pellets to ensure that the cells have formed a spherical pellet at the bottom of each 15-mL tube.
4. Aspirate the medium and replace with 0.5 mL of chondrogenic medium with BMP-2 (*see* **Note 11**).
5. Culture the pellets in chondrogenic medium with BMP-2 for 7, 14, and 21 days, carefully exchanging the medium in each 15-mL conical tube with 0.5 mL of fresh chondrogenic medium with BMP-2 every other day.

3.7 Histological Assessment of Chondrogenic Matrix Production

1. Harvest the pellets, wash 1× in PBS, and fix in formalin solution for 15 min.
2. Wash the fixed pellets 2× in PBS, and dehydrate the pellets in 1.5-mL microcentrifuge tubes with sequential washes in 50, 70, 95, and 100 % ethanol, followed by ethanol/xylene solution and 100 % xylene (5 min each, 2×) (*see* **Note 12**).
3. Embed the dehydrated pellets in low-melt paraffin.
4. Section the paraffin-embedded pellets using a microtome (5–7 μm sections) and mount onto Superfrost plus slides.
5. Deparaffinize and rehydrate the sections on slides through sequential washes in 100 % xylene (10 min, 2×), 100, 95, 70, and 50 % ethanol (2 min each, 2×), and lastly distilled water (5 min).
6. For Alcian Blue staining of sulfated proteoglycan deposits that are indicative of functional chondrocytes, rinse the formalin-fixed sections of paraffin-embedded pellets in acetic acid solution for 3 min and stain with Alcian Blue solution for 30 min at room temperature. Rinse briefly in acetic acid solution and then under running tap water for 5–10 min. Counterstain the cell nuclei with Nuclear Fast Red Kernechtrot solution for 5 min at room temperature. Rinse in running tap water for 5–10 min until clear.
7. For Safranin O staining of acidic proteoglycan present in cartilage tissues, stain hydrated formalin-fixed sections of paraffin-embedded pellets in Safranin O solution for 5 min at room temperature. Rinse under running tap water for 5–10 min to remove excess stain.
8. Dehydrate slides in 70, 95, and 100 % ethanol, then 100 % xylene (2 min each, 2×), and mount coverslips using CytoSeal Mounting Medium for microscopic evaluation of proteoglycan content assessed by diffuse blue dye deposits for Alcian Blue and diffuse orange-red deposits for Safranin O.

3.8 Quantitative PCR Analyses of Cartilage Genes

RNA is extracted for PCR analysis from cell pellets harvested over a time course of chondrogenic differentiation (*see* **Note 13**).

1. Aspirate the medium and add 300 μL of TRIzol reagent to a minimum of three pellets. Transfer to a 1.5-mL sterile, RNase-, DNase-, and pyrogen-free microcentrifuge tube.
2. Pass pellets in TRIzol reagent several times through a $26^{1/2\text{-gauge}}$ needle attached to a 1-mL syringe, vortex, and incubate at room temperature for 5 min.
3. Prepare a master mix containing 1 μL of linear acrylamide solution and 29 μL of ultrapure water for every TRIzol sample. Add 30 μL of the diluted linear acrylamide to each sample lysate.
4. Vortex samples for 30 s and incubate at room temperature for 5 min.
5. Add 60 μL of chloroform, vortex for 30 s, and incubate for 5 min.
6. Centrifuge samples at 12,000 × *g* for 15 min at 4 °C for phase separation. Place on ice.
7. Transfer the upper aqueous RNA layers into new 1.5-mL tubes.
8. Add isopropanol (150 μL) to each sample, mix by inversion, and store at −20 °C overnight for RNA precipitation. Samples may be stored longer.
9. Centrifuge at 12,000 × *g* for 30 min at 4 °C. A small glassy oval/sphere will be strongly adhered to the side of the tube. Place samples on ice.
10. Wash pellets with 500 μL of ethanol/ultrapure water solution and spin at 12,000 × *g* for 5 min at 4 °C. Aspirate the supernatant and repeat 2×.
11. Remove the ethanol and air-dry the RNA pellets. Dissolve the RNA pellets in a suitable volume of ultrapure water.
12. Measure the RNA concentration for each sample using a NanoDrop™ spectrophotometer.
13. Treat with DNase I as per the manufacturer's protocol, and reverse-transcribe the RNA to cDNA with the iScript™ cDNA Synthesis Kit, or comparable reverse transcriptase kit, following the manufacturer's protocol.
14. Perform real-time quantitative PCR on the cDNA samples synthesized from total RNA using the SYBR Green Master Mix, gene-specific real-time PCR primers (Table 2), and a real-time PCR cycler as per the manufacturer's protocol.
15. Using the data obtained from quantitative RT-PCR analyses, calculate values represented as $2^{\text{delta-delta Ct}}$, with delta-delta Ct defined as the difference in crossing threshold (Ct) values between the experimental and control samples using GAPDH as an internal standard.

Table 2
Real-time PCR primers

Gene (Gen bank accession no.)	Forward primer	Reverse primer
Gapdh (NM_002046)	aattccatggcaccgtcaag	agggatctcgctcctggaag
Sox9 (NM_000346)	agacagccccctatcgactt	cggcaggtactggtcaaact
Col2a1 (NM_001844)	ggcaatagcaggttcacgtaca	cgataacagtcttgccccactt
Col10a1 (NM_000493)	caaggcaccatctccaggaa	aaagggtatttgtggcagcatatt
Alp (NM_001177520)	gacaagaagcccttcactgc	agactgcgcctggtagttgt

16. Express the data as mean ± S.E.M. of at least three independent samples. Perform statistical comparisons between untreated and growth factor-treated groups using a two-tailed Student's *t*-test. All P values < 0.05 are considered significant (*see* **Note 14**).

4 Notes

1. "Defined" FBS is filtered through serial 40-nm pore-size rated filters and tested for endotoxin, hemoglobin, and extensive biochemical profile assays. Each lot of serum is tested to ensure quality and sterility.
2. A suitable FACS staining buffer will be isotonic and buffered to neutrality, will cushion the cells against damage during centrifugation, block nonspecific staining, prevent capping of bound antibody, and block Fc receptor binding. An example would be PBS containing 2 % FBS, 2 % HEPES, and 0.1 % BSA.
3. Verification of karyotype stability and mycoplasma-free cultures should be performed routinely. Monitor cultures daily for spontaneous differentiation. Weed cultures of any colonies displaying differentiated morphology.
4. AccuMax™ (Innovative Cell Technologies, Inc.) is an optimized combination of enzymes for the dissociation of complex tissues and is inactivated at 37 °C. AccuMax™ should be thawed and/or warmed at room temperature and stored at 4 °C. Incubation for 5–10 min should be sufficient to dissociate hESC colonies; however longer incubation times will not lyse cells.
5. Alternatively, hESC may be split mechanically by cutting and pasting or with a disposable stem cell passaging tool called STEMPRO™ EZPassage (Invitrogen).

6. Over the first four passages, hESC-MSC cells grow in colonies with slightly raised centers, and following passage onto noncoated tissue culture-treated plates, hESC-MSC cells acquire a homogenous, fibroblast-like morphology and grow in monolayer.
7. Prior to staining, cells must be singularized to prevent damage to the flow cell. Straining through nylon mesh to ensure a single-cell suspension is sufficient for tissue culture cells.
8. If samples cannot be run on a flow cytometer immediately, fix in 10 % formalin for 15 min, wash 2× in FACS staining buffer, and store at 4 °C in the dark for later analysis.
9. MSC progenitor cells are positive for MSC cell surface markers CD73, CD29, CD166, HLA-ABC, CD44, CD90, and CD105, and negative for cell surface markers characteristic of endothelial and hematopoietic lineages, including CD31, and CD45 and HLA-DR, respectively.
10. Ensure the caps are loosened for gas exchange.
11. Low-pressure aspiration using sterile 9-in. Pasteur pipettes is recommended when exchanging the medium in the 15-mL conical pellet cultures.
12. To visualize the pellets during paraffin sectioning, eosin stain can be added briefly at 1:100 to the final 95 % ethanol wash to provide color prior to complete dehydration in 100 % ethanol and xylene.
13. A minimum of three pellets is recommended per culture condition to obtain a sufficient amount of high quality RNA for downstream applications.
14. The expected real-time PCR results include an induction of gene expression of the master chondrogenic transcription factor Sox9 and the early chondrogenic matrix protein Col2a1 at 7 days of culture in chondrogenic medium with BMP-2. By days 14 and 21 of cultures in chondrogenic medium with BMP-2, the expression levels of the late chondrogenic matrix protein Col10a1 and the early osteogenic enzyme Alp are expected to increase as the differentiating chondrocytes mature into hypertrophic chondrocytes.

Acknowledgements

The authors would like to acknowledge the support of the State of Connecticut Department of Public Health Stem Cell Program (grant award no. 11SCB08).

References

1. Ahmed TA, Hincke MT (2010) Strategies for articular cartilage lesion repair and functional restoration. Tissue Eng Part B Rev 16:305–329
2. Mastbergen SC, Saris DB, Lafeber FP (2013) Functional articular cartilage repair: here, near, or is the best approach not yet clear? Nat Rev Rheumatol 9:277–290
3. van Osch GJ, Brittberg M, Dennis JE, Bastiaansen-Jenniskens YM, Erben RG, Konttinen YT, Luyten FP (2009) Cartilage repair: past and future—lessons for regenerative medicine. J Cell Mol Med 13:792–810
4. Dexheimer V, Mueller S, Braatz F, Richter W (2011) Reduced reactivation from dormancy but maintained lineage choice of human mesenchymal stem cells with donor age. PLoS One 6:e22980
5. Jones EA, Kinsey SE, English A, Jones RA, Straszynski L, Meredith DM, Markham AF, Jack A, Emery P, McGonagle D (2002) Isolation and characterization of bone marrow multipotential mesenchymal progenitor cells. Arthritis Rheum 46:3349–3360
6. Pittenger MF, Mackay AM, Beck SC, Jaiswal RK, Douglas R, Mosca JD, Moorman MA, Simonetti DW, Craig S, Marshak DR (1999) Multilineage potential of adult human mesenchymal stem cells. Science 284:143–147
7. Steinert AF, Ghivizzani SC, Rethwilm A, Tuan RS, Evans CH, Noth U (2007) Major biological obstacles for persistent cell-based regeneration of articular cartilage. Arthritis Res Ther 9:213
8. Barberi T, Willis LM, Socci ND, Studer L (2005) Derivation of multipotent mesenchymal precursors from human embryonic stem cells. PLoS Med 2:e161
9. Brown SE, Tong W, Krebsbach PH (2009) The derivation of mesenchymal stem cells from human embryonic stem cells. Cells Tissues Organs 189:256–260
10. Gruenloh W, Kambal A, Sondergaard C, McGee J, Nacey C, Kalomoiris S, Pepper K, Olson S, Fierro F, Nolta JA (2011) Characterization and in vivo testing of mesenchymal stem cells derived from human embryonic stem cells. Tissue Eng Part A 17: 1517–1525
11. Hwang NS, Varghese S, Lee HJ, Zhang Z, Ye Z, Bae J, Cheng L, Elisseeff J (2008) In vivo commitment and functional tissue regeneration using human embryonic stem cell-derived mesenchymal cells. Proc Natl Acad Sci U S A 105:20641–20646
12. Olivier EN, Rybicki AC, Bouhassira EE (2006) Differentiation of human embryonic stem cells into bipotent mesenchymal stem cells. Stem Cells 24:1914–1922
13. Vodyanik MA, Yu J, Zhang X, Tian S, Stewart R, Thomson JA, Slukvin II (2010) A mesoderm-derived precursor for mesenchymal stem and endothelial cells. Cell Stem Cell 7:718–729
14. Kimbrel EA, Kouris NA, Yavanian GJ, Chu J, Qin Y, Chan A, Singh RP, McCurdy D, Gordon L, Levinson RD et al (2014) Mesenchymal stem cell population derived from human pluripotent stem cells displays potent immunomodulatory and therapeutic properties. Stem Cells Dev 23:1611–1624
15. Wang X, Kimbrel EA, Ijichi K, Paul D, Lazorchak AS, Chu J, Kouris NA, Yavanian GJ, Lu SJ, Pachter JS et al (2014) Human ESC-derived MSCs outperform bone marrow MSCs in the treatment of an EAE model of multiple sclerosis. Stem Cell Rep 3:1–16
16. Guzzo RM, Gibson J, Xu RH, Lee FY, Drissi H (2013) Efficient differentiation of human iPSC-derived mesenchymal stem cells to chondroprogenitor cells. J Cell Biochem 114:480–490
17. Lu SJ, Feng Q, Caballero S, Chen Y, Moore MA, Grant MB, Lanza (2007) Generation of functional hemagioblasts from human embryonic stem cells. Nat Methods 4:501–9
18. Lu SJ, Ivanova Y, Feng Q, Luo C, Lanza R (2009) Hemangioblasts from human embryonic stem cells generated multilayered blood vessels with functional smooth muscle cells. Regen Med 4:37–47

Chapter 6

Differentiation of Human Induced Pluripotent Stem Cells to Chondrocytes

Rosa M. Guzzo and Hicham Drissi

Abstract

Human induced pluripotent stem (iPS) cells are relevant tools for modeling human skeletal development and disease, and represent a promising source of patient-specific cells for the regeneration of skeletal tissue, such as articular cartilage. Devising efficient and reproducible strategies, which closely mimic the physiological chondrogenic differentiation process, will be necessary to generate functional chondrocytes from human iPS cells. Our previous study demonstrated the generation of chondrogenically committed human iPS cells via the enrichment of a mesenchymal-like progenitor population, application of appropriate high-density culture conditions, and stimulation with bone morphogenetic protein-2 (Bmp-2). The differentiated iPS cells showed temporal expression of cartilage genes and the accumulation of a cartilaginous extracellular matrix in vitro. In this chapter, we provide detailed methodologies for the differentiation of human iPS cells to the chondrogenic lineage and describe protocols for the analysis of chondrogenic differentiation.

Key words Human induced pluripotent stem cells, Mesenchymal-like progenitor stem cells, Chondrogenic differentiation, Micromass culture, Bmp-2

1 Introduction

The seminal discovery that an adult human cell can be reprogrammed to a pluripotent state has provided new avenues to study molecular mechanisms in skeletal development and disease, as well as tools to accelerate drug discovery [1–3]. From a regenerative medicine perspective, human induced pluripotent stem (iPS) cells represent a potential source of patient-specific cells for the replacement of musculoskeletal tissues with poor intrinsic repair capacity, such as articular cartilage. However, the inherent pluripotent nature of human iPS cells poses a significant challenge in controlling their developmental fate to become a specialized cell type, such as a chondrocyte.

Pauline M. Doran (ed.), *Cartilage Tissue Engineering: Methods and Protocols*, Methods in Molecular Biology, vol. 1340, DOI 10.1007/978-1-4939-2938-2_6, © Springer Science+Business Media New York 2015

Various in vitro strategies have been reported for inducing the differentiation of human pluripotent stem cells to chondrocytes [4–11]. Several of these approaches utilize an initial pre-differentiation step within embryoid bodies (EB), followed by their dispersal and subsequent culture at high density to promote cell-cell interactions that mimic precartilage condensation during skeletal development. Numerous studies have examined the efficacy of stage-specific administration of developmentally relevant growth factors, such as bone morphogenetic proteins (Bmps) and transforming growth factors (Tgfs), for controlling the induction of human pluripotent stem cells to the chondrogenic lineage and subsequent chondrocyte differentiation [4, 6, 10]. As a means of limiting the developmental potency of human iPS cells, alternative approaches have differentiated human iPS cells to progenitor-like cells, which exhibited the molecular and functional properties of adult mesenchymal stem cells [5, 8, 12, 13]. A readily expandable source of iPS cell-derived multipotent progenitors, exhibiting high chondrogenicity, may provide a vast supply of cells for orthopedic and tissue engineering-related applications [13].

In our previous study, we induced chondrogenic differentiation using naive human iPS cells and iPS cell-derived mesenchymal-like progenitor cells [8]. Formation of high-density micromasses from naive human iPS cells in defined medium was sufficient to induce differentiation to the chondrogenic lineage without passing through an embryoid body stage [8]. Over a 3-week culture period, Bmp-2 treatment significantly increased the expression of early and late cartilage genes in a temporally regulated manner, as well as the accumulation of a cartilaginous extracellular matrix. We further developed a direct plating strategy [8, 10] for efficient and robust iPS cell differentiation into chondrocytes through an intermediate population of multipotent mesenchymal-like progenitors [8]. Our molecular and functional analyses indicated that iPS cell-derived progenitors exhibited mesenchymal-like features, including the expression of a defined set of cell surface markers and the capacity to differentiate into osteoblasts, chondrocytes, and adipocytes [8]. The enrichment of a mesenchymal-like progenitor cell population from differentiating human iPS cells and the appropriate high-density culture environment significantly enhanced the chondrogenic capacity of fibroblast-derived iPS cells and limited heterogeneity of the differentiated progeny.

In this chapter, we provide detailed protocols for human iPS cell chondrogenic differentiation from a naïve, pluripotent state (*see* Subheading 3.1) and via the derivation of a scalable, mesenchymal-like progenitor intermediate (*see* Subheading 3.2). We also describe the histological methods (*see* Subheading 3.3) and gene expression analyses (*see* Subheading 3.4) used to evaluate chondrogenic differentiation.

2 Materials

2.1 Cell Culture

1. Human fibroblast-derived iPS cells (*see* **Note 1**).
2. Dulbecco's Ca^{2+}- and Mg^{2+}-free phosphate-buffered saline (PBS).
3. Gelatin solution: EmbryoMax 0.1 % gelatin solution.
4. Trypsin-EDTA solution: 0.25 % trypsin, 1 mM ethylenediaminetetracetic acid (EDTA).
5. Accutase™ solution.
6. DMEM high glucose: Dulbecco's Modified Eagle's Medium (DMEM) high glucose (with 4500 mg/L glucose, with L-GLUTAMINE, without HEPES).
7. Dispase II solution: 1 mg/mL dispase II (Gibco) in DMEM high glucose. Filter sterilize and store at 4 °C for up to 2 weeks.
8. Trypan Blue solution: 0.4 % Trypan Blue in 0.85 % saline.
9. Primaria™ 6-well culture dishes.

2.2 Culture Media

1. DMEM high glucose: Dulbecco's Modified Eagle's Medium (DMEM) high glucose (with 4500 mg/L glucose, L-glutamine, without HEPES).
2. Hepes modified-DMEM high glucose: Dulbecco's Modified Eagle's Medium (DMEM) HEPES Modification and High Glucose (with 4500 mg/L glucose, L-glutamine, and 25 mM HEPES, without sodium bicarbonate and pyruvate).
3. Defined FBS: Defined fetal bovine serum (HyClone) (*see* **Note 2**).
4. Knockout™ Serum Replacement (KSR) (*see* **Note 2**).
5. Ascorbic acid-2-phosphate (AA2P) stock solution: 50 mg/mL AA2P in water. Filter sterilize. Aliquot and store at ≤−20 °C.
6. L-Proline stock solution (100×): 4 mg/mL L-proline in water. Filter sterilize. Aliquot and store at ≤−20 °C.
7. Dexamethasone stock solution: 10^{-3} M dexamethasone in sterile water. Aliquot and store at ≤−20 °C.
8. Fibroblast growth factor-basic (bFgf) stock solution: 10 μg/mL of human recombinant bFgf in sterile PBS containing 0.1 % bovine serum albumin (BSA) fraction V. Apportion the stock solution into working aliquots and store at ≤−20 °C (*see* **Note 3**).
9. Bone morphogenetic protein-2 (Bmp-2) stock solution: 100 μg/mL of human recombinant Bmp-2 in sterile PBS containing 4 mM HCl and 0.1 % BSA fraction V. Aliquot and store at −80 °C (*see* **Note 3**).
10. Rho-associated kinase (ROCK) inhibitor Y-27632: 10 mM Y-27632 in sterile water. Aliquot and store at −20 °C.
11. Micromass plating medium (Medium 1): Hepes modified-DMEM high glucose, 10 % defined FBS, 10 % KSR, 1 mM sodium pyruvate, 1× non-essential amino acid solution (Gibco),

100 U/mL penicillin/streptomycin (P/S). Sterilize by filtration (0.2 μm filter pore size) and store complete medium at 4 °C.

12. Chondrogenic differentiation medium (Medium 2): Hepes modified-DMEM high glucose, 1 % ITS$^+$ (6.25 μg/mL insulin, 6.25 μg/mL transferrin, 6.25 ng/mL selenious acid, 1.25 mg/mL BSA, 5.35 μg/mL linoleic acid), 50 μg/mL AA2P, 40 μg/mL L-proline, 10^{-7} M dexamethasone, 1 mM sodium pyruvate, 1× non-essential amino acid solution (Gibco), 2 mM Glutamax, 100 U/mL penicillin/streptomycin (P/S). Filter sterilize (*see* **Note 4**). Add Bmp-2 stock solution to give 100 ng/mL Bmp-2 immediately prior to use (*see* **Note 3**).
13. Mesenchymal stem cell induction and growth medium (Medium 3): DMEM high glucose, 10 % defined FBS, 5 ng/mL bFgf, 1× non-essential amino acid solution (Gibco), 100 U/mL penicillin/streptomycin (P/S). Store the filtered medium at 4 °C for up to 2 weeks.
14. Cryopreservation medium (2×): DMEM high glucose, 20 % defined FBS, 20 % dimethyl sulfoxide (DMSO). Filter with a 0.22-μm polyvinylidene difluoride (PVDF) membrane and store aliquots at ≤–20 °C.

2.3 Gene Expression Analyses

1. Nuclease-free water: RNAse-free, DNase-free water.
2. iScript™ cDNA synthesis kit.
3. SYBR Green I Master Mix kit (Roche).
4. Human gene-specific primers (*see* Table 1).

2.4 Histochemical Analyses

1. Formalin solution: 10 % neutral buffered formalin.
2. Alcian Blue staining solution: 1 % Alcian Blue, 3 % acetic acid, pH 2.5.
3. Nuclear Fast Red Kernechtrot solution: 0.1 % Nuclear Fast Red.
4. Acetic acid solution: 3 % acetic acid in water.
5. Tissue embedding medium (i.e., Paraplast®X-Tra™).
6. Ethanol solution series: 50, 70, and 95 % ethanol in water.
7. Xylene/ethanol solution: 1:1 xylene:100 % ethanol.

2.5 Flow Cytometry

1. Antibody staining solution: 2 % human serum and 2 % BSA fraction V in PBS.
2. Polystyrene round bottom tubes with cell strainer cap (BD Falcon).
3. Phycoerythrin- (PE-) and fluorescein isothiocyanate- (FITC-) conjugated antibodies against mesenchymal, endothelial, and hematopoietic cell surface proteins. A list of flow cytometry optimized antibodies, including their source and optimal dilutions, is provided in Table 2 (*see* **Note 5**).

Table 1
Real-time PCR primers [8]

Gene (accession No.)	Forward primer	Reverse primer
Gapdh (NM_002046)	aattccatggcaccgtcaag	agggatctcgctcctggaag
Oct3/4 (NM_203289)	tgtactcctcggtccctttc	tccaggttttctttccctagc
Nanog (NM_024865)	cagtctggacactggctgaa	ctcgctgattaggctccaac
Klf4 (NM_004235)	tatgacccacactgccagaa	tgggaacttgaccatgattg
ALP (NM_001177520)	gacaagaagcccttcactgc	agactgcgcctggtagttgt
L-Sox5 (NM_006940)	atcccaactaccatggcagct	tgcagttggagtgggccta
Sox6 (NM_017508)	gcagtgatcaacatgtggcct	cgctgtcccagtcagcatct
Sox9 (NM_000346)	agacagcccccctatcgactt	cggcaggtactggtcaaact
Aggrecan (NM_013227)	tcgaggacagcgaggcc	tcgagggtgtagcgtgtagaga
Col2a1 (NM_001844)	ggcaatagcaggttcacgtaca	cgataacagtcttgccccactt
Col2B (NM_033150)	agggccaggatgtccggca	gggtcccaggttctccatct
Col1a1 (NM_000088)	gtgctaaaggtgccaatggt	accaggttcaccgctgttac
ColXa1 (NM_000493)	caaggcaccatctccaggaa	aaagggtatttgtggcagcatatt
Runx2 (NM_004348)	gccttcaaggtggtagccc	cgttacccgccatgacagta
Runx1 (NM_001754)	aaccctcagcctcagagtca	caatggatcccaggtattgg

Table 2
Flow cytometry antibodies (BD Pharmingen) (*see* Note 5)

Antibody	Clone #	Vol (μL)/100 μL of cell suspension
FITC mouse IgG2b isotype control	clone 27-35	20
FITC mouse anti-human CD31	WM59	20
FITC mouse anti-human CD44	C26	20
FITC mouse anti-human CD45	H130	10
FITC mouse anti-human CD105	266	5
FITC mouse anti-human HLA-ABC	G46-2.6	20
PE mouse IgG1	MOPC-21	20
PE mouse anti-human CD29	MAR4	20
PE mouse anti-human CD73	AD2	20
PE mouse anti-human CD90	5E10	10
PE mouse anti-human CD166	3A6	10
PE mouse anti-human HLA-DR	G46-6	20

PE phycoerythrin, *FITC* fluorescein isothiocyanate

3 Methods

Human fibroblast-derived iPS cells are routinely propagated under serum-free culture conditions on a feeder layer of irradiated murine embryonic fibroblasts (MEFs). Detailed methods for the routine expansion and maintenance of undifferentiated human pluripotent stem cells can be found in [14]. Routine testing of human iPS cell cultures for chromosomal stability and mycoplasma contamination, as well as daily monitoring to prevent spontaneous differentiation, is advised.

3.1 Chondrogenic Differentiation of Naïve Pluripotent Stem Cells

1. Expand human iPS cells in 6-well plates at 37 °C, 5 % CO_2. Harvest cells at 75 % confluency.
2. Aspirate the medium from 6 to 12 wells containing undifferentiated human iPS cells. Wash once with 2 mL of PBS and apply 1 mL of Accutase solution to each well (*see* **Note 6**).
3. Return the plate to 37 °C and incubate for approximately 5 min.
4. Once the cells have detached, disperse iPS cell colonies into single cells by gentle, repeated pipetting using a 1-mL pipettor.
5. Transfer the cell suspension into a 15-mL sterile conical polypropylene tube containing micromass plating medium (Medium 1).
6. Centrifuge the cells at $300 \times g$ for 5 min and discard the supernatant.
7. Resuspend the cell pellet in 5 mL of Medium 1. Gently pipet up and down several times using a 1-mL pipettor.
8. Pass the cell suspension once through a 22-G needle attached to a 10-mL syringe, then pass through a 40-μm nylon cell strainer.
9. Count the cells using the Trypan Blue exclusion method with a hemocytometer and Trypan Blue solution, or using an automated cell counter.
10. Dilute the cells in Medium 1 to a final concentration of 2×10^6 cells/mL (*see* **Note 7**).
11. Seed the cells by applying 10-μL drops onto 6-well Primaria plates (*see* **Note 8**). Up to three high-density cell spots may be added per well of a 6-well dish. Ensure sufficient spacing between the drops (*see* **Note 9**).
12. Allow cells to attach for 2 h at 37 °C in a humidified 5 % CO_2 incubator.

13. To each well, carefully apply 1.5 mL of Medium 1 containing 10 μM of ROCK inhibitor Y-27632. Return the plate to the cell culture incubator overnight.
14. After 24 h, aspirate the medium and add 2 mL of chondrogenic differentiation medium (Medium 2, without growth factor). Incubate at 37 °C in a humidified 5 % CO_2 incubator.
15. On the following day (day 2), add fresh chondrogenic differentiation media (Medium 2) containing Bmp-2. Replace the growth factor-supplemented medium every other day.
16. Perform matrix staining (Subheading 3.3) and gene expression analyses (Subheading 3.4) of iPS cell micromass cultures over a time course of differentiation (i.e., days 5, 7, 10, 14, 21). By 7 days of differentiation, Bmp-2 stimulated micromasses typically display a dense Alcian Blue stained central core surrounded by a diffuse, cellular layer (*see* Fig. 1). With progressive differentiation, Alcian Blue positive cellular outgrowths and

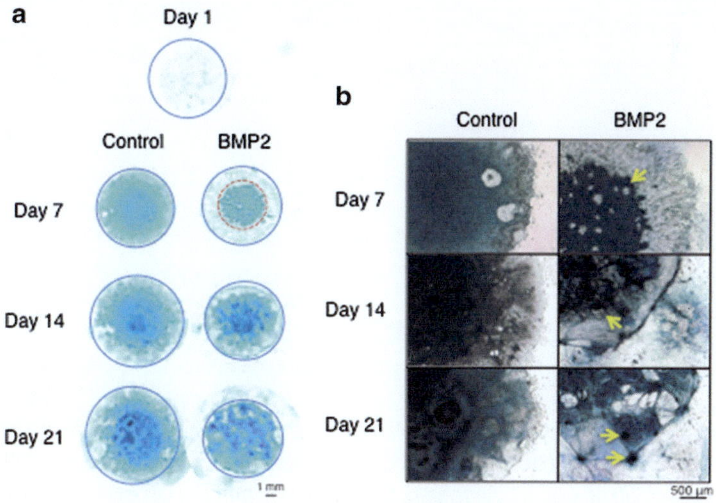

Fig. 1 Proteoglycan-rich matrix accumulation in chondrogenic iPS cell micromass cultures. (**a**) Alcian Blue staining of chondrogenic micromasses formed from human dermal fibroblast-derived iPS cells on days 1, 7, 14, and 21 of differentiation, treated with and without human recombinant Bmp-2 (100 ng/mL). Staining showed temporal accumulation of sulfated proteoglycans in control and Bmp-2 treated micromasses. Enhanced compaction of iPS cells within the central micromass core was observed in Bmp-2 treated cultures. (**b**) Higher magnification images of Alcian Blue staining in control and Bmp-2 treated iPS cell micromasses. *Arrows* show enhanced cellular compaction and nodule formation in Bmp-2 treated micromasses. Scale bar, 500 μm. Figure reproduced from Guzzo et al. [8]

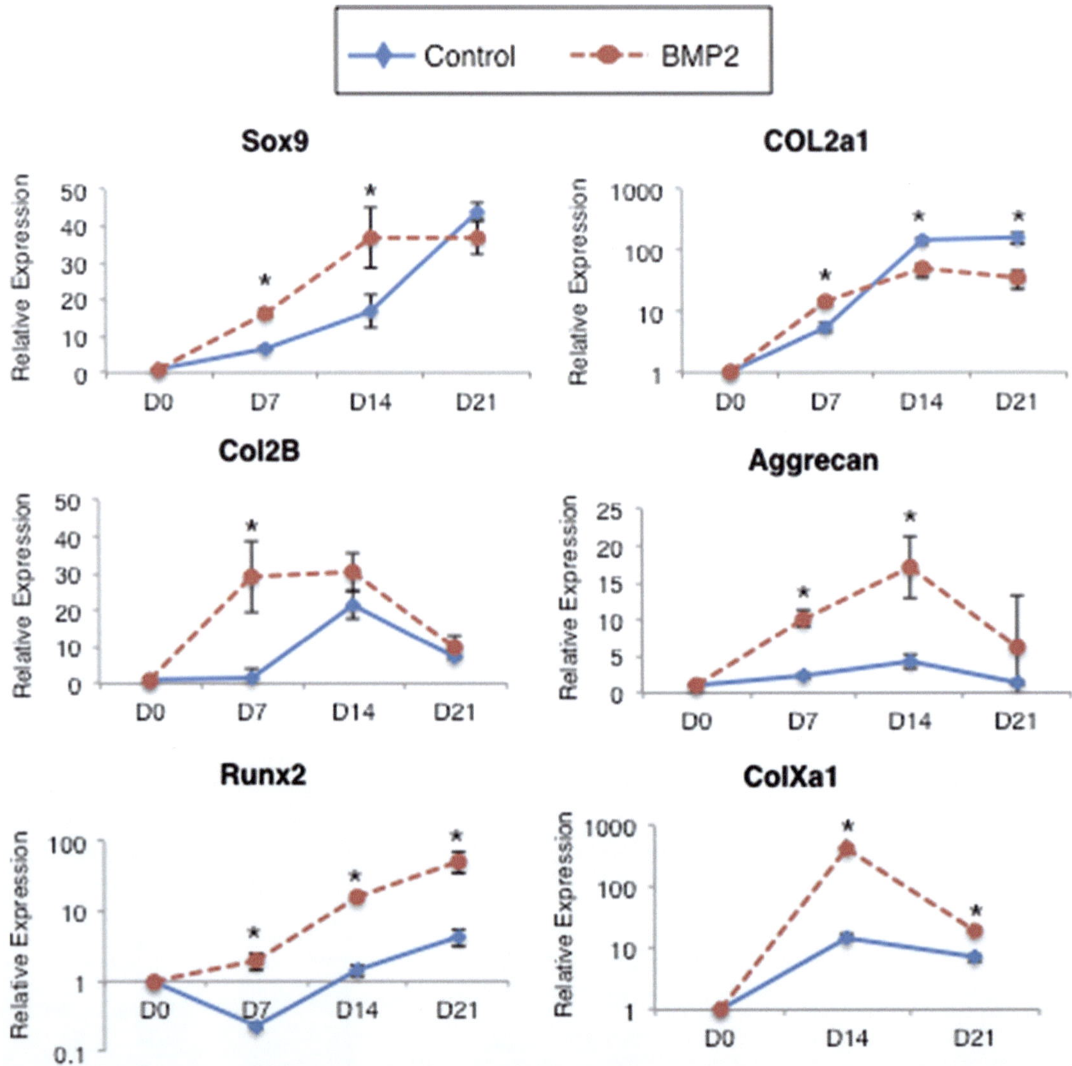

Fig. 2 BMP-2 treatment enhanced chondrogenic differentiation of human iPS cells. qRT-PCR based expression analyses of *Sox9, Col2a1, Aggrecan, Col2b, Runx2, and ColXa1* in iPS cell micromasses treated with or without Bmp-2 for 7, 14, and 21 days. The results represent the mean magnitude of transcript levels normalized to *Gapdh*, and expressed relative to undifferentiated iPS cells (day 0). Error bars indicate S.E.M. ($n=5$–9). *Asterisks* indicate statistical differences ($P<0.05$) from control (untreated) samples at each timepoint. Figure reproduced from Guzzo et al. [8]

cartilaginous nodules are typically observed in Bmp-2 treated iPS cell micromasses (*see* Fig. 1). Analyses of cartilage-specific genes can be expected to show increased expression of Sox9, Col2a1, aggrecan, and Col2B in Bmp-2 stimulated iPS cell micromasses (*see* Fig. 2).

3.2 Chondrogenic Differentiation of Pluripotent Stem Cells Through a Mesenchymal-Like Progenitor Intermediate

3.2.1 Derivation of Mesenchymal-Like Progenitors

1. Propagate human iPS cells in 6-well plates using standard procedures.
2. Aspirate iPS cell growth medium from each well of a 6-well plate and wash the cells with sterile PBS.
3. Apply 1 mL of dispase solution to each well and incubate at 37 °C for 15–20 min (*see* **Note 10**).
4. Gently dislodge the colonies from the well surface and transfer the colonies from multiple wells into a 15-mL polypropylene conical tube containing Medium 1. Allow the cell clumps to settle to the bottom of the tube for 3–5 min, and then aspirate the medium without disturbing the sediment clumps. Wash three to four times in Medium 1, with a final wash in PBS for thorough removal of dispase solution.
5. Apply 1 mL of Accutase solution to each tube and gently resuspend the cell clumps. Incubate for 2–2.5 min at 37 °C (*see* **Note 11**).
6. Add an equal volume of mesenchymal stem cell induction and growth medium (Medium 3) to each tube.
7. Centrifuge (300 × *g*) for 5 min at room temperature. Discard the supernatant and resuspend the cell pellet in Medium 3.
8. Plate the iPS cells onto 6-well plates precoated with gelatin solution. Use a cell culture split ratio of 1:1 or 1:2 (i.e., cells harvested from one well are plated onto one or two wells). Culture the cells in a humidified atmosphere with 5 % CO_2 at 37 °C. This passage is designated as iPS cell-MSC passage 0 (p0).
9. Change the medium every 2–3 days. Within 7–10 days, cultures exhibit a mixed population of flattened cuboidal and elongated spindle-shaped cells (*see* **Note 12**).
10. Once a confluent monolayer is reached, passage the cells. Apply 1 mL of Accutase solution to each well of a 6-well dish. Incubate at 37 °C for 5 min, then singularize the cells by repeated pipetting using a 1-mL pipettor. Transfer the cell suspension to a 15-mL conical tube containing Medium 3.
11. Centrifuge the cells at 300 × *g* for 5 min. Discard the supernatant.
12. Resuspend the cell pellet in Medium 3 and count the cells.
13. Prepare a dilution of 2.5×10^5 cells/mL. Plate 1 mL of the cell suspension onto each well of a 6-well culture plate precoated with gelatin solution. This passage is designated iPS cell-MSC passage 1.
14. Propagate the cells in a humidified atmosphere with 5 % CO_2 at 37 °C, with medium changes every 2–3 days.

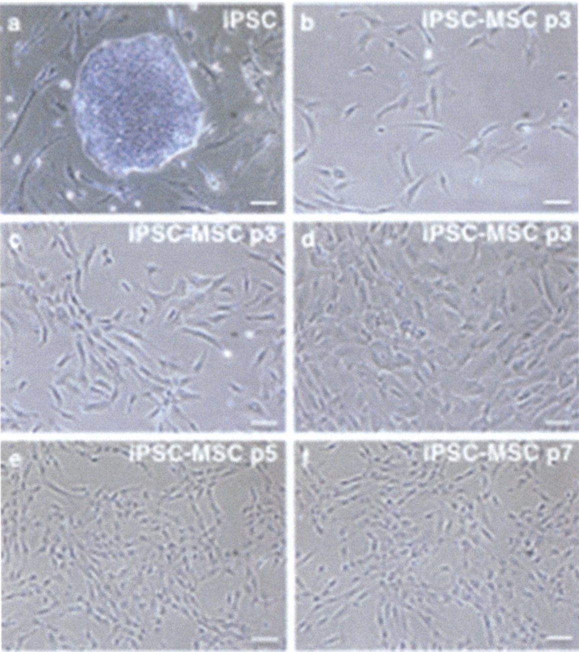

Fig. 3 Generation of mesenchymal-like progenitors from human dermal fibroblast-derived iPS cells. Image in (**a**) shows morphology of an undifferentiated human dermal fibroblast-derived iPS cell [8] colony cultured on a feeder layer of irradiated mouse embryonic fibroblasts (MEFs). iPS cell-derived mesenchymal-like progenitors (iPSC-MSC) in (**b–f**) exhibit a spindle-shaped morphology. Representative images of iPSC-MSC-like cells (passage 3) at low density (**b**), medium density (**c**), and high density (**d**) are shown. iPSC-MSC-like cells maintain mesenchymal-like morphology at higher passages (**e**, passage 5) and (**f**, passage 7). Scale bar, 100 μm. Figure reproduced from Guzzo et al. [8]

15. Within 2 weeks, cells adopt a fibroblastic, spindle-like morphology, resembling mesenchymal progenitor cells (*see* Fig. 3) (*see* **Note 13**). Once cultures reach 90 % confluency, split using trypsin-EDTA solution and seed (1×10^4 cells/cm^2) onto plates precoated with gelatin solution.
16. Change the medium every 2–3 days.
17. Assay the expression of stem cell genes (i.e., *Oct4*, *Nanog*, *alkaline phosphatase*, *Klf4*) and gene markers associated with the mesenchymal lineage (i.e., *Twist1*, *Col1a1*) by quantitative RT-PCR (as described in Subheading 3.4). Stem cell genes are suppressed in the mesenchymal-like population, whereas mesenchymal markers are significantly induced.
18. For routine expansion, seed cells at a density of 1×10^4 cell/cm^2 and maintain in Medium 3 (*see* **Note 14**). Monitor passage numbers.

19. Establish a cell bank by freezing batches at each passage. At harvest, resuspend singularized cells in 1× cryopreservation medium and aliquot 1×10^6 cells per vial. Freeze vials at −80 °C and transfer the vials to liquid nitrogen for long-term storage (*see* **Note 15**).

3.2.2 Flow Cytometry

Characterization of cell surface marker expression in iPS cell-derived progenitors is performed to validate the mesenchymal-like properties of the progenitor population.

1. Culture the iPS cell-derived mesenchymal-like progenitors (passage 4–6) to 90 % confluency and harvest the cells using trypsin-EDTA solution.
2. Transfer the cells to 15-mL conical tubes and centrifuge for 5 min at 300 × *g*.
3. Wash the cells with PBS. Strain the cells through a nylon mesh (40 μm) to ensure a single-cell suspension.
4. Count the cells. Adjust the concentration of the cell suspension (5×10^6 cells/mL) in ice-cold antibody staining solution. Incubate the cells on ice for 20–30 min to block nonspecific binding sites.
5. Perform cell staining in 5-mL polystyrene round bottom tubes with a cell strainer cap. Aliquot 5×10^5–1×10^6 cells/tube and add the appropriate volume of antibody per 100 μL of cell suspension (refer to Table 2) (*see* **Note 5**). Incubate on ice for 30 min in the dark (*see* **Note 16**).
6. Add 500 μL of ice-cold antibody staining solution to each sample and centrifuge at 300 × *g* for 5 min at 4–8 °C. Wash the samples three times with 500 μL of ice-cold antibody staining solution. Following the last wash, resuspend the cells in 300 μL of antibody staining solution and strain the cells through the caps of the tubes (*see* **Notes 17** and **18**).
7. Collect a minimum of 30,000–50,000 events on a flow cytometer, e.g., Becton-Dickinson LSR II (BD Biosciences) using FACS Diva software (Becton Dickinson). Data are analyzed using FloJo Software (Tree Star, Inc.). By flow cytometry analyses, it is expected that expression of mesenchymal surface markers (CD29, CD44, CD73, CD90, CD105, CD166, HLA-ABC) will be detected in the progenitor population derived from human iPS cells. Approximately 95–100 % of iPS cell-derived mesenchymal-like cells are likely to be positive for the aforementioned markers. The cells largely lack expression of surface endothelial markers (i.e., CD31), hematopoietic markers (i.e., CD45), as well as the MHC class II cell surface receptor HLA-DR [8].

3.2.3 Chondrogenic Differentiation of Human iPS Cell-Derived Mesenchymal-Like Progenitors

Chondrogenic differentiation of human iPS cell-derived mesenchymal-like progenitors is induced by culturing the cells in micromass according to the protocol described in Subheading 3.1.

3.3 Alcian Blue Staining of Wholemount Chondrogenic Micromasses

3.3.1 Method 1

1. Aspirate the medium from the culture dishes and wash with PBS.
2. Fix the micromasses in formalin solution for 20 min at room temperature (*see* **Note 19**).
3. Wash three times with PBS.
4. Apply 1 mL of Alcian Blue staining solution and incubate at room temperature for 2 h.
5. Wash the cells with 70 % ethanol solution to remove residual Alcian Blue stain. Repeat ethanol washes 3×.
6. Wash with distilled water 3×.
7. Aspirate the water and dry the plates overnight at room temperature (*see* **Note 20**).

3.3.2 Method 2

1. Fix micromasses as described above (*see* Subheading 3.3.1, **steps 1** and **2**).
2. Carefully detach the adherent micromasses from the plate surface using a disposable cell lifter. Transfer the detached micromasses to 15-mL polypropylene conical tubes. Pool several micromasses per tube.
3. Spin briefly (300 × *g*, 2 min) to sediment the micromasses.
4. Dehydrate the micromasses through sequential washes in 50 % ethanol (15 min), 70 % ethanol (15 min), 95 % ethanol (15 min), 100 % ethanol (15 min), xylene/ethanol solution (15 min), and 100 % xylene (15 min).
5. Embed the micromasses in tissue embedding medium (i.e., paraffin or Paraplast®X-Tra™).
6. Generate 5-μm sections using a microtome and mount onto Superfrost Plus slides.
7. Deparaffinize and rehydrate the sections through sequential washes in 100 % xylene (10 min, two times), 100 % ethanol (two washes, 2 min each wash), 95 % ethanol (two washes, 2 min each wash), 70 % ethanol (two washes, 2 min each wash), 50 % ethanol (two washes, 2 min each wash), then distilled water (5 min).
8. Place slides in Alcian Blue staining solution for 20–30 min at 37 °C.
9. Rinse briefly in acetic acid solution, then wash with distilled water until clear.

10. Counterstain in Nuclear Fast Red Kernechtrot solution for 5 min at room temperature, then wash with copious amounts of distilled water.
11. Dehydrate sections through sequential washes with 70, 95, and 100 % ethanol. Perform a final wash in xylene. Apply mounting media and glass coverslip.
12. View by light microscopy.

3.4 Analysis of Cartilage Gene Expression

Several methods are used to extract high quality total RNA, including TRIzol reagent and commercial RNA isolation kits, for the analyses of cartilage gene expression in differentiating cultures of iPS cells. The methods pertaining to RNA extraction using TRIzol reagent are described here.

1. Aspirate the medium and add 1 mL of TRIzol Reagent per well of a 6-well plate containing three micromasses. Scrape the adherent micromasses using a cell scraper.
2. Transfer the cell lysate in TRIzol to a 1.5-mL microfuge tube.
3. Pass the cell lysate several times through a 26-G needle attached to a 1-mL syringe. Incubate for 5 min at room temperature to dissociate the nucleoprotein complex.
4. Add 0.2 mL of molecular-grade chloroform and shake the tubes for 15 s. Incubate for 2 min at room temperature.
5. Centrifuge the samples at 12,000 ×*g* for 15 min at 4 °C.
6. Transfer the aqueous phase into new 1.5-mL microfuge tubes.
7. Add 0.5 mL of molecular-grade isopropyl alcohol (*see* **Note 21**).
8. Centrifuge at 12,000 ×*g* for 15 min at 4 °C.
9. Remove the supernatant and wash the RNA pellet 2× with 1 mL of 75 % ethanol solution in nuclease-free water. Centrifuge the tubes at 7500 ×*g* for 5 min at 4 °C.
10. Remove the ethanol and air-dry the pellets.
11. Add an appropriate volume (i.e., 10 μL) of nuclease-free water to the pellets. Incubate at 55 °C for 10 min to dissolve the pellet.
12. Determine RNA quantity and quality by measuring absorbance at 260 and 280 nm using a Nanodrop 2000c spectrophotometer. Store the RNA samples at −80 °C if cDNA synthesis is not being performed on the same day.
13. Eliminate genomic DNA by treating 1 μg of RNA with amplification-grade DNAse I, according to the manufacturer's instructions.
14. Perform first strand cDNA synthesis using the iScript cDNA synthesis kit, according to the manufacturer's instructions. Reactions are performed in 0.2-mL PCR tubes using 1000 ng

of total RNA (DNase I-treated) per 20 μL of reaction volume. Place the reaction tubes in a thermocycler and run with the following settings: 5 min at 25 °C, 30 min at 42 °C, 5 min at 85 °C, then hold at 4 °C.

15. Dilute with nuclease-free water (80 μL).
16. Analyze the expression of *Sox9*, *Col2a1*, *Col2B*, *Aggrecan*, *Runx2*, and *ColXa1* transcripts by qRT-PCR. Use an equivalent amount of cDNA (20 ng) per reaction, the SYBR Green I Master Mix kit, and a Real-Time PCR system.
17. Use the endogenous housekeeping gene, glyceraldehyde-3-phosphate dehydrogenase (*Gapdh*), as the internal control for normalization. Gene-specific oligonucleotide primer sequences are given in Table 1.
18. Apply the following qRT-PCR settings: 95 °C for 10 min, followed by 40 cycles of amplification (denaturation step at 95 °C for 15 s and extension step at 60 °C for 1 min). Perform melt curve analyses to validate gene-specific amplification. The level of each gene is calculated as 2^DeltaDeltaCt, with DeltaDeltaCt defined as the difference in crossing threshold (Ct) values between experimental and control samples, using *Gapdh* as an internal standard [8]. Typical results are shown in Figs. 2 and 4.

4 Notes

1. The capacity of human iPS cells to differentiate into chondrocytes may vary among different sources of iPS cells. These protocols have been successfully used to differentiate multiple lines of human iPS cells to chondrocytes.
2. Prepare working aliquots of FBS and KSR. Store at −20 °C.
3. Avoid repeated freeze-thaw cycles to ensure growth factor stability.
4. When stored at 4 °C in the dark, chondrogenic differentiation medium is generally stable for up to 2 weeks. Since medium components such as AA2P degrade over time, discard any unused medium after 2 weeks of preparation.
5. Volumes listed in Table 2 were determined for specific antibody clones purchased from BD Pharmingen. Further optimization may be necessary when antibodies from different suppliers are used.
6. Prior to use, culture medium is warmed to 37 °C in a water bath. Enzyme solutions (trypsin-EDTA, dispase, Accutase) and PBS should be used at room temperature.

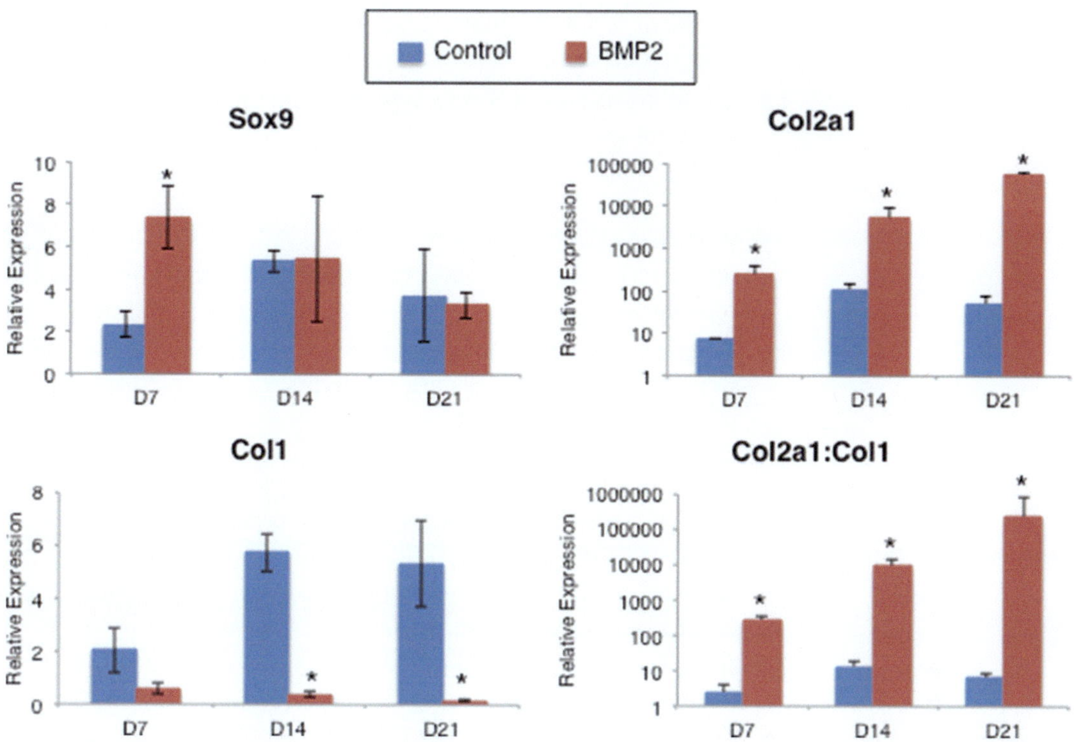

Fig. 4 High-density micromass conditions influence the chondrogenic potential of iPSC-MSC-like cells. Profile of temporal gene expression in control and BMP-2 treated iPSC-MSC-like micromasses, as determined by quantitative real-time PCR. Genes analyzed include *Sox9*, *Col2a1*, and *Col1a1*. The ratio of *Col2a1* to *Col1a1* transcript expression is also shown. *Gapdh* served as the housekeeping gene and internal control. Gene expression is represented as fold induction relative to undifferentiated iPSC-MSC-like cells (day 0), set at 1.0. *Asterisks* denote significance relative to untreated iPSC-MSC-like cells for each timepoint at $P<0.05$. "D" refers to days of differentiation. Figure reproduced from Guzzo et al. [8]

7. Add 10 μM of ROCK inhibitor Y27632 to the medium to promote survival of human iPS cells. Remove inhibitor from the medium within 24 h of cell seeding.
8. For optimal micromass attachment, we use Primaria™ 6-well culture plates (without gelatin precoating). When using other sources, it may be necessary to precoat the plates with 0.1 % gelatin. Thorough drying of the plates prior to micromass formation prevents cell spreading.
9. Avoid drying of micromasses during the 2-h incubation period by applying PBS within the reservoir between wells.
10. Monitor the detachment of iPS cell colonies from the plate surface. Intact colonies will partially peel off and exhibit rounded or rolled-up edges.
11. Do not expose cells to Accutase for longer than 2.5 min. The goal is to generate small clumps of cells rather than a single-cell suspension. To promote cell survival, 10 μM of ROCK

inhibitor Y-27632 may be added at the time of seeding and maintained in the plating medium for up to 24 h.

12. Undifferentiated iPS cells typically pile up and do not flatten. Mechanically excise the undifferentiated regions from the cultures prior to the first passage using a dissecting microscope within a laminar flow cabinet.
13. With increased passaging onto gelatin-coated tissue culture plates (p1–p2), cells develop a homogenous, fibroblast-like morphology.
14. Cells are cultured on gelatin-coated plates up to passage 2. Beyond passage 2, cells are seeded onto tissue culture plates without gelatin coating.
15. This approach has been used in our lab to generate mesenchymal-like progenitors from multiple sources of human iPS cells [8] (and unpublished data). We established the multilineage differentiation potential of human iPSC-MSC-like cells by their capacity for osteogenesis, adipogenesis, and chondrogenesis in vitro [8].
16. Include a "no stain" control and negative (similarly labeled isotype-matched antibody such as IgG1-PE and IgG2b-FITC).
17. Cells must be singularized to prevent damage to the flow cell instrument.
18. Keep samples on ice and protected from light prior to flow cytometry.
19. Formalin is toxic. Use gloves when handling and work in a fume hood.
20. Alcian blue is used as a histological stain for proteoglycan deposition. The stained micromasses may be viewed by light microscopy or the plates scanned for densitometric analyses of proteoglycan deposition.
21. GlycoBlue™ Coprecipitant added to isopropyl alcohol increases the visibility of the RNA pellet. Dilute GlycoBlue™ Coprecipitant to 50–60 μg/mL in molecular-grade isopropyl alcohol.

Acknowledgements

RMG acknowledges the support of the State of Connecticut Stem Cell Program/Department of Public Health (grants nos. 10SCA36, 13-SCA-UCHC-11). HD acknowledges the support of the State of Connecticut Stem Cell Program/Department of Public Health (Grant #11SCB08). The University of Connecticut Stem Cell Core Facility and the Flow Cytometry Center provided technical support.

References

1. Takahashi K, Tanabe K, Ohnuki M, Narita M, Ichisaka T, Tomoda K, Yamanaka S (2007) Induction of pluripotent stem cells from adult human fibroblasts by defined factors. Cell 131(5):861–872. doi:10.1016/j.cell.2007.11.019
2. Park IH, Arora N, Huo H, Maherali N, Ahfeldt T, Shimamura A, Lensch MW, Cowan C, Hochedlinger K, Daley GQ (2008) Disease-specific induced pluripotent stem cells. Cell 134(5):877–886. doi:10.1016/j.cell.2008.07.041
3. Lee J, Kim Y, Yi H, Diecke S, Kim J, Jung H, Rim YA, Jung SM, Kim M, Kim YG, Park SH, Kim HY, Ju JH (2014) Generation of disease-specific induced pluripotent stem cells from patients with rheumatoid arthritis and osteoarthritis. Arthritis Res Ther 16(1):R41. doi:10.1186/ar4470
4. Oldershaw RA, Baxter MA, Lowe ET, Bates N, Grady LM, Soncin F, Brison DR, Hardingham TE, Kimber SJ (2010) Directed differentiation of human embryonic stem cells toward chondrocytes. Nat Biotechnol 28(11):1187–1194
5. Koyama N, Miura M, Nakao K, Kondo E, Fujii T, Taura D, Kanamoto N, Sone M, Yasoda A, Arai H, Bessho K (2013) Human induced pluripotent stem cells differentiated into chondrogenic lineage via generation of mesenchymal progenitor cells. Stem Cells Dev 22(1):102–113. doi:10.1089/scd.2012.0127
6. Toh WS, Guo XM, Choo AB, Lu K, Lee EH, Cao T (2009) Differentiation and enrichment of expandable chondrogenic cells from human embryonic stem cells in vitro. J Cell Mol Med13(9B):3570–3590.doi:10.1111/j.1582-4934.2009.00762.x
7. Gong G, Ferrari D, Dealy CN, Kosher RA (2010) Direct and progressive differentiation of human embryonic stem cells into the chondrogenic lineage. J Cell Physiol 224(3):664–671. doi:10.1002/jcp.22166
8. Guzzo RM, Gibson J, Xu RH, Lee FY, Drissi H (2013) Efficient differentiation of human iPSC-derived mesenchymal stem cells to chondroprogenitor cells. J Cell Biochem 114(2):480–490. doi:10.1002/jcb.24388
9. Hwang NS, Varghese S, Elisseeff J (2008) Derivation of chondrogenically-committed cells from human embryonic cells for cartilage tissue regeneration. PLoS One 3(6):e2498. doi:10.1371/journal.pone.0002498
10. Nakagawa T, Lee SY, Reddi AH (2009) Induction of chondrogenesis from human embryonic stem cells without embryoid body formation by bone morphogenetic protein 7 and transforming growth factor beta1. Arthritis Rheum 60(12):3686–3692
11. Ko JY, Kim KI, Park S, Im GI (2014) In vitro chondrogenesis and in vivo repair of osteochondral defect with human induced pluripotent stem cells. Biomaterials 35(11):3571–3581. doi:10.1016/j.biomaterials.2014.01.009
12. Liu Y, Goldberg AJ, Dennis JE, Gronowicz GA, Kuhn LT (2012) One-step derivation of mesenchymal stem cell (MSC)-like cells from human pluripotent stem cells on a fibrillar collagen coating. PLoS One 7(3):e33225. doi:10.1371/journal.pone.0033225
13. Hynes K, Menicanin D, Mrozik K, Gronthos S, Bartold PM (2014) Generation of functional mesenchymal stem cells from different induced pluripotent stem cell lines. Stem Cells Dev 23:1084–1096. doi:10.1089/scd.2013.0111
14. Leonardo TR, Schnell JP, Nickey KS, Tran HT (2012) Culturing human pluripotent stem cells on a feeder layer. In: Peterson S, Loring JF (eds) Human stem cell manual: a laboratory guide, 2nd edn. Elsevier Inc., New York, pp 3–14

Chapter 7

Gene Transfer and Gene Silencing in Stem Cells to Promote Chondrogenesis

Feng Zhang and Dong-An Wang

Abstract

In stem cell-based chondrogenesis for articular cartilage regeneration, TGF-β3 is dosed to the stem cells to drive differentiation into chondrocytic cells. Meanwhile, type I collagen, which is endogenously expressed in some stem cells (e.g., synovium-derived mesenchymal stem cells) and upregulated by TGF-β3, poses a threat to chondrogenesis, as type I collagen may alter the components and stiffness of articular cartilage. Therefore, a wiser strategy would be to feed the cells with TGF-β3 while at the same time silencing the expression of type I collagen. In this chapter, methods for construction of adenoviral vectors and lentiviral vectors having both of the above functions are given. Their transduction into synovium-derived mesenchymal stem cells for articular cartilage engineering and following characterizations are also described.

Key words Chondrogenesis, Type I collagen, TGF-β3, Adenovirus, Lentivirus

1 Introduction

Hyaline articular cartilage trauma and degeneration are major causes of suffering and pose great threat to the life quality of human beings worldwide [1, 2]. However, there is a limited capacity for healing of cartilage itself due to its avascular nature [3]. In cell-based therapy, chondrocytes remain the major choice for cartilage regeneration, as they are the single type of cells within cartilage [4, 5]. However, chondrocytes undergo dedifferentiation during in vitro monolayer expansion, in which process the cells lose their phenotype by decreasing expression of chondrocytic markers—mainly type II collagen (Col II) and aggrecan—and upregulating type I collagen (Col I) expression [6, 7]. Alternatively, synovial mesenchymal stem cells (SMSCs) are emerging as a promising source for chondrogenesis due to the relative ease of their derivation and differentiation into chondrocytes [8]. Transforming growth factor β3 (TGF-β3) could be

Pauline M. Doran (ed.), *Cartilage Tissue Engineering: Methods and Protocols*, Methods in Molecular Biology, vol. 1340, DOI 10.1007/978-1-4939-2938-2_7, © Springer Science+Business Media New York 2015

dosed to source cells to induce differentiation in SMSCs or redifferentiation in chondrocytes due to its ability to maintain chondrocyte morphology, promote collagen synthesis, and induce chondrogenic differentiation [9, 10]. 3D alginate hydrogel is utilized as it is easy in manipulation and, more importantly, compatible with chondrogenesis [11–13]. However, a serious drawback lies in the fact that Col I is intrinsically expressed in SMSCs and upregulated in chondrocytes during monolayer culture or by the introduction of TGF-β3 in some types of cells. The existence of Col I would result in the formation of fibrous cartilage, which is significantly different from hyaline articular cartilage and may not withstand the normal loading exerted on cartilage [14, 15]. Therefore, an RNA interference (RNAi) strategy was adopted to silence or suppress the expression of Col I by degrading Col I mRNA in a posttranscriptional pathway.

To deliver both TGF-β3 and Col I-targeting short hairpin RNA (shRNA) to the source cells, both adenoviral and lentiviral vectors are applied. Among them, the adenoviral vector is supposed to induce a transient expression due to its episomal performance, whereas the lentiviral vector would lead to a more sustained expression since the vector can integrate its genome together with transgenes into the host genome [16]. In this chapter, a dual-functioning adenoviral vector is first constructed to deliver both TGF-β3 and Col I-targeting shRNA [15]. Strategies for cloning DNA sequences into the donor vector are given in detail, followed by subcloning the genes of interest into the adenovirus backbone pLP-Adeno-X ViraTrak. Mature adenoviruses are then produced by transfecting PacI-digested recombinant pLP-Adeno-X ViraTrak into low-passage human embryonic kidney (HEK) 293 cells and harvesting within 7–10 days. Adenovirus rescued from transfected cells is then amplified (1–2 weeks) and titrated before infecting target cells with the recombinant adenovirus (*see* **Note 1**).

Subsequently, four dual-functioning lentiviral vectors that have various arrangements of TGF-β3-encoding cassette and shRNA-encoding cassette are constructed [17]. Briefly, genes of interest are cloned into lentivirus backbone pLVX first. Lentiviruses are produced by co-transfecting recombinant pLVX and packaging mix, which express all the proteins necessary for the formation of mature lentiviruses. The efficiency of Col I suppression and chondrogenic induction can be compared between the four dual-functioning lentiviral vectors to choose the optimal one. LV-1, with distant and reverse arrangement of the two cassettes, proved to be the most efficient among the four (*see* **Note 2**).

Finally, some assays for characterizing chondrogenesis in 3D alginate hydrogel culture using the recombinant viral vectors are described.

2 Materials

2.1 Production of Adenoviral Vectors Expressing TGF-β3 and shRNA

1. hTGF-β3-pCMV6-XL5 plasmid DNA (OriGene Technologies, Rockville, MD, USA).
2. pDNR donor vector.
3. 50× TAE buffer: 40 mM Tris, 20 mM acetic acid, 1 mM EDTA, pH 8.4.
4. 2 and 1 % agarose gels: For 2 % agarose gel, 2 g of agarose in 100 ml of 1× TAE (diluted from 50× TAE buffer with deionized water). Heat in a microwave oven to dissolve. Pour into the gel mould for cooling at room temperature until it fully solidifies. For 1 % agarose gel, use 1 g of agarose in 100 ml of 1× TAE.
5. T4 DNA Ligase: 1 U/μl of T4 DNA ligase in 10 mM Tris–HCl, 50 mM KCl, 1 mM dithiothreitol (DTT), 0.1 mM EDTA, 50 % glycerol.
6. 10× T4 DNA Ligase buffer: 500 mM Tris–HCl (pH 7.6), 100 mM $MgCl_2$, 10 mM ATP, 10 mM DTT, 25 % (w/v) polyethylene glycol-8000. Store at −20 °C.
7. DH5α Competent *E. coli* cells.
8. PCR purification kit.
9. Gel extraction kit.
10. TRIzol reagent (Invitrogen).
11. Electroporation apparatus and electrocompetent *E. coli* cells (2×10^{10} cfu/μg pUC19).
12. Chloramphenicol (Cm) stock solution (1000×): 30 mg/ml Cm in 100 % ethanol. Filter sterilize and store in aliquots at −20 °C.
13. Ampicillin (Amp) stock solution (1000×): 100 mg/ml Amp in H_2O. Filter sterilize and store in aliquots at −20 °C.
14. SOC medium.
15. LB liquid broth, 1×.
16. LB/Cm/sucrose agar plates: Prepare your own 100-mm LB/Cm (30 μg/ml)/sucrose (7 %) agar plates using the recipe below (total volume is 1 l).
    (a) To 750 ml of deionized H_2O, add 70 g sucrose. Stir until dissolved.
    (b) Add 10 g of tryptone, 5 g of yeast extract, and 10 g of NaCl. Stir until dissolved, then add 15 g of agar.
    (c) Adjust the volume to 1 l by adding deionized H_2O.
    (d) Sterilize by autoclaving for 20 min at 121 °C. Remove from the autoclave immediately when done.
    (e) After the solution has cooled to ~50 °C, add 1 ml of filter-sterilized chloramphenicol stock solution. Mix briefly.

(f) Pour ~30 ml per 100-mm plate. Cover the plates and allow them to dry overnight.

17. HEK 293 cells: human embryonic kidney cells immortalized with adenovirus 5 genes.
18. Human fibroblast cells.
19. Complete growth medium-1: DMEM (Dulbecco's Modified Eagle's Medium) containing 4 mM l-glutamine, 4.5 g/l glucose, 1 mM sodium pyruvate, 1.5 g/l sodium bicarbonate, and supplemented with 10 % fetal bovine serum (FBS), 100 U/ml of penicillin G sodium, and 100 μg/ml of streptomycin.
20. Adeno-X™ Virus Purification Kits (Clontech).
21. Adeno-X™ Virus Rapid Titer Kit (Clontech).
22. LB/Amp/Cm liquid broth: LB liquid broth 1× with 100 μg/ml ampicillin and 30 μg/ml chloramphenicol.
23. 95 % ethanol solution: 95 % ethanol in deionized water.
24. Ammonium acetate (NH_4Ac) solution: 10 M NH_4Ac in deionized water.
25. Glycogen solution: 20 mg/ml glycogen in deionized water.
26. 70 % ethanol solution: 70 % ethanol in deionized water.
27. Dry ice/ethanol bath: Dry ice is crushed to a powder and slowly added to the 95 % ethanol solution until a thick slurry is achieved.
28. TE buffer: 10 mM Tris, 1 mM EDTA in deionized water. Bring the pH to 8.0 with HCl.
29. 10× Bovine serum albumin (BSA) solution (New England Biolabs).
30. Adeno-X™ ViraTrak Expression System 2 kit (Clontech).
31. PacI restriction endonuclease solution: 10,000 U/ml of PacI restriction endonuclease in 10 mM Tris–HCl, 200 mM NaCl, 1 mM DTT, 0.1 mM EDTA, 200 μg/ml BSA, 50 % glycerol (New England Biolabs).
32. 10× digestion buffer: 500 mM potassium acetate, 200 mM Tris-acetate, 100 mM magnesium acetate, 1 mg/ml BSA, pH 7.9.
33. UltraPure™ Phenol:Chloroform:Isoamyl Alcohol (25:24:1, v/v): highly pure phenol, chloroform, and isoamyl alcohol mixed at 25:24:1 (v/v) ratio, saturated with Tris–HCl.

2.2 Production of Lentiviral Vectors Expressing TGF-β3 and shRNA

1. HEK 293T cell line: human embryonic kidney cells immortalized with SV40 T large antigen.
2. Complete growth medium-2: DMEM containing 4 mM l-glutamine, 4.5 g/l glucose, 1 mM sodium pyruvate, 1.5 g/l sodium bicarbonate, and supplemented with 10 % tetracycline (Tc)-free FBS, 100 U/ml of penicillin G sodium, and 100 μg/ml of streptomycin.

3. Polybrene (hexadimethrine bromide).
4. Lenti-X Expression System (Clontech).
5. Lenti-X HT Packaging Mix (Clontech).
6. Lentiphos Transfection Reagent kit (Clontech).
7. Phosphate buffered saline (PBS): 137 mM NaCl, 2.7 mM KCl, 4.3 mM Na_2HPO_4, 1.4 mM KH_2PO_4, pH 7.2.
8. Trypsin solution (Sigma).
9. 1 % agarose gel: 1 g of agarose in 100 ml of 1× TAE (diluted from 50× TAE buffer with deionized water). Heat in a microwave oven to dissolve. Pour into the gel mould for cooling at room temperature until it fully solidifies.

2.3 Three-Dimensional Culture of SMSCs

1. Washing buffer: 0.15 M NaCl and 25 mM HEPES *N*-(2 hydroxyethyl) piperazine-*N*'-(2-ethanesulfonic acid) in distilled water.
2. Sterile alginate solution: 1.2 % alginate in deionized water. The solution is sterilized at 120 °C for 15 min and stored at 4 °C after cooling.
3. Sterile $CaCl_2$ solution: 102 mM $CaCl_2$ in deionized water. Filter through a 0.22-μm filter for sterilization.
4. Complete growth medium-1.
5. Trypsin solution (Sigma).

2.4 Quantitative Real-Time PCR (qPCR)

1. TRIzol reagent (Invitrogen).
2. Superscript® First-Strand Synthesis System (Invitrogen).
3. IQ™ SYBR Green Supermix system (Bio-Rad).

2.5 Cell Viability and Proteoglycan Assays

1. Premixed WST-1 reagent (4-[3-(4-iodophenyl)-2-(4-nitrophenyl)-2*H*-5-tetrazolio]-1,3-benzene disulfonate) (Roche Diagnostics).
2. Papain solution: 3 mg/ml of papain (≥10 U/mg of protein) in 0.1 M Tris, 0.1 M sodium acetate at pH 7.3.
3. Dimethylmethylene blue dye solution: 80 % dye content in deionized water.

2.6 Histology and Immunohistochemistry

1. 4 % (w/v) neutral-buffered paraformaldehyde solution.
2. PBS.
3. 90, 80, 70, and 50 % ethanol solutions in deionized water.
4. Safranin-O solution (Sigma).
5. Masson's Trichrome Stains.
6. Glutaraldehyde solution: 2.5 % glutaraldehyde in deionized water.

7. Goat serum solution: 1 % w/v goat serum in PBS.
8. Primary antibody solution for Col I (Santa Cruz Biotechnology): 2 ng/ml in PBS.
9. Primary antibody solution for Col II (Chemicon): 2 ng/ml in PBS.
10. HRP-conjugated secondary antibodies (Invitrogen): 1 μg/ml in PBS.
11. DAB Substrate kit (Clontech).

3 Methods

3.1 Production of Adenoviral Vectors Expressing TGF-β3 and shRNA (Ad-D)

The Ad-D vector was constructed by sequentially incorporating the following fragments into pDNR: TGF-β3 coding sequence (**steps 1–6**), human U6 promoter (**steps 7–13**), and shRNA targeting type I collagen (**steps 14–20**). The structure of Ad-D is shown in Fig. 1.

3.1.1 Construction and Production of Recombinant Adenoviral Vector

1. PCR-amplify the TGF-β3 coding sequence from hTGF-β3-pCMV6-XL5 plasmid DNA using the following primers: forward ACGCGTCGACATGAAGATGCACTT; reverse TGCACTGCAGTCAGCTACATTTC, with a suitable commercial PCR kit. Run the PCR product on 2 % agarose gel and cut the exact band for gel purification using a commercial kit (*see* **Note 3**).
2. Double digest the PCR product (1 μg) using SalI and PstI restriction endonucleases. Purify the DNA using a PCR purification kit or gel purification kit.
3. Double digest the pDNR donor vector (1 μg) using SalI and PstI restriction endonucleases. Run the product on 1 % agarose gel and cut the exact band for gel purification (*see* **Note 4**).
4. Combine 200 ng of digested pDNR and 100 ng of digested TGF-β3 cDNA. Add 1 μl of T4 DNA Ligase and top up to 10 μl with double-distilled water (ddH_2O). Incubate at 18 °C overnight.
5. Transform DH5α Competent *E. coli* cells with the ligation mixture and incubate at 37 °C overnight.
6. Pick the colonies for mini- or midi-scaled cultures. Purify plasmid from the bacterial culture using any commercial plasmid extraction kit. Confirm successful recombination either by restriction digestion analysis, PCR, or sequencing.

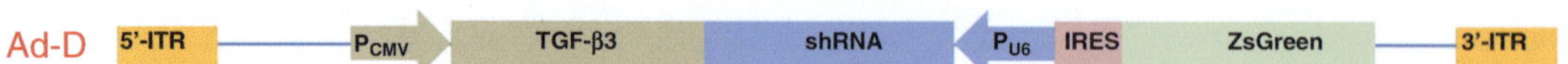

Fig. 1 Schematic illustration of the structure of Ad-D

7. Extract genomic DNA from human fibroblast cells using TRIzol reagent.
8. PCR-amplify the U6 promoter sequence from fibroblast genomic DNA using the following primers: forward ATTTGCGGGCCCGCAGGAAGAGGGCCTAT; reverse CCGCTCGAGTCGTCCTTTCCACAAG, with a suitable commercial PCR kit. Run the PCR product on 2 % agarose gel and cut the exact band for gel purification using a commercial kit.
9. Double digest the PCR product (1 μg) using ApaI and XhoI restriction endonucleases. Purify the DNA using a PCR purification kit or gel purification kit.
10. Double digest the pDNR-TGFβ3 (1 μg) using ApaI and XhoI restriction endonucleases. Run the product on 1 % agarose gel and cut the exact band for gel purification.
11. Combine 750 ng of digested pDNR-TGFβ3 and 100 ng of digested U6 promoter fragment. Add 1 μl of T4 DNA Ligase and top up to 10 μl with ddH_2O. Incubate at 18 °C overnight.
12. Transform DH5α Competent *E. coli* cells with the ligation mixture and incubate at 37 °C overnight.
13. Pick the colonies for mini- or midi-scaled cultures. Purify plasmid from the bacterial culture using any commercial plasmid extraction kit. Confirm successful recombination either by restriction digestion analysis, PCR, or sequencing.
14. The two single-stranded DNAs (ssDNA) for shRNA expression are as follows (with XhoI and PstI overhangs at the two ends):

    Sense: TCGAGCAATCACCTGCGTACAGAATTCAAGAGATTCTGTACGCAGGTGATTGTTTTTTACGCGTCTGCA;

    Antisense: GACGCGTAAAAAACAATCACCTGCGTACAGAATCTCTTGAATTCTGTACGCAGGTGATTGC.

    Use TE buffer to resuspend each purified oligonucleotide so that the final concentration is 100 μM. Mix the two oligos at a 1:1 ratio.
15. Place the mixture in a thermo cycler to undergo the following program: heat the mixture to 95 °C for 30 s; heat at 72 °C for 2 min; heat at 37 °C for 2 min; heat at 25 °C for 2 min; store on ice (*see* **Note 5**).
16. Dilute the annealed oligo with TE buffer to obtain a concentration of 0.5 μM (*see* **Note 6**).
17. Double digest the pDNR-TGFβ3-U6 with XhoI and PstI.

18. Set the following ligation reaction in a microcentrifuge tube. Incubate the reaction mixture for 3 h at room temperature (*see* **Note 7**).

    2 μl of linearized pDNR-TGFβ3-U6 (25 ng/μl).

    1 μl of diluted, annealed oligonucleotide (0.5 μM).

    1.5 μl of 10× T4 DNA Ligase buffer.

    1.5 μl of 10× BSA solution.

    8.5 μl of nuclease-free H_2O.

    0.5 μl of T4 DNA Ligase.

19. Transform DH5α Competent *E. coli* cells with the ligation mixture and incubate at 37 °C overnight.
20. Pick the colonies for mini- or midi-scaled cultures. Purify plasmid from the bacterial culture using any commercial plasmid extraction kit. Confirm successful recombination either by restriction digestion analysis, PCR, or sequencing.

3.1.2 Subclone Recombinant pDNR to pLP-Adeno-X ViraTrak Acceptor Vector

1. Measure the pDNR concentration and adjust the concentration to 200 ng/μl. Mix 200 ng (1 μl) of pDNR, 1 μl of Cre recombinase (included in the Adeno-X™ ViraTrak Expression System 2 kit), and 18 μl of Adeno-X LP Reaction Mix (included in the Adeno-X™ ViraTrak Expression System 2 kit). Tap the tube, spin briefly, and incubate at room temperature for 15 min, followed by heat inactivation at 70 °C for 5 min (*see* **Notes 8–10**).
2. Take out 1.5 μl of the reaction mixture to electroporate 40 μl of electrocompetent *E. coli* cells. Add 1 ml of SOC medium and incubate the cells at 37 °C for 60 min (*see* **Note 11**).
3. Transfer the cells to a centrifuge tube and spin at 3500 ×*g* for 1 min.
4. Resuspend the cell pellet in 100 μl of fresh SOC medium and spread it on a LB/Cm/sucrose agar plate. Leave the plate on the bench for 15–20 min and then transfer the dish to a 37 °C incubator for incubation overnight (*see* **Notes 12** and **13**).
5. The next day, pick several colonies and propagate them in 100 ml of LB/Amp/Cm liquid broth. Allow the cells to grow until they reach log phase. Purify the plasmid using commercially available kits (*see* **Note 14**).

3.1.3 Protocols for Producing Recombinant Adenovirus

1. In a sterile 1.5-ml microcentrifuge tube, combine 20 μl of sterile H_2O, 4 μl of 10× digestion buffer, 10 μl of recombinant pLP-Adeno-X DNA, 2 μl of PacI restriction endonuclease solution, and 4 μl of 10× BSA. Mix the contents and spin the tube briefly. Incubate at 37 °C for 2 h.

2. Add 60 μl of TE buffer and 100 μl of UltraPure™ Phenol:Chloroform:Isoamyl Alcohol. Vortex gently (*see* **Note 15**).
3. Spin the tube in a microcentrifuge at 15,000 × *g* for 5 min at 4 °C to separate the phases.
4. Carefully transfer the top aqueous layer to a clean sterile 1.5-ml microcentrifuge tube. Discard the interface and lower phase.
5. Add 400 μl of 95 % ethanol solution, 25 μl of 10 M NH_4Ac solution (or 2.5 μl of 3 M sodium acetate solution), and 1 μl of glycogen solution. Vortex gently.
6. Spin the tube in a microcentrifuge at 15,000 × *g* and 4 °C for 5 min. Remove and discard the supernatant.
7. Wash the pellet with 300 μl of 70 % ethanol solution.
8. Spin in a microcentrifuge at 15,000 × *g* for 2 min.
9. Carefully aspirate off the supernatant.
10. Air-dry the pellet for ~15 min at room temperature.
11. Dissolve the DNA precipitate in 10 μl of sterile TE buffer.
12. About 12–24 h before transfection, plate healthy, log-phase HEK 293 cells at a density of $1–2 \times 10^6$ cells per 60-mm culture plate. Allow the cells to grow in an incubator overnight at 37 °C with 5 % CO_2 (*see* **Note 16**).
13. Transfect HEK 293 cells in each 60-mm culture plate with 10 μl (5 μg) of PacI-digested Adeno-X DNA solution from **step 11** using transfection reagents (*see* **Note 17**).
14. Observe the cells daily for the presence of green fluorescence and cytopathic effect (CPE), which should begin 3–10 days after transfection (*see* **Note 18**).

3.1.4 Harvesting and Amplifying Adenovirus from Transfected Cells

About 7–10 days after transfection (with or without CPE present), recombinant adenoviruses are ready for harvesting by lysing the cells with multiple freeze-thaw cycles.

1. Dislodge the cells and transfer them to a sterile 15-ml conical centrifuge tube. Centrifuge the suspension at 1250 × *g* for 5 min at room temperature (*see* **Note 19**).
2. Freeze cells in a dry ice/ethanol bath, then place the tube in a 37 °C water bath until the ice is just thawed. Do not allow the suspension to reach 37 °C. Vortex the cells after thawing. Repeat for another two freeze-thaw cycles.
3. After the third cycle, centrifuge at 1250 × *g* for 5 min at room temperature and transfer the lysate to a clean, sterile centrifuge tube. The stock can now be stored at −20 °C.
4. Infect a fresh nontransfected, 50 % confluent, 60-mm culture of HEK 293 cells by adding 250 μl (50 %) of the cell lysate

from **step 3**. Add the lysate directly to complete growth medium-1 and incubate at 37 °C with 5 % CO_2. CPE should be evident within 1 week (*see* **Note 20**).

5. When infection results in 50–95 % of the cells detaching from the plate due to CPE in less than 1 week, prepare a viral stock by following **steps 1–4**. Name this stock "Primary Amplification Stock" and store it at −20 °C. This stock can be used to further amplify adenoviruses in large HEK 293 cultures. Large stocks can be purified using the Adeno-X Virus Purification Kit.
6. The infected cells can also be used to verify the presence of adenoviral vectors either by western blotting, ELISA, PCR or any biochemical assays.
7. Titrate the stock using the Adeno-X™ Virus Rapid Titer Kit.

3.1.5 Cell Culture and Transduction

1. SMSCs are maintained in culture in complete growth medium-1. Replace the medium every 3 days. Subculture the cells using a 1:3 dilution ratio when the cultures reach 90 % confluence. Resuspend the pellet in 500 μl of sterile, fresh, complete growth medium-1.
2. One day before transduction, seed a certain number of cells on dishes or flasks.
3. Calculate the volume of adenovirus stock to be added using the formula below.

$$\text{Volume of stock} = \frac{\text{cell number} \times \text{MOI}}{\text{titer of adenoviral stock from Subheading 3.1.4}}$$

   The multiplicit y of infection (MOI) is 200 in our experiment (*see* **Note 21**).

4. Add the calculated volume of adenovirus stock to the cell medium and gently swirl the dish or flask.
5. Two days after transduction, replace the medium with fresh complete growth medium-1 without adenovirus. Continue the cell culture or proceed to alginate hydrogel encapsulation for 3D culture (Subheading 3.3).

3.2 Production of Lentiviral Vectors Expressing TGF-β3 and shRNA

3.2.1 Construction of Dual-Functioning Recombinant Lentiviral Vectors (See Notes 22 and 23)

The structures of the four dual-functioning lentiviral vectors are shown in Fig. 2. Among the four vectors (Fig. 2), LV-3 is the one that has the same arrangement of TGF-β3 and shRNA-expressing cassettes in Ad-D as demonstrated in Subheading 3.1. LV-4 differs from LV-3 in that the shRNA-expressing cassette is in the same direction as the TGF-β3 expressing cassette in LV-4, whereas in LV-3 the directions for the two cassettes are reversed. In LV-1 and LV-2, the shRNA-expressing cassette is moved to elsewhere between the MfeI and FseI restriction sites, out of the multiple cloning site, to minimize the potential interference between the

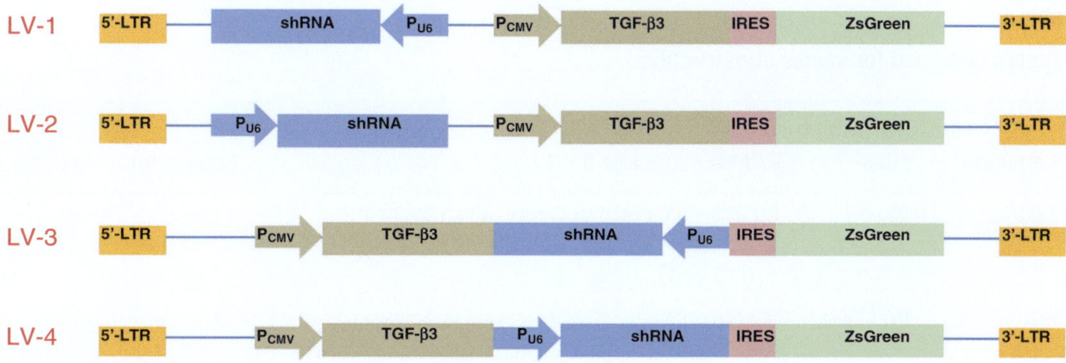

Fig. 2 Schematic structures of the recombinant lentiviral vectors

two expressing cassettes and attenuate the impact on ZsGreen expression and transcription of the whole lentiviral genome. In LV-1, the two cassettes are in reverse directions, while in LV-2 they are in the same direction.

1. To construct LV-3, digest the recombined adenoviral shuttle vector pDNR with the restriction endonuclease PspOM I. Use a PCR purification kit to purify the linearized plasmid.
2. Further digest the linearized plasmid from **step 1** with the restriction endonuclease Sal I. Run a 1 % agarose gel with digested products and extract the fragment containing the TGF-β3 cDNA sequence, U6 promoter and Col I-targeted shRNA, which is about 1600 bp long, using a gel extraction kit.
3. Meanwhile, double digest the pLVX-IRES-ZsGreen vector with restriction endonucleases Xho I and Not I, and extract the digestion product from a 1 % agarose gel after gel electrophoresis.
4. Mix the fragment from **step 2** and the linearized pLVX-IRES-ZsGreen vector from **step 3** and ligate using T4 DNA Ligase at 16 °C overnight. Amplify the recombinant plasmid in competent *E. coli* cells and extract using a plasmid purification kit.
5. To construct other lentiviral vectors, first construct the lentiviral plasmid vector carrying only the TGF-β3 expressing cassette (pLVX-T) as the backbone and also as a control in the experiments. Amplify the TGF-β3 sequence through PCR using primers as shown in Table 1. Double digest both the TGF-β3 sequence and pLVX vector with XhoI and NotI, and ligate the two linear segments with T4 DNA Ligase.
6. PCR-amplify the shRNA-expressing cassette (U6 and shRNA encoding sequence) from the recombinant dual-functioning adenoviral plasmid described in Subheading 3.1.2 as template, with primers carrying various restriction sites at the ends. Double digest the fragments with corresponding restriction

Table 1
Primers designed for vector construction

Amplicon	Restriction sites	Primer sequence 5′–3′	Length (bp)	AT (°C)
TGF-β3	XhoI/ NotI	F:CCGCTCGAGATGAAGATGCACTT R:ATAAGAATGCGGCCGCTCAGCTAC ATTTAC	1259	50
shRNA cassette-1	FseI/MfeI	F:AGACTACAATTGTGCAGGAAGAGGGCCTAT R:CATCAAGGCCGGCCACGCGTAAAAAACAA TCACCTG	~350	55
shRNA cassette-2	MfeI/FseI	F:AGACTAGGCCGGCCGACCATGTTCACTTA CCTAC R:CATCAACAATTGACGCGTAAAAAACAATC ACCTG	~400	58
shRNA cassette-4	NotI/ BamHI	F:AAGGAAAAAAGCGGCCGCATATAGCAGGA AGAGGGCCTAT R:CGGGATCCACGCGTAAAAAACAATCACCTG	~350	55

AT annealing temperature

endonucleases and ligate them with pLVX-T from **step 5** which has been digested with the same restriction endonucleases. Primers for the amplification of TGF-β3 and shRNA-expressing cassettes for LV-1, LV-2, and LV-4, respectively, are listed in Table 1.

7. Verify the recombinant lentiviral plasmids with restriction endonuclease digestion followed by gel electrophoresis and gene sequencing. Subsequently, use these plasmids to generate lentiviral vectors that express both TGF-β3 and shRNA (LV-1, LV-2, LV-3, LV-4) according to the steps given in Subheading 3.2.2.

3.2.2 Producing Lentivirus with the Lenti-X HT Packaging System

1. Plate 5×10^6 HEK 293T cells/100-mm plate in 10 ml of complete growth medium-2 containing Tc-free FBS to achieve 50–80 % confluency the next day for transfection (*see* **Notes 24** and **25**).
2. In a polystyrene tube, mix 15 μl of Lenti-X HT Packaging Mix, 3 μg of Lenti-X plasmid DNA, and sufficient sterile H_2O to achieve a final volume of 438 μl. Then add 62 μl of Lentiphos1 solution from the Lentiphos Transfection Reagent kit to the diluted DNA and vortex thoroughly.
3. While vortexing the DNA/Lentiphos1 solution, add 500 μl of Lentiphos2 from the Lentiphos Transfection Reagent kit, dropwise, into the tube.

4. Incubate at room temperature for 5–10 min to allow the DNA precipitate to form.
5. Gently vortex the transfection solution, and add the entire contents of the tube (1 ml), dropwise, to the cell culture medium.
6. Gently move the plate back and forth to distribute the transfection solution evenly. Incubate the plate at 37 °C overnight in a CO_2 incubator.
7. Replace the transfection medium with 10 ml of fresh complete growth medium-2 and incubate at 37 °C for 48–72 h.
8. Harvest the lentivirus-containing supernatants. Centrifuge briefly (500 × *g* for 10 min) or filter through a 0.45-μm filter (*see* **Notes 26** and **27**).
9. Titrate the virus stock, use the virus to transduce target cells, or freeze the stock in aliquots as described in **step 10**.
10. To store virus stocks, aliquot the cleared supernatant into single-use cryotubes to avoid multiple freeze-thaw cycles. Store the tubes at −80 °C. No cryoprotectant is required (*see* **Note 28**).

3.2.3 Titration of the Lentiviral Vector Produced from HEK 293T Cells

To titrate the produced lentiviral vector, flow cytometry is utilized to calculate the number of viable lentivirus particles that can successfully transduce HEK 293T cells, assuming that one positively transduced 293T cell represents one viable viral particle (*see* **Note 29**).

1. One day before transduction, seed HEK 293T cells in 6-well plates at a density of 1×10^6 cells per well in complete growth medium-1 for culture in the incubator at 37 °C with 5 % CO_2.
2. Make serial tenfold dilutions of the vector in complete growth medium-1 and add to the 293T cells in the presence of polybrene at a final concentration of 2 μg/ml.
3. After culturing for an additional 8 h in the incubator at 37 °C with 5 % CO_2, change the medium and incubate the cells for up to 72 h post-infection.
4. At 72 h post-infection, dislodge the 293T cells by gentle pipetting and wash twice with cold PBS. Subject the cells in each well to flow cytometry analysis to obtain the percentages of green fluorescence-positive cells against the total number of cells.
5. Use an additional well to count the total cell number in each well.
6. Select a viral dilution resulting in 10–20 % infected cells for calculation to ensure one infectious unit (IFU) per cell. Calculate the virus titer using the formula below:

$$\text{IFU / ml} = \left(\text{Number of cells / well at the time of harvest} \times \text{\% positive cells by flow cytometry}\right) / \text{volume of virus in ml}$$

3.2.4 Transduction of SMSCs with Recombinant Lentiviral Vector and Subsequent Cell Sorting

1. Plate SMSCs in complete growth medium-1 for culture in the incubator at 37 °C with 5 % CO_2 12–18 h before transduction.
2. Thaw lentiviral stocks and supplement the viral stock with polybrene to obtain the desired final polybrene concentration of 2 μg/ml during the transduction step. Add the viral supernatant to the cells and allow for transduction.
3. Eight hours post-transduction, discard the viral vector-containing transduction medium and replace with fresh complete growth medium-1. Incubate the cells in the incubator at 37 °C with 5 % CO_2 for another 3–4 days to allow the expressed protein to accumulate in the target cells.
4. After the appearance of green fluorescence observed under a fluorescence microscope, trypsinize the cells using trypsin solution and wash with PBS. Subject the cells to fluorescence-activated cell sorting (FACS) to select the ZsGreen positive cells that have been successfully transduced with recombinant lentiviral vector. Cells without transduction are used as a negative control. The excitation wavelength is 470 nm and the emission wavelength is 520 nm.
5. Plate the sorted cells in dishes for amplification, marked as P0. Label subsequent passages as P1, P2, etc.

3.3 Three-Dimensional Culture of SMSCs

1. Trypsinize cells that have been transduced with dual-functioning adenovirus or lentivirus with trypsin solution and rinse with washing buffer. Centrifuge at 1000 rpm for 5 min.
2. Resuspend the cells in sterile alginate solution at a density of 7×10^6 cells/ml.
3. Slowly drop 40 μl of the cell suspension into a beaker containing sterile $CaCl_2$ solution. Polymerize for 10 min.
4. Place four beads into each well of a 24-well plate with 1 ml of complete growth medium-1. Collect and replace the medium every 3 days for quantitative analysis of TGF-β3 using ELISA.

3.4 Quantitative Real-Time PCR (qPCR)

1. After 72 h of infection with various recombinant adenoviral vectors, lyse the cells and isolate total RNA using TRIzol reagent (*see* **Note 30**).
2. For cDNA synthesis, reverse-transcribe 1 μg of total RNA using the Superscript® First-Strand Synthesis System.
3. Perform qPCR for Col I, TGF-β3, and other chondrogenesis-related genes using the IQ™ SYBR Green Supermix system according to the manual. The reaction conditions are as follows: initial denaturation at 95 °C for 3 min, followed by 40 cycles of denaturation at 95 °C for 30 s, annealing at 53 °C for 30 s, and extension at 72 °C for 30 s. Use porcine RPL4 for

Table 2
Primers for qPCR

Gene	Forward primer (5′–3′)	Reverse primer (5′–3′)	AT (°C)
Col I	CCTGCGTGTACCCCACTCA	ACCAGACATGCCTCTTGTCCTT	58
TGF-β3	5′-GCGGAGCACAACGAACTG-3′	CTGCTCATTCCGCTTAGAG	58
Col II	GCTATGGAGATGACAACCTGGCTC	ACAACGATGGCTGTCCCTCA	58
Col X	CAGGTACCAGAGGTCCCATC	CATTGAGGCCCTTAGTTGCT	58
Comp	GGCACATTCCACGTGAACA	GGTTTGCCTGCCAGTATGTC	58
Aggrecan	CGAGGAGCAGGAGTTTGTCAAC	ATCATCACCACGCAGTCCTCTC	58
RPL4	CAAGAGTAACTACAACCTTC	GAACTCTACGATGAATCTTC	58
Human β-actin	CCTGGCACCCAGCACAAT	GGGCCGGACTCGTCATACT	58

AT annealing temperature (°C)

normalization. Calculate the relative gene expression values using the comparative $\Delta\Delta C_T$ (threshold cycle) method. Sequences of all the primers required are listed in Table 2.

3.5 Cell Viability Test

Cell viability is tested using the WST-1 assay.

1. Add 10 μl of premixed WST-1 reagent to each well containing one sample bead in 100 μl of complete growth medium-1, followed by incubation in the incubator at 37 °C with 5 % CO_2 for 2 h.
2. After 2 h incubation, determine the absorbance of the medium at 450 nm using a microplate reader. The absorbance values represent the relative cell viability in each group (*see* **Note 31**).

3.6 Quantitative Analysis of Proteoglycan Synthesis

1. After 42 days of culture, take out the constructs and wash three times with deionized water to remove the salt in the medium.
2. Freeze-dry the cell beads for 36 h in a freeze drier.
3. Digest the freeze-dried residues with 1 ml of papain solution per sample.
4. Add dimethylmethylene blue dye solution to the digested solution and measure the absorbance at 525 nm using a UV–VIS spectrophotometer (*see* **Note 32**).

3.7 Histology and Immunohistochemistry

1. Fix 3D construct samples in 4 % (w/v) neutral-buffered paraformaldehyde solution overnight.
2. Embed the fixed constructs in paraffin and section to 5 μm thick.

3. Deparaffinize the sections by incubating in three washes of xylene for 5 min each.
4. Incubate the sections in two washes of 100 % ethanol for 10 min each.
5. Wash the sections by incubating in serial washes of 90, 80, 70, and 50 % ethanol for 10 min each.
6. Rehydrate the sections by incubating in deionized H_2O for 10 min.
7. Submerge the deparaffinized sections in Safranin-O solution for 10 min and observe under a microscope for GAGs. GAGs are stained in red color and can be easily distinguished from other substances.
8. Submerge deparaffinized sections in Masson's Trichrome Stains for 30 min and observe under a microscope for total collagen, which is stained in blue color.
9. For immunohistochemistry, fix the specimens in glutaraldehyde solution for 30 min and block with goat serum solution for 1 h. Afterwards, apply primary antibody solution for Col I or Col II for 2 h at 4 °C. Following three PBS washes, incubate the sections for Col I and Col II with their respective HRP-conjugated secondary antibodies at room temperature for 1 h. Observe the presence of Col II using the DAB Substrate kit (*see* **Note 33**).

4 Notes

1. This adenoviral vector was used to transduce various cell lines, including human fibroblasts, osteoblasts, etc., and has been demonstrated to be functional for the release of TGF-β3 and suppression of Col I expression. For details regarding the release profile and inhibition efficiency, please refer to ref. 15.
2. The four LVs have distinct behaviors regarding the release of TGF-β3 and suppression of Col I expression. Some of them may be better, among others, in promoting chondrogenesis (or even in only one chondrogenic marker), while others may have better profiles in suppressing Col I expression. Therefore, it is extremely difficult to make a decisive agreement as to which one is the best. LV-1 was determined to be only suboptimal per our judgement, and it is up to readers themselves to decide which one to select, according to the results compiled in ref. 17. Yet, this chapter has, for the first time, brought up the idea that while chondrogenesis using stem cells is of great interest in recent research, Col I upregulation should be circumvented. This would promote the development of more

clinically suitable engineered cartilage tissue constructs and ameliorate the problems associated with existing approaches.

3. A *pfu* high-fidelity DNA polymerase is recommended to minimize mismatch during PCR reaction.

4. Double digestion is recommended to maximize reaction efficiency and minimize DNA loss, compared to sequential digestion and DNA purification. Follow the double digestion conditions from the vendor's instructions. Also, as the DNA product in this step is thousands of base pairs long, 1 % or even lower concentration of agarose gel is to be used, while in **step 1** where PCR products are only a few hundred base pairs long, 2 % agarose gel can be applied.

5. The internal hairpin of the oligonucleotide is disrupted by heating to 95 °C, which would promote intermolecular annealing between the two strands.

6. An excess of oligos will block ligation efficiency. Therefore, it is better to dilute the oligos to appropriate concentrations.

7. The reaction should not last over 3 h.

8. The recombination reaction requires precise donor to acceptor DNA ratios. Confirm the concentrations before mixing.

9. The incubation time should not exceed 15 min, as overtime reaction will decrease the yield.

10. Upon first use, Cre recombinase should be thawed rapidly from –70 °C in fingers until most of the ice has been thawed. The recombinase should then be placed on ice for complete melting. Once thawed, the Cre recombinase should be stored at –20 °C in a non-frost-free freezer.

11. As the pAdeno-X vector is large in size (~36 kb), transformation should be carried out in electrocompetent *E. coli* cells using electroporation to increase the transformation efficiency. The transformation efficiency of the competent cells should be $>5\times10^9$ cfu/μg. If not, replace with a fresh sample of cells.

12. Cre-loxP recombination can directly transfer the sequence of interest (up to 3.5 kb) from the donor vector into the loxP sites in the acceptor vector. During recombination, a chloramphenicol resistance gene (Cm^r) is transferred along with the gene of interest into the adenoviral Acceptor Vector, to enable positive selection of recombinants with chloramphenicol. The *B. subtilis* sucrase gene (*SacB*) present in the Donor Vector allows negative selection against bacterial cells containing residual Donor Vectors and by-products of the recombination reaction.

13. As the growth condition is stringent and the acceptor plasmid is large in size, the colonies grow very slowly. It may take over 24 h before colonies appear. On the other hand, work should

be carried out rapidly once the pLP-Adeno-X vector is in the bacteria, in order to minimize its recombination and modification in the bacterial cells. Therefore, pick up the colonies as soon as they appear, i.e., within 30 h.

14. For plasmid extraction, clarify the bacterial lysate by filtration rather than centrifugation. Quantify the DNA by OD_{260}.
15. Use UltraPure™ Phenol:Chloroform:Isoamyl Alcohol (25:24:1) that is commercially available, instead of preparing the mixture in your own laboratory, as this step requires reagents of high purity.
16. Cells should be 50–70 % confluent with good adherence and flat morphology before transfection in order to obtain optimal transfection efficiency and virus production.
17. Both calcium phosphate and lipofectamine can be used for transfection in this step.
18. Transfection efficiency can be assessed directly by observing green fluorescence under a fluorescence microscope. The efficiency is ~20 % due to the large size. Fluorescent viral plaques can also be observed well before the appearance of CPE, indicating the production of viable viruses in the cells.
19. Trypsinization should not be exercised in this step. Directly dislodge the cells by gentle pipetting.
20. CPE is expected to be evident within 1 week after infection. If no CPE is observed more than 7 days after infection, it is possible that the virus lysates from previous steps are of low titer, and transduction should be carried out from the beginning.
21. MOI stands for multiplicity of infection. Explicitly, it refers to the average number of viral particles applied to one target cell. In application, it depends on the aggressiveness of the virus as well as the vulnerability of the target cells.
22. The inserted gene sequence should contain an ATG start codon in order to express the desired gene product. Meanwhile, the inserted gene should not contain any polyadenylation signal, as this signal would lead to premature polyadenylation during virus transcription and production of undesired virions.
23. Precautions should be exercised as the produced lentiviruses can infect human cells. All the procedures should be performed in a Biosafety Level 2 tissue culture hood that has been approved for use with lentiviruses.
24. In all the procedures, extreme care is necessary in order not to disturb the HEK 293T cells, as cell adherence is loose for this cell type. Collagen-coated dishes or plates are recommended for the culture of 293T cells. Other commercial coated plates,

i.e., Corning CellBIND plates, can also be used for this purpose.

25. Tet System Approved FBS (Tc-free) should be used in lentivirus production with HEK 293T cells (e.g., using complete growth medium-2). The presence of tetracycline would affect the performance of the packaging mix and reduce the production of lentiviruses. Yet, for other procedures using HEK 293T cells, complete growth medium-1 with conventional FBS can be used.
26. Lentivirus can be collected at both 48 and 72 h. Generally, virus titer is highest at 48 h.
27. For filtering unwanted cell fragments, a cellulose acetate or polyethersulfone (PES) filter should be used instead of nitrocellulose. Nitrocellulose can bind proteins on the membrane of the lentivirus and compromise the integrity of the virus.
28. Lentivirus stock should be stored at −80 °C and repeated freeze-thaw cycles avoided. Each freeze-thaw cycle may reduce the virus titer by two- to fourfold.
29. An alternative method can be used to measure the number of RNA genomic copies of the viral stock. Briefly, lentiviral genomic RNA is purified using a NucleoSpin® RNA Virus Kit. Residual plasmid DNA is removed by treatment with DNase I. The purified RNA is subjected to qPCR and the threshold cycle (C_T) is obtained for the sample using a Lenti-X qRT-PCR Titration Kit. Meanwhile, standard samples with known numbers of genomic RNA copies are also tested and a standard curve of C_T versus lg (genomic copies) is derived. The genomic RNA copy number for the sample lentiviral vector is obtained from the standard curve and compared with the infectious units obtained through flow cytometry. A ratio of genomic RNA copy number to IFU is established so that, in subsequent titration experiments, the Lenti-X qRT-PCR Titration Kit can be easily used indirectly to determine infectious units.
30. To obtain enough RNA for real-time PCR from the cells encapsulated in polysaccharide hydrogels, an alternative could be the use of plant RNA isolation kits, as such kits are advantageous in removing the interference of polysaccharide from the crude materials [18, 19].
31. The WST assay is a colorimetric assay based on conversion of the tetrazolium salt of WST-1 into a colored dye by mitochondrial dehydrogenase enzymes and subsequent release of soluble dye into the medium. Within a certain time and range, the absorbance of the colored dye is proportional to the amount of mitochondrial dehydrogenase and can reflect cell number and viability.

32. Dimethylmethylene is able to react with proteoglycans, the product of which can be detected by measuring the absorbance at 450 nm. Higher absorbance values indicate higher contents of proteoglycans.
33. 3,3′ Diaminobenzidine (DAB) is a widely used chromogen for immunohistochemical staining. In the presence of peroxidase enzyme (HRP conjugated to secondary antibody), DAB produces a brown precipitate that is insoluble in alcohol and can be easily detected under a microscope.

Acknowledgements

This work was supported by AcRF Tier 1 RG 36/12 and AcRF Tier 2 ARC 1/13, Ministry of Education, Singapore.

References

1. Flugge LA, Miller-Deist LA, Petillo PA (1999) Towards a molecular understanding of arthritis. Chem Biol 6:R157–R166
2. Bader RMSA (2007) Cartilage tissue engineering and bioreactor systems for the cultivation and stimulation of chondrocytes. Eur Biophys J 36:539–568
3. Chancellor JV, Hunsche E, de Cruz E, Sarasin FP (2001) Economic evaluation of celecoxib, a new cyclo-oxygenase 2 specific inhibitor, in Switzerland. Pharmacoeconomics 19:59–75
4. Darling EM, Hu JC, Athanasiou KA (2004) Zonal and topographical differences in articular cartilage gene expression. J Orthop Res 22:1182–1187
5. Hidaka C, Cheng C, Alexandre D, Bhargava M, Torzilli PA (2006) Maturational differences in superficial and deep zone articular chondrocytes. Cell Tissue Res 323:127–135
6. Goessler UR, Bugert P, Bieback K, Baisch A, Sadick H, Verse T, Klüter H, Hörmann K, Riedel F (2004) Expression of collagen and fiber-associated proteins in human septal cartilage during in vitro dedifferentiation. Int J Mol Med 14:1015–1022
7. Goessler UR, Bugert P, Bieback K, Sadick H, Baisch A, Hormann K, Riedel F (2006) In vitro analysis of differential expression of collagens, integrins, and growth factors in cultured human chondrocytes. Otolaryngol Head Neck Surg 134:510–515
8. Yoshimura H, Muneta T, Nimura A, Yokoyama A, Koga H, Sekiya I (2007) Comparison of rat mesenchymal stem cells derived from bone marrow, synovium, periosteum, adipose tissue, and muscle. Cell Tissue Res 327:449–642
9. Na K, Kim S, Woo DG, Sun BK, Yang HN, Chung HM, Park KH (2007) Synergistic effect of TGFbeta-3 on chondrogenic differentiation of rabbit chondrocytes in thermo-reversible hydrogel constructs blended with hyaluronic acid by in vivo test. J Biotechnol 128:412–422
10. Yun K, Moon HT (2008) Inducing chondrogenic differentiation in injectable hydrogels embedded with rabbit chondrocytes and growth factor for neocartilage formation. J Biosci Bioeng 105:122–126
11. Awad HA, Wickham MQ, Leddy HA, Gimble JM, Guilak F (2004) Chondrogenic differentiation of adipose-derived adult stem cells in agarose, alginate, and gelatin scaffolds. Biomaterials 25:3211–3222
12. Steinert A, Weber M, Dimmler A, Julius C, Schütze N, Nöth U, Cramer H, Eulert J, Zimmermann U, Hendrich C (2003) Chondrogenic differentiation of mesenchymal progenitor cells encapsulated in ultrahigh-viscosity alginate. J Orthop Res 21:1090–1097
13. Park H, Kang SW, Kim BS, Mooney DJ, Lee KY (2009) Shear-reversibly crosslinked alginate hydrogels for tissue engineering. Macromol Biosci 9:895–901
14. Yao Y, Zhang F, Zhou RJ, Li M, Wang DA (2010) Continuous supply of TGFbeta3 via adenoviral vector promotes type I collagen and viability of fibroblasts in alginate hydrogel. J Tissue Eng Regen Med 4:497–504

15. Zhang F, Yao Y, Hao J, Zhou R, Liu C, Gong Y, Wang DA (2010) A dual-functioning adenoviral vector encoding both transforming growth factor-beta3 and shRNA silencing type I collagen: construction and controlled release for chondrogenesis. J Control Release 142:70–77
16. Oberholzer A, John T, Kohl B, Gust T, Müller RD, La Face D, Hutchins B, Zreiqat H, Ertel W, Schulze-Tanzil G (2007) Adenoviral transduction is more efficient in alginate-derived chondrocytes than in monolayer chondrocytes. Cell Tissue Res 328:383–390
17. Zhang F, Yao YC, Zhou RJ, Su K, Citra F, Wang DA (2011) Optimal construction and delivery of dual-functioning lentiviral vectors for type I collagen-suppressed chondrogenesis in synovium-derived mesenchymal stem cells. Pharm Res 28:1338–1348
18. Wang CM, Hao JH, Zhang F, Su K, Wang DA (2008) RNA extraction from polysaccharide-based cell-laden hydrogel scaffolds. Anal Biochem 380:333–334
19. Wang CM, Li X, Yao YC, Wang DA (2009) Concurrent extraction of proteins and RNA from cell-laden hydrogel scaffold free of polysaccharide interference. J Chromatogr B 877:3762–3766

Part III

Biomaterials and Scaffolds

Chapter 8

Hydrogels with Tunable Properties

Peggy P.Y. Chan

Abstract

This chapter describes the preparation of tissue engineered constructs by immobilizing chondrocytes in hydrogel with independently tunable porosity and mechanical properties. This chapter also presents the methods to characterize these tissue engineered constructs. The resulting tissue engineered constructs can be useful for the generation of cartilage tissue both in vitro and in vivo.

Key words Injectable hydrogel, Tunable property, Gelation time, Porosity, Tissue scaffolds, Cartilage tissue engineering

1 Introduction

Hydrogels are hydrated networks formed by crosslinking hydrophilic polymers. Hydrogels can absorb water up to thousands of times the dry weight of the polymer [1]. Due to the highly hydrated and soft-tissue-like biomechanical properties of hydrogels, these materials provide a physiologically relevant environment for cell cultivation [2]. The use of injectable hydrogels as tissue scaffolds offers many advantages over preformed hydrogels. Injectable hydrogels can be administered using a syringe to fill any shape or defect; they are therefore suitable for laparoscopic surgery applications. Three-dimensional injectable hydrogel networks are capable of entrapping proteins and releasing them subsequently, and are often used as protein delivery depots [3, 4]. Cells and proteins such as growth factors are often co-immobilized in hydrogels [5–7]: the growth factors slowly release from the hydrogel network and regulate the growth and differentiation of the surrounding cells. Hydrogels are therefore attractive tissue scaffold materials and offer great promise for cartilage repair [8, 9].

The preparation of cell-immobilized gelatin-hydroxyphenylpropionic acid/carboxymethylcellulose-tyramine (GTN-HPA/CMC-TYR) hydrogel is described in this chapter. The method to tune the porosity of the GTN-HPA/CMC-TYR

Pauline M. Doran (ed.), *Cartilage Tissue Engineering: Methods and Protocols*, Methods in Molecular Biology, vol. 1340, DOI 10.1007/978-1-4939-2938-2_8, © Springer Science+Business Media New York 2015

hydrogel is also described. GTN-HPA/CMC-TYR is an injectable porous hydrogel comprised of gelatin and carboxymethylcellulose polymer backbones. Gelatin is a protein derived from collagen hydrolysis [4]. Gelatin contains arginine-glycine-aspartate (RGD) motifs that are known to provide high affinity binding sites for cell adhesion. Gelatin is biocompatible, biodegradable, and available abundantly, and has therefore been widely used for many pharmaceutical and medical applications [10]. Gelatin-based hydrogels have been extensively investigated for cartilage regeneration [11–17]. CMC is a Food and Drug Administration (FDA) approved cellulose derivative that has been widely employed for pharmaceutics manufacturing [10]. To prepare the GTN-HPA/CMC-TYR hydrogel, horseradish peroxidase (HRP) enzyme is used to catalyze the oxidative coupling of the phenol moieties on GTN-HPA and CMC-TYR, while hydrogen peroxide (H_2O_2) is used as an oxidant for the formation of linkages at the C-C and C-O positions of the phenols [18]. The stiffness of the hydrogel is dependent on the number of linkages in the hydrogel network, which can be controlled by adjusting the amount of oxidant used for crosslinking [19].

HRP-mediated crosslinking can occur in aqueous environments and at room temperature, and is thus suitable for in situ immobilization [18], where cells are premixed with hydrogel precursor prior to crosslinking and are fixed inside the hydrogel upon crosslinking. Gelation time (crosslinking time) is an important parameter that determines the success of an injectable tissue engineered construct. A fast gelation time (5–15 min) is usually preferred, as cells can be immobilized in the hydrogel uniformly during crosslinking [20]. If the crosslinking is not undertaken rapidly, cells will settle at the bottom of the precursor solution before crosslinking takes place, thus resulting in uneven cell distribution inside the hydrogel. The gelation time of the GTN-HPA/CMC-TYR hydrogel can be tuned independently by varying the amount of HRP enzyme used for crosslinking. The HRP-mediated crosslinking reaction occurs rapidly, taking between seconds and 25 min [4].

Porosity is an important parameter that controls cell binding, cell migration, intracellular signaling, extracellular matrix (ECM) production, nutrient and gas transport, and waste removal, thus determining the success of a tissue engineered construct [21]. Smaller pore sizes are preferred during early stages of tissue development as they allow better cell adhesion. Larger pore sizes are preferred during later stages of tissue development as they allow better mass transfer, thus facilitating faster cell growth [18]. The CMC component of the GTN-HPA/CMC-TYR hydrogel can be selectively digested by cellulase derived from *Trichoderma longibrachiatum*, an enzyme that has been recognized as safe (GRAS) by the FDA. Post-fabrication, the pore size of the GTN-HPA/CMC-TYR hydrogel can be tuned by injecting cellulase into the scaffolds, thus leaving behind the GTN-HPA as the supporting structure of the tissue scaffold. This system allows users to tailor the porous structure of the hydrogel to match the growth rate of a tissue [18].

2 Materials

Prepare all solutions using analytical grade reagents unless indicated otherwise. Waste disposal regulations should be followed when disposing of waste materials.

2.1 Cell Culture

1. Cell culture medium: Dulbecco's modified Eagle medium (DMEM) with 10 % serum and other additives [22]. Under a sterile laminar flow hood and using aseptic handling techniques, remove 50 mL of medium from a 500-mL bottle of DMEM. To this bottle, add 50 mL of heat-inactivated fetal bovine serum, 5 mL of 10,000 U/mL penicillin-streptomycin, 5 mL of Minimum Essential Medium (MEM) nonessential amino acids (100×), 5 mL of 1 M HEPES, 1 mL of 250 μg/mL Fungizone® antimycotic (Gibco), and 2.3 mg of L-proline. Invert the bottle a few times to mix the solution. Immediately, sterilize the mixture using a 0.22-μm vacuum filter system. Store at 4 °C. Discard unused cell culture medium after 6 months.
2. Chondrocytes, e.g., from bovine articular cartilage (*see* **Note 1**).

2.2 Dialysis Solutions

1. Ultrapure water: Prepare by purifying deionized water to attain a resistivity of 18 MΩ cm at 25 °C using a water purification system.
2. NaCl solution: 100 mM NaCl in ultrapure water. Weigh 29.2 g of NaCl and transfer to a 5-L beaker. Use a measuring cylinder to measure 5 L of ultrapure water and add to the beaker. Mix the solution.
3. 25 % ethanol solution: 25 % (v/v) ethanol in ultrapure water. Use a measuring cylinder to measure 3.75 L of ultrapure water and transfer to a 5-L beaker. Use a measuring cylinder to measure 1.25 L of absolute ethanol and add to the beaker. Mix the solution.

2.3 Hydrogel Stock Solutions

1. PBS solution: 1× phosphate buffered saline, pH 7.4. Under a sterile laminar flow hood and using aseptic handling techniques, sterilize the solution using a 0.22-μm vacuum filter system.
2. Horseradish peroxidase (HRP) stock solution: 150 units/mL of HRP in PBS. Weigh 0.15 g of HRP powder (100 units/mg) and transfer into a 100-mL screw-top bottle. Add 100 mL of PBS solution to the bottle and mix until the HRP powder has

fully dissolved. Under a sterile laminar flow hood and using aseptic handling techniques, sterilize the solution using a 0.2-μm syringe filter, collecting the filtrate in a sterile 100-mL screw-top bottle (sterilize the bottle beforehand by autoclaving at 121 °C for at least 15 min) (*see* **Note 2**).

3. H_2O_2 stock solution: 1 % (v/v) H_2O_2 in PBS. Transfer 11.6 mL of cold PBS solution (at 4 °C) to a 15-mL conical centrifuge tube. Add 400 μL of cold 30 % H_2O_2 solution to the centrifuge tube (*see* **Note 3**). Close the centrifuge tube and vortex the solution for 10 s. Under a sterile laminar flow hood and using aseptic handling techniques, sterilize the solution using a 0.2-μm syringe filter, collecting the filtrate in a sterile 15-mL conical centrifuge tube (*see* **Note 4**).
4. Cellulase stock solution: 1 mg/mL of cellulase in cell culture medium. Sterilize two 100-mL screw-top bottles by autoclaving them at 121 °C for at least 15 min. Weigh 0.1 g of cellulase powder from *Trichoderma longibrachiatum* (≥1.0 unit/mg, Sigma-Aldrich). Transfer the cellulase powder into one of the sterile bottles. Add 100 mL of cell culture medium. Mix the solution until the powder has fully dissolved. Immediately, sterilize the solution using a 0.2-μm syringe filter under a sterile laminar flow hood and using aseptic handling techniques. Collect the filtrate in the second sterile 100-mL screw-top bottle (*see* **Note 5**).

2.4 Analytical and Hydrogel Degradation Solutions

1. Papain solution: 1.875 units/mL of papain [22]. Dissolve 1420 mg of disodium phosphate, 2922 mg of ethylenediaminetetraacetic acid (EDTA), and 309 mg of dithiothreitol (DTT) in 1 L of ultrapure water. Adjust the pH to 6.8 with HCl and NaOH. Into 100 mL of this solution, dissolve 187.5 mg of papain type III from *Carica papaya* and 121.2 mg of L-cysteine. Immediately, sterilize the mixture using a 0.22-μm vacuum filter system under a sterile laminar flow hood and using aseptic handling techniques.
2. Blyscan™ GAG assay kit (Bicolor).
3. Trypsin solution: 2.5 % trypsin solution (10×), gamma irradiated.
4. Hydrogel degradation solution: 5 mg/mL cellulase and 0.25 % trypsin in PBS. Weigh 0.25 g of cellulase powder from *Trichoderma longibrachiatum* (≥1.0 unit/mg, Sigma-Aldrich). Transfer the cellulase into a 50-mL conical centrifuge tube. Add 45 mL of PBS solution. Mix the solution until the powder has fully dissolved. Immediately, sterilize the solution using a 0.2-μm syringe filter under a sterile laminar flow hood and using aseptic handling techniques. Collect the filtrate in a new sterile 50-mL conical centrifuge tube. Using aseptic handling techniques, transfer 5 mL of trypsin solution into the centrifuge tube. Invert the solution to mix.

3 Methods

Sterile conditions and aseptic handling techniques must be used for all steps of hydrogel synthesis after filter sterilization of the CMC-TYR and GTN-HPA conjugate solutions, and for all procedures involving cells.

3.1 Prepare the CMC-TYR Conjugate [23]

1. Weigh 5 g of carboxymethyl cellulose (CMC, $M_w = 90$ kDa).
2. Insert a clean magnetic stirrer bar into a 1-L flask. Add 250 mL of ultrapure water and stir using a magnetic stirrer.
3. Add the CMC powder slowly to the ultrapure water. Continue stirring until the powder has fully dissolved.
4. Weigh separately 0.8648 g of tyramine hydrochloride (TYR), 0.5732 g of *N*-hydroxysuccinimide (NHS), and 0.9547 g of *N*-(3-dimethylaminopropyl)-*N*-ethylcarbodiimide hydrochloride (EDC.HCl).
5. Add the TYR and NHS, followed by the EDC.HCl, to the solution. Adjust the pH to 4.7 using HCl and NaOH. Stir the solution for 24 h at room temperature.
6. To this solution, add 200 mL of ultrapure water.
7. Cut four cellulose membrane tubes (molecular weight cut-off 3.5 kDa, 29 mm diameter) so that each is approximately 30 cm in length. Rinse and rub the membrane tubes under running water.
8. Close one end of the membrane tube using a dialysis tubing closure. Transfer the solution to the membrane tube using a funnel. Close the other end of the membrane tube using a dialysis tubing closure.
9. Place the membrane tubes in a 5-L beaker with 5 L of NaCl solution. Add a magnetic stirrer bar to the beaker.
10. Cover the beaker with aluminum foil and stir the NaCl solution. Replace spent solution with fresh NaCl solution every 8 h until the CMC-TYR has dialyzed against NaCl for 2 days.
11. Place the membrane tubes in a 5-L beaker containing 5 L of 25 % ethanol solution.
12. Dialyze the CMC-TYR against the 25 % ethanol solution and ultrapure water in sequence for 2 days each. Replace spent solution with fresh solution every 8 h.
13. Remove the membrane tubes from the beaker. Carefully open the membrane tubes and transfer the solution to several 50-mL conical centrifuge tubes.
14. Transfer the centrifuge tubes into a freezer and freeze at −20 °C for 48 h.
15. Lyophilize the CMC-TYR for 72 h using a freeze dryer.
16. Store the lyophilized CMC-TYR conjugate at room temperature.

3.2 Prepare the GTN-HPA Conjugate [23]

1. Insert a clean magnetic stirrer bar into a 1-L flask. Add 150 mL of ultrapure water and 100 mL of *N*,*N*-dimethylformamide (DMF). Stir the solution using a magnetic stirrer.
2. Weigh separately 3.32 g of 3,4-hydroxyphenylpropionic acid (HPA), 3.2 g of NHS, and 3.82 g of EDC.HCl.
3. Add the HPA and NHS, followed by the EDC.HCl, to the solution. Adjust the pH to 4.7 using HCl and NaOH. Stir the solution for 5 h at room temperature.
4. Weigh 10 g of gelatin.
5. Add the gelatin to 150 mL of ultrapure water. Continue stirring at 60 °C until the powder has fully dissolved.
6. Add the gelatin solution to the mixture containing HPA, NHS, and EDC.HCl.
7. Stir the solution overnight at room temperature.
8. Dialyze the product according to Subheading 3.1, **steps 7–14**.
9. Lyophilize the GTN-HPA for 72 h using a freeze dryer.
10. Store the lyophilized GTN-HPA conjugate at room temperature.

3.3 Prepare the Hydrogel Precursor Solution

1. Weigh 0.65 g of lyophilized GTN-HPA and transfer into a 15-mL conical centrifuge tube.
2. Weigh 0.65 g of lyophilized CMC-TYR and transfer into a separate 15-mL conical centrifuge tube.
3. Add 13 mL of PBS solution (or cell culture medium) to each tube.
4. Close the tubes and vortex the solutions until the GTN-HPA and CMC-TYR conjugates are fully dissolved (*see* **Note 6**).
5. Sterilize the GTN-HPA solution using a 0.2-μm syringe filter. Collect the filtrate in a sterile 15-mL conical centrifuge tube. Apply gentle pressure to filter the solution through the syringe filter membrane (*see* **Note 6**).
6. Repeat the filtration step with a fresh syringe filter to sterilize the CMC-TYR solution. Collect the filtrate in a separate sterile 15-mL conical centrifuge tube (*see* **Note 6**).
7. Transfer 8 mL of the sterilized GTN-HPA solution into a sterile 15-mL conical centrifuge tube. Transfer 2 mL of the sterilized CMC-TYR solution into the same tube. Close the cap of the centrifuge tube and mix the solution by vortexing for 30 s.

3.4 In Situ Hydrogel Gelation

1. Suspend chondrocytes in cell culture medium to make a cell suspension with final concentration of 1×10^6 cells/mL.
2. Transfer 0.7 mL of cell suspension to a sterile 2-mL boil-proof microcentrifuge tube. Spin down the cells to form a cell pellet by centrifuging the tube at $500 \times g$ for 5 min.

3. Gently open the cap of the microcentrifuge tube. Decant the supernatant carefully. Do not disturb the cell pellet at the bottom of the tube. Discard the supernatant as liquid biohazard waste.
4. Transfer 0.7 mL of hydrogel precursor solution to the microcentrifuge tube (*see* **Note 7**).
5. Dilute 1 part of HRP stock solution with 99 parts of sterile PBS solution to obtain 1.5 units/mL HRP (*see* **Note 8**). Transfer 3.5 μL of the 1.5 units/mL HRP solution to the microcentrifuge tube.
6. Transfer 3.5 μL of fresh H_2O_2 stock solution to the microcentrifuge tube (*see* **Note 9**). Immediately mix the solution for 2 s using a vortex.
7. Immediately, transfer 500 μL of the cell-hydrogel solution into a well of a sterile 24-well plate (*see* **Notes 10** and **11**) without introducing bubbles (*see* **Note 12**). Cells will be immobilized inside the hydrogel as the GTN-HPA and CMC-TYR are crosslinked enzymatically (*see* **Notes 13** and **14**). Discard the microcentrifuge tube as biohazard waste. Multiple cell-immobilized hydrogels can be prepared by repeating the above procedure.
8. Cover the well plate with its lid. Transfer the well plate into a 5 % CO_2 humidified incubator and allow the cell-immobilized hydrogels to incubate at 37 °C overnight.
9. Add 1 mL of sterile cell culture medium to every well that contains cell-immobilized hydrogel. Allow the cells to incubate further in the CO_2 incubator.
10. Exchange the cell culture medium once a day. Gently withdraw most of the spent cell culture medium and transfer 1 mL of fresh cell culture medium into each well.
11. Incubate at 37 °C in a 5 % CO_2 humidified incubator with daily medium exchange for the desired period of time.

3.5 Tuning the Pore Size Post-gelation

1. Withdraw most of the spent cell culture medium from the well plate containing the hydrogels.
2. Transfer 1 mL of cellulase stock solution into each well (*see* **Note 15**).
3. Cover the well plate with its lid. Transfer the well plate to a 5 % CO_2 humidified incubator and allow the hydrogel to incubate in cellulase solution at 37 °C overnight.
4. Replace the spent cellulase solution daily. To increase the rate of pore degradation, spent cellulase solution can be replaced every 2 h (*see* **Note 16**).
5. At the end of the digestion period, gently withdraw most of the spent cellulase enzyme solution from the well plate.

6. Transfer 1 mL of sterile PBS solution into each well to rinse the hydrogel. Allow the hydrogels to incubate in PBS for 10 min.
7. Rinse the hydrogel three times by repeating the above steps.
8. Add 1 mL of cell culture medium to every well that contains cell-immobilized hydrogel. Allow the cells to continue culture at 37 °C in a 5 % CO_2 humidified incubator.
9. Exchange the cell culture medium once a day. Gently withdraw most of the spent cell culture medium and transfer 1 mL of fresh cell culture medium into each well.
10. Incubate at the incubator with daily cell culture medium exchange for the desired period of time.

3.6 Pore Size and Porosity Measurement

The hydrogel pore size and porosity can be measured at any time during cell culture and/or after the pore size has been tuned post-gelation.

1. Gently withdraw most of the spent cell culture medium from the well plate containing the cell-immobilized hydrogels.
2. Transfer 1 mL of sterile PBS solution into each well to rinse the hydrogel. Allow the hydrogels to incubate in PBS for 10 min.
3. Rinse the hydrogel three times by repeating the above steps.
4. Transfer the well plate into a freezer and freeze the hydrogel at −20 °C for 24 h.
5. Lyophilize the hydrogel for 48 h using a freeze dryer.
6. Measure the pore sizes and porosities of the lyophilized hydrogel using a mercury porosimeter or other suitable method.

3.7 Biochemical Assays of Cartilage Matrix

1. Withdraw most of the spent cell culture medium from the wells that contain cell-immobilized hydrogel.
2. Transfer 1 mL of sterile PBS solution into each well. Allow the hydrogel to incubate in PBS for 10 min.
3. Rinse the cell-immobilized hydrogel three times by repeating the above steps.
4. Transfer the well plate into a freezer and freeze the hydrogel at −20 °C overnight.
5. Lyophilize the hydrogel for 48 h using a freeze dryer.
6. Remove the hydrogel from the well plate using a sterile spatula. Transfer the hydrogel into a 5-mL conical centrifuge tube.
7. Crush the lyophilized hydrogel inside the centrifuge tube.
8. Transfer 5 mL of papain solution into each centrifuge tube containing the crushed hydrogel. Incubate at 60 °C for 16 h on an orbital shaker.
9. Centrifuge the tube at 10,000 × g for 10 min to spin down the cell debris.

10. Decant the supernatant to a new 15-mL conical centrifuge tube. Discard the centrifuge tube containing cell debris as biohazard waste.
11. Use 50 μL of the supernatant for evaluation of the glycosaminoglycan (GAG) content. Evaluate GAG using a Blyscan™ GAG assay kit according to the manufacturer's instructions. Calculate the GAG content in the hydrogels using a calibration curve generated from standards prepared from known amounts of chondroitin sulfate.
12. To evaluate the collagen content, prepare crushed hydrogel by repeating **steps 1** to 7. Quantify the collagen content using ELISA detection of collagen types I and II or other suitable method (*see* **Note 17**).

3.8 Retrieving Cells from Hydrogels for Cellular Analysis

1. Remove the cell-immobilized hydrogels from the well plate carefully using a sterile spatula.
2. Transfer each of the hydrogels into a sterile 15-mL conical centrifuge tube.
3. Transfer 10 mL of hydrogel degradation solution to each centrifuge tube. Shake the centrifuge tube on an orbital shaker at 37 °C and 100 rpm until the hydrogel has fully degraded (*see* **Note 16**).
4. Transfer the centrifuge tube to a bench-top centrifuge. Centrifuge at 1200 × *g* for 10 min at room temperature.
5. Discard the supernatant and add 10 mL of sterile PBS solution. Vortex to resuspend the cell pellet and rinse the cells.
6. Centrifuge at 1200 × *g* for 10 min at room temperature. Rinse the cells another two times.
7. Remove the supernatant. The collected cell pellet can be used for further analyses such as cell proliferation (*see* **Note 18**), polymerase chain reaction (PCR) (*see* **Note 19**), and flow cytometry (FACS).

4 Notes

1. Isolate primary chondrocyte cells according to Ragan et al. [24] or obtain the cells from a supplier. Mesenchymal stem cells (MSCs) can be used instead of chondrocytes. Isolate MSCs according to Park et al. [25].
2. Proteins lose their bioactivity and degrade within hours when they are stored in solution at room temperature. The HRP stock solution can be prepared in large batches, and then frozen as 300-μL aliquots and stored at −20 °C for better day-to-day reproducibility. Remove the required amount from the freezer

and allow it to warm to room temperature before adding to the hydrogel precursor solution.

3. H_2O_2 oxidizes to form O_2 and water if it is left at room temperature for a long time. To maintain H_2O_2 in a non-oxidized state, it should be stored at the highest concentration available (usually 30 %) at −20 °C. H_2O_2 (30 % solution) can be frozen in 500-μL aliquots and stored at −20 °C for better day-to-day reproducibility. Remove the required amount from the freezer and allow it to warm to room temperature before preparation of the 1 % (v/v) H_2O_2 stock solution.
4. We found that it is best to prepare H_2O_2 stock solution (1 %) fresh each time.
5. The cellulase stock solution can be prepared in large batches and then frozen as 10-mL aliquots. Store at −20 °C for better day-to-day reproducibility. Remove the required amount from the freezer and allow it to warm to room temperature before adding it to the hydrogel.
6. Both GTN-HPA and CMC-TYR are polysaccharides that are susceptible to bacterial contamination once they are hydrated. After dissolving lyophilized GTN-HPA and/or CMC-TYR, the solution should be filter sterilized immediately to prevent bacterial contamination. We found that it is best to prepare GTN-HPA solution and/or CMC-TYR solution fresh each time.
7. The volume of hydrogel can be scaled up or scaled down. We recommend the following hydrogel volumes for well plates of different size: 2 mL per well for a 6-well plate, 1 mL per well for a 12-well plate, and 250 μL per well for a 48-well plate.
8. Discard the HRP stock solution if it has been placed at room temperature for more than 4 h.
9. Discard the H_2O_2 stock solution if it has been placed at room temperature for more than 2 h.
10. As an alternative to well plates, the hydrogel may be injected to fill irregular defects or to fill a mold with a desired shape.
11. The hydrogel may be implanted into an animal by subcutaneous injection. All animal studies should be carried out in compliance with the necessary regulations.
12. To avoid air bubbles, immerse the pipette tip just below the meniscus during aspiration. Use consistent plunger pressure and speed.
13. Increasing the concentration of HRP can increase the gelation time of the hydrogel.
14. Increasing the concentration of H_2O_2 can increase the stiffness of the hydrogel. We recommend using up to 10 μL of H_2O_2 stock solution to produce hydrogel with higher stiffness.

15. Alternatively, the cellulase stock solution may be injected into hydrogels in moulds or well plates. For in vivo study, the cellulase stock solution may be injected into hydrogels implanted in animals.
16. Hydrogel degradation time is dependent on the concentration of trypsin and cellulase as well as the size of the hydrogel. For larger hydrogels, use a higher concentration and higher volume of cellulase or trypsin.
17. Collagen content may be quantified according to Yamaoka et al. [26].
18. Perform a cell proliferation assay by measuring the DNA content using commerically available DNA quantitation kit.
19. Perform PCR assay according to Li et al. [27].

Acknowledgements

Funding for this work was partly provided through Australian Research Council Discovery Grant DP 120102570. This work was performed in part at the Melbourne Centre for Nanofabrication (MCN) which comprises the Victorian Node of the Australian National Fabrication Facility (ANFF). P.P.Y. Chan is grateful for an MCN Technology Fellowship.

References

1. Hoffman AS (2002) Hydrogels for biomedical applications. Adv Drug Deliv Rev 54:3–12
2. Hoo SP, Sarvi F, Li WH, Chan PPY, Yue Z (2013) Thermoresponsive cellulosic hydrogels with cell-releasing behavior. ACS Appl Mater Interfaces 5:5592–5600
3. Zhang R, Huang Z, Xue M, Yang J, Tan T (2011) Detailed characterization of an injectable hyaluronic acid-polyaspartylhydrazide hydrogel for protein delivery. Carbohydr Polym 85: 717–725
4. Fon D, Al-Abboodi A, Chan PPY, Zhou K, Crack P, Finkelstein DI, Forsythe JS (2014) Effects of GDNF-loaded injectable gelatin-based hydrogels on endogenous neural progenitor cell migration. Adv Healthc Mater 3:761–774
5. Wang F, Li Z, Khan M, Tamama K, Kuppusamy P, Wagner WR, Sen CK, Guan J (2010) Injectable, rapid gelling and highly flexible hydrogel composites as growth factor and cell carriers. Acta Biomater 6:1978–1991
6. Hwang JH, Kim IG, Piao S, Jung AR, Lee JY, Park KD, Lee JY (2013) Combination therapy of human adipose-derived stem cells and basic fibroblast growth factor hydrogel in muscle regeneration. Biomaterials 34:6037–6045
7. Li H, Koenig AM, Sloan P, Leipzig ND (2014) In vivo assessment of guided neural stem cell differentiation in growth factor immobilized chitosan-based hydrogel scaffolds. Biomaterials 35:9049–9057
8. Kim IL, Mauck RL, Burdick JA (2011) Hydrogel design for cartilage tissue engineering: a case study with hyaluronic acid. Biomaterials 32:8771–8782
9. Bhattacharjee M, Coburn J, Centola M, Murab S, Barbero A, Kaplan DL, Martin I, Ghosh S (2014) Tissue engineering strategies to study cartilage development, degeneration and regeneration. Adv Drug Deliv Rev. doi:10.1016/j.addr.2014.08.010
10. Sarvi F, Yue Z, Hourigan K, Thompson MC, Chan PPY (2013) Surface-functionalization of PDMS for potential micro-bioreactor and embryonic stem cell culture applications. J Mater Chem B 1:987–996
11. Balakrishnan B, Joshi N, Jayakrishnan A, Banerjee R (2014) Self-crosslinked oxidized

alginate/gelatin hydrogel as injectable, adhesive biomimetic scaffolds for cartilage regeneration. Acta Biomater 10:3650–3663

12. Xia W, Liu W, Cui L, Liu Y, Zhong W, Liu D, Wu J, Chua K, Cao Y (2004) Tissue engineering of cartilage with the use of chitosan-gelatin complex scaffolds. J Biomed Mater Res B Appl Biomater 71B:373–380
13. Hu X, Ma L, Wang C, Gao C (2009) Gelatin hydrogel prepared by photo-initiated polymerization and loaded with TGF-β1 for cartilage tissue engineering. Macromol Biosci 9:1194–1201
14. Awad HA, Quinn WM, Leddy HA, Gimble JM, Guilak F (2004) Chondrogenic differentiation of adipose-derived adult stem cells in agarose, alginate, and gelatin scaffolds. Biomaterials 25:3211–3222
15. Sarem M, Moztarzadeh F, Mozafari M (2013) How can genipin assist gelatin/carbohydrate chitosan scaffolds to act as replacements of load-bearing soft tissues? Carbohydr Polym 93:635–643
16. Chen F-M, Zhao Y-M, Sun H-H, Jin T, Wang Q-T, Zhou W, Wu Z-F, Jin Y (2007) Novel glycidyl methacrylated dextran (Dex-GMA)/gelatin hydrogel scaffolds containing microspheres loaded with bone morphogenetic proteins: formulation and characteristics. J Control Release 118:65–77
17. Wang L-S, Du C, Toh WS, Wan ACA, Gao SJ, Kurisawa M (2014) Modulation of chondrocyte functions and stiffness-dependent cartilage repair using an injectable enzymatically crosslinked hydrogel with tunable mechanical properties. Biomaterials 35:2207–2217
18. Al-Abboodi A, Fu J, Doran PM, Tan TTY, Chan PPY (2014) Injectable 3D hydrogel scaffold with tailorable porosity post-implantation. Adv Healthc Mater 3:725–736
19. Wang L-S, Chung JE, Pui-Yik CP, Kurisawa M (2010) Injectable biodegradable hydrogels with tunable mechanical properties for the stimulation of neurogenesic differentiation of human mesenchymal stem cells in 3D culture. Biomaterials 31:1148–1157
20. Al-Abboodi A, Tjeung R, Doran PM, Yeo LY, Friend J, Yik Chan PP (2014) In situ generation of tunable porosity gradients in hydrogel-based scaffolds for microfluidic cell culture. Adv Healthc Mater 3(10):1655–1670
21. Hoo SP, Loh QL, Yue Z, Fu J, Tan TTY, Choong C, Chan PPY (2013) Preparation of a soft and interconnected macroporous hydroxypropyl cellulose methacrylate scaffold for adipose tissue engineering. J Mater Chem B 1:3107–3117
22. Martens PJ, Bryant SJ, Anseth KS (2003) Tailoring the degradation of hydrogels formed from multivinyl poly(ethylene glycol) and Poly(vinyl alcohol) macromers for cartilage tissue engineering. Biomacromolecules 4:283–292
23. Al-Abboodi A, Fu J, Doran PM, Chan PPY (2013) Three-dimensional nanocharacterization of porous hydrogel with ion and electron beams. Biotechnol Bioeng 110:318–326
24. Ragan PM, Chin VI, Hung H-HK, Masuda K, Thonar EJMA, Arner EC, Grodzinsky AJ, Sandy JD (2000) Chondrocyte extracellular matrix synthesis and turnover are influenced by static compression in a new alginate disk culture system. Arch Biochem Biophys 383: 256–264
25. Park H, Temenoff JS, Tabata Y, Caplan AI, Mikos AG (2007) Injectable biodegradable hydrogel composites for rabbit marrow mesenchymal stem cell and growth factor delivery for cartilage tissue engineering. Biomaterials 28:3217–3227
26. Yamaoka H, Asato H, Ogasawara T, Nishizawa S, Takahashi T, Nakatsuka T, Koshima I, Nakamura K, Kawaguchi H, Chung U-i, Takato T, Hoshi K (2006) Cartilage tissue engineering using human auricular chondrocytes embedded in different hydrogel materials. J Biomed Mater Res A 78A:1–11
27. Li W-J, Tuli R, Okafor C, Derfoul A, Danielson KG, Hall DJ, Tuan RS (2005) A three-dimensional nanofibrous scaffold for cartilage tissue engineering using human mesenchymal stem cells. Biomaterials 26:599–609

Chapter 9

Decellularized Extracellular Matrix Scaffolds for Cartilage Regeneration

Shraddha Thakkar, Hugo Fernandes, and Lorenzo Moroni

Abstract

Decellularized extracellular matrix (ECM) is gaining a lot of attention as a biomaterial for tissue engineering applications. This chapter describes the processing techniques for decellularization of cell-derived ECM and protocols for the fabrication of ECM-based scaffolds in the form of hydrogels or fibrous polymer meshes by electrospinning. It describes the protocols to analyze the morphology and presence of collagen in fabricated scaffolds using scanning electron microscope and Picrosirius Red staining respectively. Methods to evaluate the metabolic activity and proliferation of cells (resazurin-based assay and DNA assay, respectively) and gene expression are also presented. Furthermore, histological techniques to analyze the presence of sulfated glycosaminoglycans are also described.

Key words Mesenchymal stromal cells, Extracellular matrix, Decellularization, Lyophilization, Chondrocytes, Electrospinning, Gene expression, Hydrogel

1 Introduction

Cartilage is a connective tissue composed of chondrocytes trapped in extracellular matrix (ECM). Damage in articular cartilage leads to degeneration of the joint which progresses to the development of osteoarthritis (OA) [1–3]. The lack of vascular supply and low matrix turnover limits its repairability [4]. Cell-based cartilage tissue engineering has emerged as a promising technique to regenerate or restore cartilage [5–9].

In native tissues, the ECM plays an important role in maintenance and renewal of tissues [10, 11]. Interaction of the ECM proteins with growth factors modulates cell behavior affecting growth, differentiation, cell migration, as well as viability [12, 13]. In recent years, ECM has been used for different tissue engineering applications [14–17]. One of the major challenges of cartilage tissue engineering is controlling the fate of chondrocytes outside of their natural three-dimensional (3D) environment. It is known that the native ECM comprises various types of collagens,

Pauline M. Doran (ed.), *Cartilage Tissue Engineering: Methods and Protocols*, Methods in Molecular Biology, vol. 1340, DOI 10.1007/978-1-4939-2938-2_9, © Springer Science+Business Media New York 2015

Table 1
Commercially available extracellular matrix (biological scaffolding materials)

	Commercial products	Type	Source	Company
1	AlloDerm®	Dermis	Human	Lifecell
2	AlloPatch®	Fascia lata	Human	Musculoskeletal Transplant Foundation
3	Restore™	Small intestinal submucosa (SIS)	Porcine	DePuy
4	TissueMend®	Dermis	Bovine	TEI Biosciences
5	Zimmer collagen Patch®	Dermis	Porcine	Tissue Science Laboratories
6	SurgiMend™	Dermis	Bovine	TEI Biosciences
7	Oasis®	Small intestinal submucosa (SIS)	Porcine	Healthpoint
8	DurADAPT™	Pericardium	Horse	Pegasus Biologicals
9	Durasis®	Small intestinal submucosa (SIS)	Porcine	Cook SIS

proteoglycans, and other functional molecules vital for proper functioning of the tissues. This complex composition is difficult to mimic, promoting researchers to use the native ECM after decellularization treatment as an alternative. Decellularization would ideally remove all the cells and the antigens while retaining all the functional cues that reside in the ECM [18]. Decellularized ECM from a variety of tissues including heart [19–21], heart valves [22–24], blood vessels [25, 26], and small intestinal sub mucosa [27–29] has been studied. Some of the commercially available ECM scaffolds [30] are presented in Table 1.

ECM has been utilized to improve the biological property of synthetic materials. Titanium fiber meshes deposited with ECM derived from rat marrow stromal cells (rMSC) showed an increase in osteogenic differentiation of rMSCs and calcium deposition compared to titanium meshes without ECM [31]. Another study showed that poly (lactic-co-glycolic acid) (PLGA) meshes consisting of ECM secreted by mesenchymal stromal cells (MSCs) showed higher chondrogenesis compared to standard pellet culture [32]. Chen et al. cultured MSCs on bone marrow-derived ECM and showed that marrow ECM enhanced proliferation of the mesenchymal progenitors, which retained their stem cell characteristics [33]. Furthermore, Postovit et al. observed the ability of the stem cell microenvironment to reprogram the fate of melanoma cells to normal pigment cells [34]. It is widely recognized that ECM is a rich source of biological signals but lacks the mechanical properties required to make a scaffold. To circumvent this, ECM has been combined with synthetic materials to improve its mechanical properties while maintaining its bioactivity. Additionally, to mimic the natural 3D environment found in ECM, scaffolds have been prepared in the form of a hydrogel and of fibrous mesh using electrospinning techniques [35–44].

A wide range of techniques are available for fabrication of 3D ECM scaffolds, depending on the intended application. The aim of the present chapter is to describe the experimental procedures used to fabricate electrospun and hydrogel scaffolds containing ECM. We describe the steps required to fabricate randomly oriented fibrous meshes as well as hydrogels. In vitro studies have already been performed to evaluate the effect of human mesenchymal stromal cell (hMSC) ECM on the fate of osteoarthritic (OA) cells [45]. We provide complete protocols for expansion, culturing, and seeding of cells on scaffolds, including ECM production using hMSCs, and its processing using decellularization of the cell layers and subsequent lyophilization. We focus on the fabrication of ECM scaffolds as hydrogels and electrospun meshes, in both cases with and without ECM. Poly-caprolactone (PCL)–chloroform solution is electrospun to form fibrous meshes, while gelation of sodium alginate in contact with calcium chloride is used to form hydrogel. Electrospun scaffolds with and without ECM are characterized using scanning electron microscopy. Incorporation of ECM (i.e., collagen content) in the scaffolds is assessed by Picrosirius Red staining. The metabolic activity of cells seeded on electrospun scaffolds is measured by a resazurin-based assay (Alamar Blue assay), and alterations in gene expression are monitored using PCR. Histological staining and DNA and glycosaminoglycan (GAG) analyses are performed on hydrogel samples.

2 Materials

2.1 Cell Culture and Production of Extracellular Matrix

1. Bone marrow aspirates obtained from human patients with informed consent.
2. Osteoarthritic (OA) and healthy (HL) cartilage obtained from human donors with informed consent, e.g., during knee surgery.
3. Proliferation medium: α-minimum essential medium (α-MEM), 10 % fetal bovine serum, 0.2 mM ascorbic acid, 2 mM l-glutamine, 100 units/mL penicillin, 100 μg/mL streptomycin, and 1 ng/mL basic fibroblast factor-2 (FGF-2).
4. Basic medium: α-MEM, 10 % fetal bovine serum, 0.2 mM ascorbic acid, 2 mMl-glutamine, 100 units/mL penicillin, and 100 μg/mL streptomycin.
5. Chondrogenic medium: Dulbecco's Modified Eagle Medium (DMEM), 1 % GlutaMAX (100X, Gibco), 0.2 mM ascorbic acid, 100 units/mL penicillin, 100 μg/mL streptomycin, 0.4 mM proline, 100 μg/mL sodium pyruvate, and 50 μg/mL

insulin transferrin selenium-premix. Add 10 ng/mL transforming growth factor-beta3 (TGF-β3) and 1×10^{-7} M dexamethasone shortly before use.

6. Collagenase type II solution (for cell isolation): 0.15 % collagenase type II (Worthington) in DMEM with 100 units/mL penicillin and 100 μg/mL streptomycin.
7. Phosphate buffered saline (PBS) supplemented with antibiotics: PBS containing 100 units/mL penicillin and 100 units/mL streptomycin.
8. Chondrocyte culture medium: DMEM with 10 % fetal bovine serum, 1× nonessential amino acids, 10 mM HEPES buffer, 0.2 mM ascorbic acid 2-phosphate, 0.4 mM proline, 100 units/mL penicillin, 100 μg/mL streptomycin.
9. Trypsin solution: 0.05 % trypsin (Sigma-Aldrich) for cell detachment.
10. Ammonium hydroxide (NH_4OH) solution (for disruption of cell membranes): 20 mM NH_4OH in distilled water.
11. Cell scraper (Corning®).
12. Deoxyribonuclease (DNase) 1 solution: 10 units/mL of DNase 1 (Invitrogen) in storage buffer provided. Store at −20 °C.

2.2 Electrospun Scaffolds

1. PCL/HFIP solution: 1998.75 mg of poly-caprolactone (PCL, Mw 42,500 kDa) in 10 mL of 1, 1, 1, 3, 3, 3-hexafluoro-2-propanol (HFIP, 99.9 % w/v) in a glass vial. To avoid evaporation, seal the lid of the vial with parafilm. Stir the PCL/HFIP solution overnight on a magnetic stirrer at room temperature.
2. PBS.
3. Picrosirius Red staining kit (Polysciences), for staining collagen.
4. 70 % ethanol solution.
5. Alamar Blue solution: 10 % v/v Alamar Blue reagent in 1 mL of basic medium.
6. NucleoSpin RNA II isolation kit (Macherey-Nagel).
7. TRIzol reagent (Invitrogen).
8. Nanodrop 1000 spectrophotometer, for quantifying RNA.
9. iScript cDNA synthesis kit (Bio-Rad). Store at −20 °C at constant freezer temperature; components remain stable for a year.
10. MJ mini gradient thermal cycler (Bio-Rad).
11. SYBR Green supermix (iQ SYBR) (Bio-Rad).
12. Roche light cycler, for qPCR.
13. Bio-Rad software, for data analysis.

2.3 Hydrogel Scaffolds

1. Tris/EDTA buffer: 6.05 g Tris in 950 mL of distilled water. Adjust the pH to 7.6 using HCl and add 0.372 g of EDTA. Add 1000 mL of distilled water.
2. Digestion buffer: 18.5 μg/mL of iodoacetamide (BioUltra, Sigma) and 1 μg/mL of pepstatin A (Sigma) in Tris/EDTA buffer.
3. Proteinase K solution: 1 mg/mL of Proteinase K (Sigma, from *Tritirachium album*, lyophilized powder, >30 units/mg of protein) in digestion buffer.
4. Bicinchoninic acid (BCA)-based protein assay kit (Thermo Fisher).
5. CyQUANT Cell Proliferation Assay kit (Life Technologies), for DNA quantification.
6. Sodium alginate solution: 2 % w/v sodium alginate (Pronatal, LF10/60FT) in PBS. Alginate was sterilized using a UV lamp (wavelength 234 nm) for 15 min.
7. Calcium chloride ($CaCl_2$) solution: 100 mM $CaCl_2$ anhydrous in distilled water.
8. PBS.
9. Lysis buffer: Concentrated cell lysis buffer (component B provided in the CyQUANT Cell Proliferation Assay kit) diluted 20-fold in distilled water containing 180 mM NaCl and 1 mM EDTA.
10. Lysis buffer with RNase: 1.35 kuntz/mL of RNase in lysis buffer.
11. PBE buffer: 7.1 g of Na_2HPO_4 and 1.86 g of Na_2EDTA in 495 mL of double-distilled H_2O (d_2H_2O). Adjust the pH to 6.5 ± 0.1 using concentrated HCl solution. Adjust the volume of the buffer to 500 mL and sterilize by filtration. Store at 4 °C.
12. PBE-cysteine solution: 17.5 mg of cysteine-HCl in 10 mL of PBE buffer.
13. Dimethyl methylene blue (DMMB) solution: 9.5 mL of 0.1 M HCl solution, 90.5 mL of d_2H_2O, 0.304 g of glycine, and 0.237 g of NaCl. Adjust the pH to 3 and, while stirring, add 16 μg of DMMB/mL into the glycine/NaCl solution. Store in the dark at room temperature. The solution remains stable for 3 months. Sterilize by filtering just prior to use.
14. NaCl solution: 13.4 g of NaCl in 100 mL of d_2H_2O. Store at room temperature.
15. Formalin solution: 10 % formalin in distilled water.
16. Ethanol solution series: 70, 80, and 90 % ethanol in distilled water.
17. 3 % v/v acetic acid solution: 3 mL of glacial acetic acid and 97 mL of distilled water.

18. Alcian Blue solution: 1 g of Alcian Blue 8 GX in 100 mL of 3 % v/v acetic acid solution. Stir the solution and adjust the pH to 2.5.
19. Fast Green solution: 0.01 g of Fast Green FCF concentrate in 1000 mL of distilled water.
20. 1 % v/v acetic acid solution: 1 mL of glacial acetic acid and 99 mL of distilled water.
21. Safranin O solution: 0.1 g of Safranin O in 100 mL of distilled water.
22. Nuclear Fast Red solution: 0.25 g of Nuclear Fast Red and 12.5 g of aluminum sulfate in 250 mL of distilled water.
23. Picrosirius Red staining kit (Polysciences).

3 Methods

3.1 Cell Culture and Production of Extracellular Matrix

3.1.1 Isolation of hMSCs from Bone Marrow

Isolation of hMSCs from bone marrow has been described previously [46].

1. Suspend the bone marrow aspirates using 20-gauge needles and filter through a 70-μm cell strainer (*see* **Note 1**).
2. Prepare T175 cm^2 culture flasks with 15 mL of proliferation medium and seed 5×10^5 mononuclear cells per flask (Passage 0). Incubate at 37 °C in a humid atmosphere with 5 % CO_2.
3. Refresh the medium twice a week.
4. To freeze the hMSC cells, add 10 % fetal bovine serum (FBS) and 10 % dimethyl sulfoxide (DMSO) to the proliferation medium in a cryovial. Add 5×10^6 cells per vial and store in liquid nitrogen. Passages 2–4 are needed to obtain cells for cryopreservation.

3.1.2 Isolation and Proliferation of Chondrocytes

1. Incubate dissected OA and HL cartilage in collagenase type II solution for 20–22 h at 37 ° C.
2. Filter the cell suspension through a 100-μm cell strainer.
3. Wash the cells twice with PBS supplemented with antibiotics.
4. Centrifuge the cells and remove the supernatant. Resuspend the pellet in basic medium and count the cells with a hemocytometer. Adjust the cell concentration to 1×10^6 cells/mL.
5. Prepare T175 cm^2 culture flasks with 15 mL of chondrocyte culture medium. Seed the cells at 5000 cells/cm^2 (Passage 0) and incubate at 37 °C in a humid atmosphere of 5 % CO_2 for 3 weeks.
6. Change the medium every 3 days.

7. When confluence reaches 80 %, trypsinize the cells by adding 1 mL of trypsin solution to the culture flask for 3 min. Gently tap the flask to ensure detachment of the cells (*see* **Note 2**).
8. Confirm cell detachment from the flask by observing under a microscope.
9. Transfer the suspension to a tube and centrifuge at $300 \times g$ for 3 min.
10. Discard the supernatant and resuspend the pellet with basic medium.
11. Perform a cell count with a hemocytometer and adjust to 1×10^6 cells/mL of medium.

3.1.3 Cell Culture to Obtain Extracellular Matrix (ECM)

1. Seed cryopreserved hMSCs in T175 cm^2 culture flasks at a seeding density of 5000 cells/cm^2. To compare the effects of media on cell culture, two different media including basic medium and chondrogenic medium are used for culturing. Triplicate flasks are used for each medium to obtain ECM from hMSCs and differentiated hMSCs in chondrogenic medium.
2. Culture the cells for 3 weeks. Initially refresh the medium after 10 days and then every 4 days during subsequent weeks.
3. Let cells reach confluence to maximize ECM deposition. Follow decellularization methods to recover the ECM.

3.1.4 Decellularization of ECM

1. Isolate ECM by adding 5 mL of NH_4OH solution to the tissue culture flasks for 3–4 min.
2. Upon confirmation of complete cell lysis under a microscope (cells start to burst), add distilled water to dilute the NH_4OH. Wash and remove the remaining cell debris.
3. Cautiously aspirate the NH_4OH solution (*see* **Note 3**). Add DNase 1 solution to the flask and incubate for 30 min.
4. Wash carefully with distilled water without disturbing the thin layer of ECM at the bottom of the flask.
5. Use a cell scraper to detach the layer of ECM from the flask. ECM thus obtained looks transparent with jelly-like structure.

3.2 Electrospun Scaffolds

3.2.1 Lyophilization of ECM

1. Measure the wet weight of the jelly-like ECM in an Eppendorf tube and note it.
2. Seal the tube containing the matrix. Freeze the sample to begin lyophilization of the ECM by dipping the tube in liquid nitrogen for about 2–3 min.
3. Place the frozen tube in a gas freeze-drying vessel (*see* **Note 4**).
4. Attach the vessel to the freeze-dryer and turn on the vacuum.

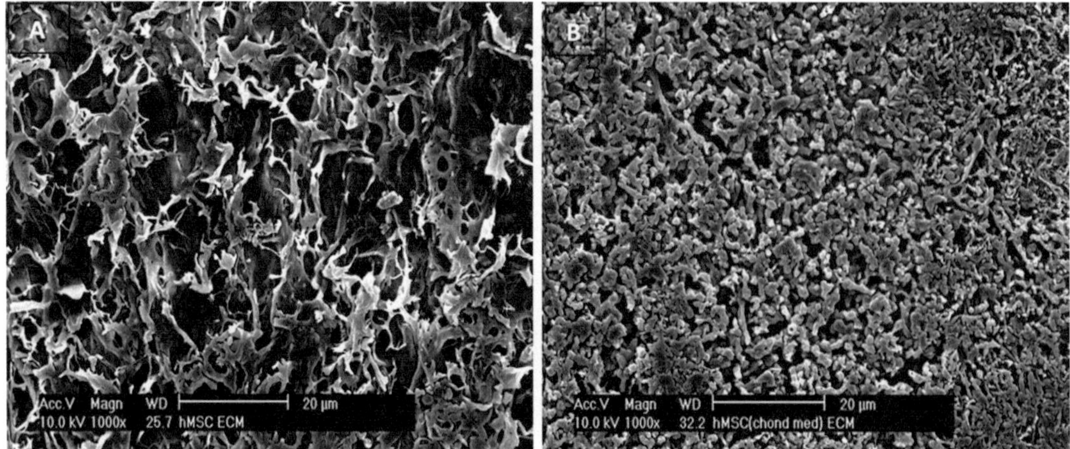

Fig. 1 Scanning electron microscopy images (magnification 1000×) of (**a**) ECM produced by hMSCs cultured in basic medium (*left*); and (**b**) ECM produced by hMSCs cultured in chondrogenic medium (*right*). The observed difference in morphology could be the effect of culture medium on the cells, leading to a difference in protein composition expressed by the cells

5. It takes about 3–4 days to remove about 300 μL of water from 1 mg of sample. Upon completion of the process, measure the dry weight of the sample. As protein quantification of lyophilized ECM is not possible, all of the lyophilized ECM is assumed to be proteins (Fig. 1).

3.2.2 Preparation of Solutions for Electrospinning

1. Add 0.125 μg/mL of lyophilized ECM to the PCL/HFIP solution to prepare solutions with each type of ECM (i.e., ECM produced by hMSCs cultured in basic medium and ECM produced by hMSCs in chondrogenic medium).
2. Stir the PCL/HFIP solutions containing ECM for 8 h on a magnetic stirrer for homogenous distribution of proteins in the solution (*see* **Note 5**).

3.2.3 Electrospinning Procedure

1. Set up electrospinning apparatus consisting of a syringe pump, voltage supply, capillary tube (0.8 mm diameter, to transport solution from the syringe to a metallic needle), and aluminum foil of thickness 0.2 mm.
2. Place the collector or aluminum foil at a distance of 20 cm from the needle.
3. Attach the positive end of the voltage supply to the needle and ground the collector.
4. Maintain the temperature at 20 °C and humidity between 33 % and 35 %.
5. Fill a 5-mL syringe with polymeric solution from Subheading 3.2.2. Remove all air bubbles and then attach a needle to the syringe.

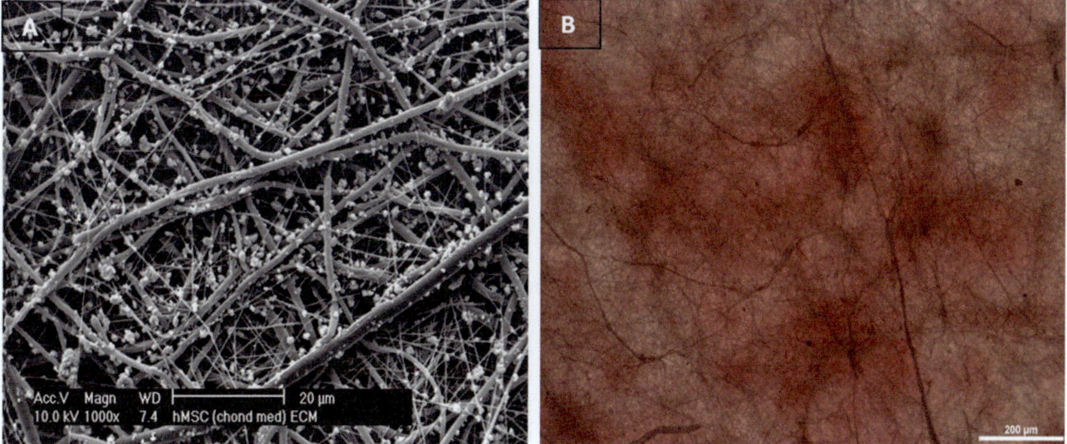

Fig. 2 Electrospun scaffold consisting of ECM produced by hMSCs cultured in chondrogenic medium. (**a**) SEM image; and (**b**) Polarized light microscopic image. The red staining indicates the presence and distribution of collagen and hence ECM in the scaffold

6. Position the syringe needle on the syringe pump and set the flow rate to the desired value (e.g., 2.5 mL/h).
7. Apply a potential difference of 15 kV to create an electric field to form fibers.
8. Collect fibers on the aluminum foil (*see* **Note 6**).

3.2.4 Scanning Electron Microscopy

1. Attach 25-mm adhesive carbon tabs onto aluminum SEM specimen stubs.
2. Cut the electrospun scaffold using a sharp blade.
3. Place the cut samples on the specimen stubs and sputter with gold for 200 s at 60 mA current and 10^{-1} mbar vacuum.
4. Use SEM with a beam of 10 kV at different magnifications for capturing images.
5. Apply Image J software to calculate the fiber diameter (Fig. 2a). Measure at least 20 fibers per condition.

3.2.5 Picrosirius Red Staining

1. Prior to staining, wash the electrospun scaffold samples twice in PBS.
2. Place the samples in solution A from the Picrosirius Red staining kit for 2 min and wash with distilled water.
3. Then, place samples in solution B for 110 min.
4. Last, immerse the samples in solution C for 2 min.
5. Dehydrate by treating with 70 % ethanol solution for 45 s.
6. Dry the samples at room temperature for 20 min after peeling off the aluminum foil. Mount on glass microscope slides. Collagen type I and type III fibrils are stained in red when observed under a polarized light microscope (Fig. 2b) (*see* **Note 7**).

3.2.6 Cell Seeding and Culture on Electrospun Scaffolds

1. Prepare electrospun discs (15 mm in diameter) with and without ECM.
2. Soak the discs in 100 % ethanol to sterilize for 30 min and dry in a flow cabinet for 15 min.
3. Wash the scaffolds twice with PBS and transfer to a 24-well nontreated plate.
4. Secure the scaffolds in place using O-rings, size 14 × 1.78.
5. Soak the scaffolds along with the O-rings in basic medium and incubate for 4 h.
6. Seed each scaffold with 50,000 OA or HL cells in 50 μL of chondrogenic culture medium and incubate for 1 h to allow the cells to attach to the scaffold.
7. To understand the influence of media on cell culture, add 1 mL of either basic or chondrogenic medium to each well.
8. Culture the electrospun scaffolds in an incubator at 37 °C and 5 % CO_2 for 21 days, changing the medium every 3 days.

3.2.7 Alamar Blue Staining

1. Remove the culture medium and add 1 mL of Alamar Blue solution to each well in the 24-well plate.
2. Incubate the plate for 4 h at 37 °C. Cover with aluminum foil to avoid light.
3. Prepare negative controls without cells using basic medium or chondrogenic medium in a 96-well black-bottom plate.
4. Read the fluorescence using an excitation wavelength of 530–560 nm and emission wavelength of 580–610 nm. The fluorescence measured is the response to chemical reduction of the medium due to cell growth (*see* **Note 8**).

3.2.8 RNA Extraction

1. Combine two samples containing cells that have been cultured in the same medium. This is to ensure a sufficient quantity of RNA. Add 500 μL of TRIzol reagent to each well and transfer the contents to Eppendorf tubes.
2. Add 200 μL of chloroform (at room temperature) to the samples, vigorously agitate the tubes for 15 s, and incubate at room temperature for 2 min.
3. Centrifuge the tubes at 7245 × *g* for 15 min at 4 °C.
4. Remove the aqueous phase containing the RNA by inclining the tube at 45° and carefully pipetting out the solution to a new Eppendorf tube.
5. Precipitate the RNA (approximately 500 μL) at room temperature using isopropyl alcohol.

6. Follow the instructions of the RNA isolation kit from Macherey-Nagel, starting from **step 4**. The kit includes rDNase and RA2 and RA3 buffers.
   (a) Prepare the rDNase stock solution by adding 55 μL of RNase-free H_2O to the rDNase vial (10 preps) and incubate for 1 min at room temperature. For homogenous mixing, gently swirl the vials (*see* **Note 9**).
   (b) Prepare the wash buffer RA3 by adding 8 mL of 100 % ethanol to 2 mL of wash buffer RA3 concentrate (*see* **Note 10**).
7. Filter the lysate by placing the violet ring filter on 2-mL collection tubes. Add the mixture and centrifuge at 11,000 × g for 30 s.
8. Discard the filter and add 100 μL of 70 % ethanol solution to homogenize the lysate. Mix the solution well by pipetting up and down a few times.
9. For binding of RNA, place a Nucleospin RNA XS column on the collection tubes and add the lysate to it. Centrifuge the tubes at 11,000 × g for 30 s.
10. Discard the fluid and place the columns on a new collection tube.
11. Add 350 μL of MDB (membrane desalting buffer) from the RNA isolation kit to the column and centrifuge at 11,000 × g for 1 min.
12. Prepare the rDNase reaction mixture in a sterile microcentrifuge tube. To make a stock solution, add 10 μL of rDNase stock solution to 90 μL of Reaction Buffer for rDNase from the RNA isolation kit, times the number of samples. For homogenous mixing of components, flick the tube.
13. Place 95 μL of rDNase reaction mixture at the center of the silica membrane.
14. Cover the tube with a lid and incubate for 15 min at room temperature.
15. Wash and dry the silica membrane by adding 200 μL of RA2 buffer to the column and centrifuge at 11,000 × g for 30 s. Place the column on a new collection tube.
16. Wash a second time in a similar way; add 600 μL of RA3 buffer and centrifuge at 11,000 × g for 30 s.
17. For the last wash, add 250 μL of RA3 buffer to the column and centrifuge in a similar way for 2 min. Discard the fluid and place the column on a nuclease-free collection tube.
18. Elute RNA by adding 40 μL of RNase-free water and centrifuge at 11,000 × g for 30 s.
19. Place the isolated RNA on ice to maintain optimal stability.

20. Quantify RNA using Nanodrop spectrophotometry.

3.2.9 Nanodrop Quantification of RNA

1. Select the Nanodrop 1000 software module and further choose the nucleic acid application module.
2. Clean the pedestal surface and perform a blank measurement.
3. Load the blank with 1 μL of RNAse-free water onto the pedestal and lower the sampling arm carefully.
4. Select the blank option to run the measurement and save it.
5. Clean the pedestal, load 1 μL of sample, and measure its absorbance value.
6. Use an absorbance ratio of 260/280 to assess the purity and concentration of RNA (*see* **Note 11**).

3.2.10 First-Strand cDNA Synthesis Using Reverse Transcriptase

Follow the iScript instruction manual for cDNA synthesis.

1. To ensure all samples use an equal amount of RNA for cDNA synthesis, calculate the amount of RNA for each sample. For cDNA synthesis, an average 240 ng of RNA is required for a total working volume of 20 μL. The volume of master mix is 5 μL of the total working volume; the remaining volume of 15 μL is available for nuclease-free water and RNA. Based on the RNA values obtained from Nanodrop quantification, use the lowest/minimum value to calculate the RNA amount for each sample.
2. To attain the remaining volume of 15 μL for each sample, add nuclease-free water to the calculated amount of RNA (in ng/μL).
3. Place the reagents and tubes on ice while preparing the master mix.
4. Defrost the RNA samples on ice.
5. Prepare a master mix by adding 4 μL of 1× iScript select reaction mix to 1 μL of 1× iScript reverse transcriptase.
6. Label 0.5-mL Eppendorf tubes and add RNA and nuclease-free water as per calculation.
7. To these tubes, add 5 μL of master mix and vortex for 15 s to ensure homogenous mixing.
8. Centrifuge the tubes for 15 s and arrange in the MJ mini gradient thermal cycler.
9. Select the program for a volume of 20 μL; the thermal cycler incubates the reactions at 25 °C for 5 min.
10. Further, it mixes gently and incubates at 42 °C for 30 min.
11. Last, it incubates the reaction at 85 °C for 5 min to inactivate the reverse transcriptase and terminate the cDNA synthesis.
12. Use the obtained cDNA directly for PCR or store it at −20 °C for up to 6 months.

3.2.11 Quantitative PCR (qPCR)

1. Use 1 μL of cDNA to perform the qPCR for each condition.
2. Prepare the master mix of 20 μL with the following components: 10 μL of iQ SYBR supermix, 1 μL of optimized upstream primers and downstream primers, 1 μL of cDNA, and the remaining amount of nuclease-free water to obtain a total volume of 20 μL. Use water as a control.
3. Mix the components gently and avoid bubble formation in the mixture.
4. Centrifuge the tubes at 600 rpm for 1 min.
5. Arrange 20-μL glass capillary tubes and fill the capillary tubes with 20 μL of master mix. Seal the capillary tubes with good sealing lids and place them in the Roche light cycler and select the program for specific genes.
6. Use the primer sequences for selected target genes (e.g., collagen I, collagen II, collagen X, aggrecan, Sox 9) and housekeeping gene (e.g., GAPDH).
7. Analyze the data using Bio-Rad software by adjusting the noise band to the exponential phase.
8. Quantify the target gene by normalizing the target gene to the housekeeping gene, i.e., the cycle threshold (Ct value) of GAPDH.

3.3 Hydrogel Scaffolds

Digest the decellularized ECM from Subheading 3.1.4 in proteinase K according to the manufacturer's protocol.

3.3.1 Digestion of Decellularized ECM in Proteinase K

1. Weigh the decellularized ECM: approximately 350 mg of ECM is obtained by culturing 5000 hMSCs/cm^2 for 3 weeks as described in Subheading 3.1. Centrifuge the decellularized ECM and extract the water to obtain a pellet of ECM.
2. Add 1 mL of proteinase K solution to the decellularized ECM. Allow digestion to proceed at 56 °C for 16 h.
3. Pass the solution through a 0.22-μm filter and quantify the total protein content using a bicinchoninic acid (BCA) assay.

3.3.2 BCA Assay for ECM Protein

Using the BCA protein assay kit, proceed according to the manufacturer's protocol to analyze the protein content in digested ECM samples.

1. Add 10 μL of sample or standards to plates containing 200 μL of working reagent from the BCA assay kit, incubate for 30 min at 37 °C, and allow cooling at room temperature.
2. Measure the absorbance of the resultant purple-colored samples by spectrometry at 562 nm.
3. Plot the results from standards to determine a standard curve and use the equation to calculate the amount of protein (μg) in each sample (*see* **Note 12**).

3.3.3 Cell Encapsulation and Cell Culture in Hydrogel Scaffolds

The hydrogel scaffolds are formed by crosslinking sodium alginate solution containing cells in $CaCl_2$ solution.

1. Sodium alginate solution (2 % w/v alginate in PBS) is supplemented with ECM (125 μg/mL of PBS) to form an alginate-ECM solution.
2. Suspend OA and HL cells in basic medium at a concentration of a million cells per mL of medium. Add the cell suspension to the alginate-ECM solution with density of 2×10^6 cells/mL of alginate-ECM solution.
3. For alginate-ECM gelation, pipette 400 μL of $CaCl_2$ solution to each well of a well plate, dropwise add 100 μL of cell alginate-ECM suspension (using a pipette), and incubate for 30 min at room temperature (*see* **Note 13**). The size of the alginate beads formed is around 1–3 mm diameter.
4. To understand the effect of media on cell culture, add 1 mL of either basic medium or chondrogenic medium to the wells after removal of excess $CaCl_2$.
5. Repeat the cell suspension protocol for encapsulation of OA and HL cells in alginate solution without ECM.
6. Culture all cell-hydrogel constructs for 3 weeks in either basic or chondrogenic medium in an incubator at 37 °C and 5 % CO_2. Refresh the medium every 3 days.

3.3.4 DNA Assay

1. Store the cultured cell-hydrogel constructs at −80 °C after discarding the medium and washing twice in PBS.
2. Prior to the assay, digest the scaffolds overnight at 56 °C in proteinase K as mentioned in Subheading 3.3.1, using cultured cell-hydrogel constructs instead of decellularized ECM.
3. Add 400 μL of lysis buffer with RNase to 100 μL of sample in proteinase K and incubate for 1 h at room temperature.
4. Add an equal amount (100 μL) of CyQUANT GR dye (available in the CyQUANT kit) to each well and incubate at 37 °C for 15 min in the dark.
5. Measure the fluorescence emission at 520 nm by spectrometry at an excitation wavelength of 480 nm. The measured fluorescence is used to plot a standard curve and obtain a linear equation, which is used to calculate the amount of DNA in the sample.

3.3.5 GAG Assay

1. Store the cultured cell-hydrogel constructs at −80 °C after discarding the medium and washing twice in PBS.
2. Prior to the assay, digest the scaffolds overnight at 56 °C in proteinase K as described in Subheading 3.3.1, using cultured cell-hydrogel constructs instead of decellularized ECM.

3. Add 5 μL of NaCl solution to 25 μL of sample or standards in a 96-well plate.
4. To this, add 150 μL of DMMB solution and measure the absorbance at 544 nm in a plate reader.
5. Plot the results from known amounts of chondroitin sulfate B (standards) to determine a standard curve. Use the equation to calculate the amount of sulfated GAG present in each sample.

3.3.6 Paraffin Embedding for Histological Staining

1. Fix the cultured cell-hydrogel constructs in formalin solution for 1 h, using 500 μL per sample.
2. Dehydrate the samples using an ethanol solution series with increasing order (70–100 %) and xylene.
3. Embed the samples in paraffin by placing the sample in a plastic block holder and carefully adding paraffin (56–58 °C) onto it (*see* **Note 14**).
4. Refresh the paraffin on the following day.
5. Trim the paraffin blocks before mounting on a microtome and cut 7-μm sections of the embedded sample.
6. Place the paraffin slides in a water bath at 40 °C to avoid wrinkles.
7. Use glass slides to fish out the paraffin sections. Dry the sections at 37 °C.
8. Lastly, rehydrate the samples by washing in xylene and then through an ethanol solution series (100 % to 70 %). Place the slides in distilled water for 10 min.
9. Use the glass slides with paraffin sections for Safranin O, Alcian Blue, and Picrosirius Red staining.

3.3.7 Safranin O/Fast Green Staining

1. Add Fast Green solution to the paraffin sections and incubate for 3 min.
2. Rinse the sections with 1 % v/v acetic acid solution for 15 s.
3. Add Safranin O solution to the sections and incubate for 5 min.
4. Dehydrate the slides in 100 % ethanol for 2 min.
5. Mount the sections and observe using a light microscope. Cartilage will be stained orange to red and nuclei in black (Fig. 3a).

3.3.8 Alcian Blue Staining

1. Clean the paraffin sections with 3 % v/v acetic acid for 3 min.
2. Add Alcian Blue solution to the sections and incubate for 30 min.
3. Wash in running water for 2 min and then rinse in distilled water.
4. To provide a good contrast to Alcian Blue staining, counterstain the nuclei with Nuclear Fast Red solution for 5 min.

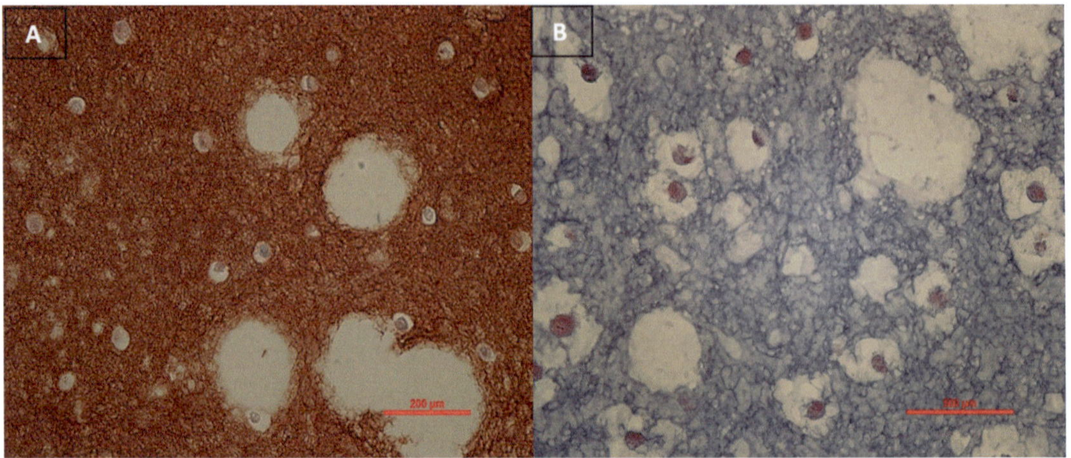

Fig. 3 Staining images of OA cells encapsulated in an alginate-ECM construct consisting of ECM made by hMSC culture. (**a**) Safranin O staining showing nuclei stained orange. Embedded cells are seen in isolated pockets. A lot of background staining is observed. However cell-alginate constructs without ECM show less nuclei stained in orange than constructs with ECM. (**b**) Alcian Blue staining indicating the distribution of proteoglycans within the hydrogel. Embedded chondrocytes are seen in lacunae, with nuclei stained in red and proteoglycans in blue

5. Wash in running water for 1 min, dehydrate the samples in 100 % ethanol for 2 min, and mount the slide using resinous medium.
6. Observe using a light microscope. Cells are located in lacunae and nuclei are stained in red and GAGs are stained blue (Fig. 3b).

3.3.9 Picrosirius Red Staining

1. Place the paraffin sections in solution A from the Picrosirius Red staining kit for 2 min and wash with distilled water.
2. Then, place the sections in solution B for 110 min.
3. Last, immerse the sections in solution C for 2 min.
4. Incubate by treating with 70 % ethanol for 45 s.
5. Dehydrate in 95 and 100 % ethanol for 2 min in each solution.
6. Observe the paraffin sections under a polarized light microscope. Collagen is stained red.

4 Notes

1. hMSCs are known for their self-renewal and multilineage differentiation. Classical or density gradient centrifugation can be used for isolation of hMSCs from bone marrow. FACS analysis is typically used for characterization of hMSCs.

2. After trypsinization and counting, the cells in suspension can be stored for a maximum of 2 h at 4 °C.
3. During decellularization of ECM, aspirate the NH_4OH solution carefully to ensure that the ECM is not detached from the flask and aspirated out with the NH_4OH.
4. During the lyophilization process, open the lid of the Eppendorf tube containing ECM before placing it in the gas freeze-drying vessel.
5. ECM-incorporated polymeric solution should be spun within 10 h after adding the ECM, to avoid protein degradation by HFIP.
6. When aluminum foil is used as the electrospinning target, the collected electrospun fibers stick to the aluminum foil and over time form an electrospun mesh.
7. A large amount of collagen is seen in scaffolds containing ECM [47].
8. Absorbance measurement instead of fluorescence is also possible, since the reduced potential maintained by the living cells converts the Alamar Blue reagent into a detectable fluorescent or absorbent product. An increase in metabolic activity is expected during initial time point measurements. A significant increase in metabolic activity has been observed during the initial 2 weeks for all conditions. A drop in activity was then observed as the cells reached a plateau.
9. As rDNase is sensitive to mechanical agitation, avoid shaking the vial vigorously. The aliquots can be stored at –20 °C for 6 months.
10. The wash buffer is stable for a year when kept at room temperature (18–25 °C).
11. Purity of RNA is assessed by the ratio of absorbance at 260 nm and 280 nm. A ratio of approximately 2.0 is generally accepted as "pure" RNA. A lower ratio may indicate the presence of protein or other contaminations absorbed at 280 nm.
12. The amount of protein may vary depending on the amino acid structures, cell type, culture duration, and ECM digestion protocol.
13. For gelation of sodium alginate, ensure that contact between $CaCl_2$ and the alginate solution occurs for no more than 30 min. Longer time causes loosening of the hydrogel.
14. While embedding the samples in paraffin, ensure that the sample is placed at the center of the plastic block holder so that the complete sample is covered during sectioning.

References

1. Felson DT, Zhang YQ (1998) An update on the epidemiology of knee and hip osteoarthritis with a view to prevention. Arthritis Rheum 41(8):1343–1355
2. van der Kraan PM (2012) Osteoarthritis year 2012 in review: biology. Osteoarthritis Cartilage 20(12):1447–1450
3. Lohmander S (2012) Osteoarthritis year 2012 in review. Osteoarthritis Cartilage 20(12):1439
4. Pearle AD, Warren RF, Rodeo SA (2005) Basic science of articular cartilage and osteoarthritis. Clin Sports Med 24(1):1
5. Temenoff JS, Mikos AG (2000) Review: tissue engineering for regeneration of articular cartilage. Biomaterials 21(5):431–440
6. Klein TJ et al (2009) Tissue engineering of articular cartilage with biomimetic zones. Tissue Eng Part B Rev 15(2):143–157
7. Tur K (2009) Biomaterials and tissue engineering for regenerative repair of articular cartilage defects. Turkish J Rheumatol—Turk Romatoloji Dergisi 24(4):206–217
8. Bian LM et al (2009) Functional tissue engineering of articular cartilage with adult chondrocytes. Proceedings of the ASME summer bioengineering conference—2009, Part A and B, pp. 317–318
9. Thiede RM, Lu Y, Markel MD (2012) A review of the treatment methods for cartilage defects. Vet Comp Orthop Traumatol 25(4):263–272
10. Badylak SE (2002) The extracellular matrix as a scaffold for tissue reconstruction. Semin Cell Dev Biol 13(5):377–383
11. Badylak SF (2007) The extracellular matrix as a scaffold for regenerative medicine. Faseb Journal 21(5):A140
12. Giancotti FG, Ruoslahti E (1999) Transduction—integrin signaling. Science 285(5430):1028–1032
13. Taipale J, KeskiOja J (1997) Growth factors in the extracellular matrix. Faseb Journal 11(1):51–59
14. Badylak SF, Freytes DO, Gilbert TW (2009) Extracellular matrix as a biological scaffold material: Structure and function. Acta Biomater 5(1):1–13
15. Hoshiba T et al (2011) Effects of extracellular matrices derived from different cell sources on chondrocyte functions. Biotechnol Prog 27(3):788–795
16. Eslaminejad MB, Bagheri F, Zomorodian E (2010) Matrigel enhances in vitro bone differentiation of human marrow-derived mesenchymal stem cells. Iran J Basic Med Sci 13(1):187–194
17. Hoshiba T, Mochitate K, Akaike T (2007) Hepatocytes maintain their function on basement membrane formed by epithelial cells. Biochem Biophys Res Commun 359(1): 151–156
18. Hoshiba T et al (2010) Decellularized matrices for tissue engineering. Expert Opin Biol Ther 10(12):1717–1728
19. Horn MA et al (2012) The cardiac extracellular matrix is remodelled divergently with age in heart failure: a role for altered collagen degradation in an ovine rapid pacing model. Heart 98:A2
20. Ott HC et al (2008) Perfusion-decellularized matrix: using nature's platform to engineer a bioartificial heart. Nat Med 14(2):213–221
21. Sarig U et al (2012) Thick acellular heart extracellular matrix with inherent vasculature: a potential platform for myocardial tissue regeneration. Tissue Eng Part A 18(19–20): 2125–2137
22. Steinhoff G et al (2000) Tissue engineering of pulmonary heart valves on allogenic acellular matrix conduits—in vivo restoration of valve tissue. Circulation 102(19):50–55
23. Schornik D et al (2012) Extracellular matrix structure of autologous human pericardium and significance for heart valve tissue engineering. J Tissue Eng Regen Med 6:113–113
24. Schenke-Layland K et al (2004) Comparative study of cellular and extracellular matrix composition of native and tissue engineered heart valves. Matrix Biol 23(2):113–125
25. McFetridge PS et al (2004) Preparation of porcine carotid arteries for vascular tissue engineering applications. J Biomed Mater Res A 70A(2):224–234
26. Rhodes JM, Simons M (2007) The extracellular matrix and blood vessel formation: not just a scaffold. J Cell Mol Med 11(2):176–205
27. Voytik-Harbin SL et al (1998) Small intestinal submucosa: a tissue-derived extracellular matrix that promotes tissue-specific growth and differentiation of cells in vitro. Tissue Eng 4(2):157–174
28. Badylak SF et al (1989) Small intestinal submucosa as a large diameter vascular graft in the dog. J Surg Res 47(1):74–80
29. Nihsen ES, Johnson CE (2007) Small intestinal submucosa (SIS*) provides a natural extracellular matrix microarchitecture for difficult to heal and chronic wounds. J Wound Ostomy Continence Nurs 34(3):S65
30. Aurora A et al (2007) Commercially available extracellular matrix materials for rotator cuff

repairs: State of the art and future trends. J Shoulder Elbow Surg 16(5):171s–178s

31. Datta N et al (2006) In vitro generated extracellular matrix and fluid shear stress synergistically enhance 3D osteoblastic differentiation. Proc Natl Acad Sci U S A 103(8): 2488–2493
32. Lu HX et al (2011) Cultured cell-derived extracellular matrix scaffolds for tissue engineering. Biomaterials 32(36):9658–9666
33. Chen XD et al (2007) Extracellular matrix made by bone marrow cells facilitates expansion of marrow-derived mesenchymal progenitor cells and prevents their differentiation into osteoblasts. J Bone Miner Res 22(12): 1943–1956
34. Postovit LM et al (2006) A three-dimensional model to study the epigenetic effects induced by the microenvironment of human embryonic stem cells. Stem Cells 24(3):501–505
35. Jagur-Grodzinski J (2010) Polymeric gels and hydrogels for biomedical and pharmaceutical applications. Polym Adv Technol 21(1): 27–47
36. Agarwal S, Wendorff JH, Greiner A (2008) Use of electrospinning technique for biomedical applications. Polymer 49(26):5603–5621
37. Agarwal S, Wendorff JH, Greiner A (2009) Progress in the field of electrospinning for tissue engineering applications. Adv Mater 21(32–33):3343–3351
38. Kumbar SG et al (2008) Electrospun nanofiber scaffolds: engineering soft tissues. Biomed Mat 3(3):034002
39. Li WJ et al (2002) Electrospun nanofibrous structure: a novel scaffold for tissue engineering. J Biomed Mater Res 60(4):613–621
40. Lannutti J et al (2007) Electrospinning for tissue engineering scaffolds. Mater Sci Eng C 27(3):504–509
41. Pham QP, Sharma U, Mikos AG (2006) Electrospinning of polymeric nanofibers for tissue engineering applications: a review. Tissue Eng 12(5):1197–1211
42. Shen MX et al (2009) Biocompatible polymeric hydrogels with tunable adhesion to both hydrophobic and hydrophilic surfaces. 2009 3rd international conference on bioinformatics and biomedical engineering, Vols 1-11, pp 1074-1077
43. Rosiak JM, Ulanski P, Rzeznicki A (1995) Hydrogels for biomedical purposes. Nucl Instrum Meth B 105(1–4):335–339
44. Hoffman AS (2001) Hydrogels for biomedical applications. Bioartificial organs iii: tissue sourcing, immunoisolation, and clinical trials. Annals of the New York Academy of Science 944: 62-73
45. Thakkar S et al (2013) Mesenchymal stromal cell-derived extracellular matrix influences gene expression of chondrocytes. Biofabrication 5(2):025003
46. Both SK et al (2007) A rapid and efficient method for expansion of human mesenchymal stem cells. Tissue Eng 13(1):3–9
47. Choi JS et al (2010) Fabrication of porous extracellular matrix scaffolds from human adipose tissue. Tissue Eng Part C Methods 16(3):387–396

Chapter 10

Use of Interim Scaffolding and Neotissue Development to Produce a Scaffold-Free Living Hyaline Cartilage Graft

Ting Ting Lau, Wenyan Leong, Yvonne Peck, Kai Su, and Dong-An Wang

Abstract

The fabrication of three-dimensional (3D) constructs relies heavily on the use of biomaterial-based scaffolds. These are required as mechanical supports as well as to translate two-dimensional cultures to 3D cultures for clinical applications. Regardless of the choice of scaffold, timely degradation of scaffolds is difficult to achieve and undegraded scaffold material can lead to interference in further tissue development or morphogenesis. In cartilage tissue engineering, hydrogel is the highly preferred scaffold material as it shares many similar characteristics with native cartilaginous matrix. Hence, we employed gelatin microspheres as porogens to create a microcavitary alginate hydrogel as an interim scaffold to facilitate initial chondrocyte 3D culture and to establish a final scaffold-free living hyaline cartilaginous graft (LhCG) for cartilage tissue engineering.

Key words Scaffold-free, Cartilage, Hydrogel, Biomaterials, Tissue engineering

1 Introduction

Three-dimensional (3D) cell culture systems often make use of biomaterial-based scaffolds as mechanical supports during the early stages of cell growth and development. As the culture progresses, cells proliferate and secrete sufficient extracellular matrix (ECM) proteins to establish their own niche in the microenvironment. Ideally at this time, the biomaterial-based scaffold should degrade accordingly, making room for protein deposition and also to facilitate further tissue development or morphogenesis. However, given that degradation profiles of scaffolds generally do not synchronize well with cell proliferation and tissue development, scaffolds become barriers for further growth. Thus, to circumvent this issue, we employed hydrogel as a transient mechanical support during the early stages of culture, before removing it at later stages to establish a scaffold-free hyaline cartilaginous graft (LhCG) for cartilage tissue engineering purposes.

Pauline M. Doran (ed.), *Cartilage Tissue Engineering: Methods and Protocols*, Methods in Molecular Biology, vol. 1340, DOI 10.1007/978-1-4939-2938-2_10, © Springer Science+Business Media New York 2015

To set up a macro-scaled 3D LhCG construct, porcine hyaline chondrocytes were first co-encapsulated with gelatin microspheres in alginate to create microcavitary (with cavities hundreds of microns in diameter) hydrogel constructs. The cells were guided to outgrow the gel phase gradually and fill up the cavities, forming scattered pieces of pure micro-tissues (in the form of isogenous groups), using a previously reported strategy named "phase transfer cell culture" (PTCC) [1, 2]. With further culture, an integrated 3D macro-network consisting of inter-connecting pure micro-tissues is created in the hydrogel matrix. It is only then that the structural integrity of the tissue construct is considered stable and no longer relies on the alginate scaffold. Therefore, this alginate hydrogel is no longer necessary and thus is completely and noninvasively removed by simple citric leaching treatment [3]. As a result, a pure cartilaginous ECM and chondrocyte-based 3D construct, namely a living cartilaginous graft, LhCG, is created.

2 Materials

2.1 Chondrocyte Extraction and Culture

1. 100× Antibiotic-antimycotic solution (Gibco).
2. Chondrocyte construct (CC) medium: DMEM high glucose with Glutamax, 20 % heat-inactivated fetal bovine serum (HI FBS), 0.01 M 4-(2-hydroxyethyl)-piperazine-1-ethanesulfonic acid (HEPES), 1× nonessential amino acids (Gibco), 0.4 mM proline, 0.05 mg/ml vitamin C, 100 units/ml penicillin, and 100 μg/ml streptomycin (*see* **Note 1**).
3. DMEM working medium: DMEM supplemented with 10 % v/v FBS, and 1× antibiotic-antimycotic solution.
4. Chondrocyte monolayer (CM) medium: 1 part CC medium to 4 parts DMEM working medium.
5. Collagenase medium: 1 mg/ml of collagenase type II in CM medium. Filter with a 0.2-μm membrane before use.
6. Porcine femur.
7. 1× PBS solution.

2.2 Acellular Gelatin Microspheres

1. Gelatin type B solution: 30 ml of 10 wt% bovine skin gelatin type B solution in deionized water. Warm the solution in a 70 °C oven to completely dissolve the gelatin.
2. Soya oil.
3. Dioxane:acetone solution: 80 ml of acetone and 400 ml of dioxane.
4. Ethyl acetate.

5. Cold 100 % ethanol: Pre-chill to −20 °C.
6. Ice-water bath.
7. Sieve for size 165–198 μm.
8. 100× Antibiotic-antimycotic solution (Gibco).
9. 100-μm nylon mesh cell strainer.

2.3 Cell-Laden Gelatin Microspheres

1. Gelatin type A solution: 5 % w/v Type A gelatin in 1:1 PBS (1×) and CC medium. Boil and then keep warm in a 37 °C water bath.
2. Filtered soya oil: Filter soya oil through a 0.2-μm membrane and keep warm in a 37 °C water bath.
3. Cold 1× PBS: Pre-chill PBS at 4 °C before use.
4. Trypsin solution: 0.25 % trypsin-EDTA (1×), phenol red.
5. Ice-water bath.
6. 40-μm nylon mesh cell strainer.

2.4 Microcavitary Hydrogel for 3D Chondrocyte Cell Culture

1. 1.5 wt% alginate precursor solution: 1.5 g of alginic acid sodium salt from brown algae in 100 ml of 0.15 M NaCl solution. To fully dissolve the alginate, autoclave the solution. Store at 4 °C before use.
2. 102 mM calcium chloride solution: 0.5661 g of $CaCl_2$ in 50 ml of deionized water. Filter with a 0.2-μm membrane and store at 4 °C.
3. 30-mm Petri dish coated with gelatin substrate at the bottom: Dissolve 1.5 g of gelatin and 0.1132 g of $CaCl_2$ in 10 ml of deionized water. Heat in an oven to fully dissolve the gelatin. Place 1.5 ml of gelatin solution onto a 30-mm Petri dish and spread evenly across the entire dish. Keep at 4 °C until use (*see* **Note 2**).
4. 6-Well tissue culture plate filled with CC medium: Coat the bottom of the wells with 1.2 wt% agarose in PBS. After the agarose is cooled and gelated, add 6–8 ml of CC medium to each well of the 6-well tissue culture plate (*see* **Note 3**).
5. Trypsin solution: 0.25 % trypsin-EDTA (1×), phenol red.

2.5 Scaffold-Free Living Hyaline Cartilaginous Graft

1. Sodium citrate solution: 1.62 g of sodium citrate in 100 ml of 0.15 M NaCl. Filter with a 0.2-μm filter membrane.
2. 6-Well tissue culture plate filled with CC medium: Coat the bottom of the wells with 1.2 wt% agarose in PBS. After the agarose is cooled and gelated, add 6–8 ml of CC medium to each well of the 6-well tissue culture plate (*see* **Note 3**).

3 Methods

3.1 Chondrocyte Harvesting and Culture

1. Isolate the femur from a 6-month-old pig and expose the patella region.
2. Using a scalpel and tweezers, harvest the cartilage from the patella groove.
3. Prepare 10× antibiotic-antimycotic solution by diluting 100× antibiotic-antimycotic stock with 1× PBS.
4. Immerse all isolated cartilage slices in 10× antibiotic-antimycotic solution in a 50-ml tube. Shake vigorously for 1 min.
5. Remove antibiotic-antimycotic solution and replace with 1× PBS solution.
6. Transfer all cartilage slices onto a 100-mm Petri dish and dice them into small pieces (~2 mm × 2 mm).
7. Incubate cartilage chips in collagenase medium for 14 h on an orbital shaker (50 rpm) at 37 °C.
8. Collect supernatant and centrifuge at 205 × *g* (*see* **Note 4**).
9. Remove collagenase medium and resuspend the cell pellet in CM medium.
10. Seed the chondrocytes into T175 flasks and culture them in CM medium until 80 % confluent.

3.2 Preparation of Acellular Gelatin Microspheres

1. Pour 25–30 ml of warm gelatin type B solution into a 100-ml beaker containing 10 ml of ethyl acetate.
2. Obtain a gelatin/ethyl acetate emulsion by stirring at 700 rpm for 1 min.
3. Transfer the emulsion into another 100-ml beaker containing 60 ml of soya oil. After stirring at 350 rpm for 1.5 min, place the beaker into an ice-water bath and continue stirring at 350 rpm for 15 min.
4. Transfer the emulsion to a 1000-ml beaker containing 300 ml of cold ethanol. Stir gently using a gentle swirling motion for 10 min for further cooling (*see* **Note 5**).
5. Use a spatula to transfer the gelatin microspheres into a 250-ml beaker containing 120 ml of dioxane:acetone solution to remove the soya oil. Leave the microspheres in the mixture for 10 min.
6. Repeat **step 5** for another three times to ensure that all soya oil has been removed.
7. Transfer all microspheres to a 250-ml beaker containing 100 ml of 100 % ethanol and wash two more times.

8. Pour away excess ethanol and dry the gelatin microspheres by placing them in a 70 °C oven for about 3 h or as appropriate.
9. Sieve the gelatin microspheres into appropriate sizes (165–198 μm).
10. Transfer the collected gelatin microspheres to a 50-ml tube containing 50 ml of 10× antibiotic-antimycotic solution to prevent contamination and leave at 4 °C overnight.
11. For long-term storage, strain the microspheres using a 100-μm nylon mesh cell strainer and transfer them to a new tube containing 50 ml of DMEM with 10 % FBS and keep them at 4 °C before use.

3.3 Preparation of Cell-Laden Gelatin Microspheres [4]

1. When chondrocytes reach 80 % confluence, detach them using trypsin solution.
2. Suspend 1×10^7 chondrocytes in 1 ml of gelatin type A solution.
3. At room temperature, add 15 ml of filtered soya oil into a 50-ml beaker containing a magnetic stirrer bar and stir at 500 rpm.
4. Add the cell suspension slowly (dropwise) to the oil and continue to stir at 500 rpm for 2 min.
5. Transfer the beaker to an ice-water bath. Stir at 300 rpm for 10 min.
6. Transfer the emulsion to a centrifuge tube. Centrifuge at 700 rpm for 3 min.
7. Aspirate the supernatant and add 3 ml of cold PBS.
8. Mix the microspheres using a 1-ml micropipette by pipetting up and down (*see* **Note 6**).
9. Add another 5 ml of cold PBS and mix again until no clumps are observed.
10. Top up the mixture with cold PBS to 25 ml.
11. Centrifuge at 700 rpm for 3 min.
12. Repeat washing **steps 7–11** for another three times.
13. After centrifuging, resuspend the cell-laden microspheres with 10 ml of cold PBS.
14. Collect the cell-laden microspheres by passing the mixture through a 40-μm nylon mesh cell strainer (Fig. 1a).

3.4 Microcavitary Hydrogel for 3D Chondrocyte Cell Culture

1. When chondrocytes reach 80 % confluence, detach them using trypsin solution and suspend them at a density of 1×10^7 cells/ml in 1.5 wt% alginate precursor solution at 4 °C.
2. Add this cell suspension to the gelatin microspheres (from either Subheading 3.2 or 3.3) using 0.3 g of microspheres/ml

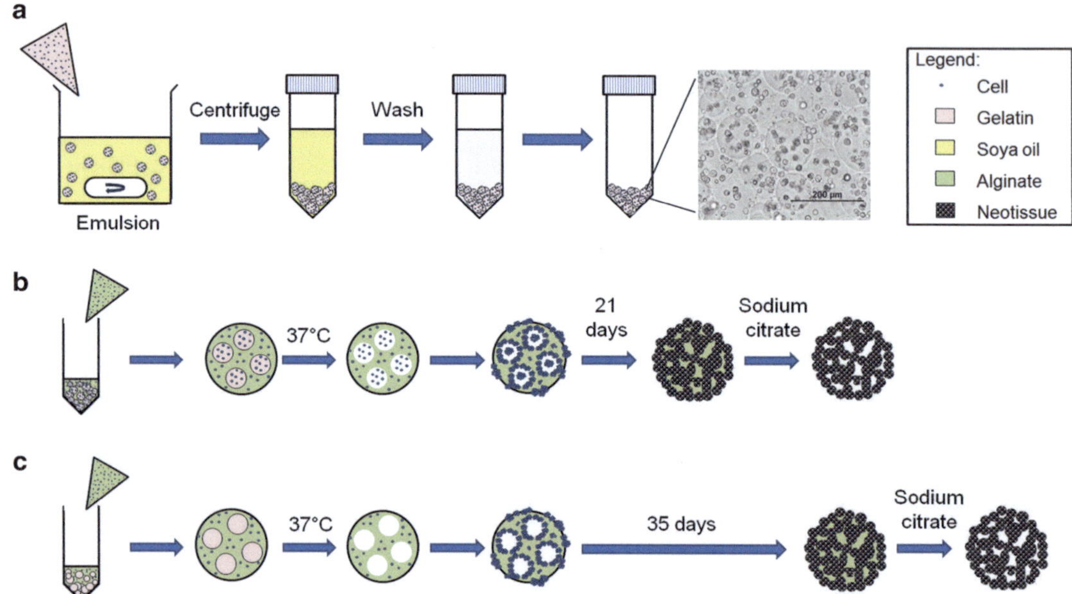

Fig. 1 Schematic diagram of fabrication processes of microcavitary hydrogel for 3D chondrocyte cell culture. (**a**) Fabrication process of cell-laden microspheres. Use of (**b**) cell-laden microspheres or (**c**) acellular microspheres to generate microcavitary hydrogel constructs for scaffold-free living hyaline cartilaginous graft establishment. A suspension of chondrocytes and microspheres is mixed with alginate precursor solution. Upon gelation, the construct is exposed to a culture condition of 37 °C that causes dissolution of the microspheres, creating microcavities of corresponding sizes. Cells from the bulk alginate infiltrate the cavities left behind by the gelatin microspheres and neotissue develops. After 35 days of culture, alginate is removed via sodium citrate treatment to yield a scaffold-free living hyaline cartilaginous graft establishment (reproduced from ref. [4] with permission from Elsevier)

of alginate suspension and mix well (Fig. 1b and c, respectively) (*see* **Note 6**).

3. Transfer 1.5 ml of the mixture to a 30-mm Petri dish precoated with gelatin. Spread evenly across the dish and incubate at 4 °C for 4 min.
4. Gently introduce 1 ml of cold 102 mM $CaCl_2$ solution to the mixture from the edge of the Petri dish. Incubate at 4 °C for another 4 min.
5. The cell-laden hydrogel construct should fully gelate by now and should easily be transferred to a new Petri dish using a spatula.
6. Cut the hydrogel construct into pieces approximately 5 × 5 mm in size and transfer them to the 6-well tissue culture plate containing CC medium. Each well should have no more than three hydrogel pieces.
7. Culture the constructs in CC medium at 37 °C and 5 % CO_2 for 30–35 days on an orbital shaker operated at 50 rpm

(*see* **Note 7**). Microcavitary hydrogel should form spontaneously upon exposure to 37 °C (*see* **Note 8**). Medium should be changed every 2–3 days.

3.5 Establishment of Scaffold-Free Living Hyaline Cartilaginous Graft

1. After 30–35 days of culture, transfer the constructs to a 15-ml tube containing 8 ml of sodium citrate solution. Each tube can contain up to three constructs.
2. Incubate the constructs at room temperature for 10 min. Gently rock the tube every 2–3 min.
3. Pour the contents onto a clean 100-mm Petri dish and transfer the constructs to a newly prepared 6-well tissue culture plate filled with CC medium (*see* **Note 9**).
4. Alginate is removed by citrate treatment and the constructs are now named as living hyaline cartilaginous grafts (LhCG).

4 Notes

1. Heat-inactivated FBS is a necessary component in CC medium as chondrocytes thrive better in HI FBS than FBS when cultured in a 3D environment. HI FBS from Life Technologies (certified, US origin) is highly recommended.
2. This gelatin substrate serves as a source of calcium ions for gelation of alginate at the bottom of the dish.
3. Often, cells from the construct fall out, attach to the bottom of the well, and compete with cells within the construct for nutrients. By coating the bottom of the well with agarose to prevent cell adhesion, any cells that fall out can be removed during medium change.
4. If the cartilage chips are not fully digested, they could be further digested again to maximize yield. Incubate the chips in collagenase medium for an additional 5–8 h on an orbital shaker (50 rpm) at 37 °C to harvest another round of chondrocytes [5].
5. Gently stir using a stirring rod or spatula to prevent clumping of gelatin microspheres.
6. Cut the tip of the pipette to facilitate pipetting. The microspheres will clog up typical pipette tips.
7. Culturing on an orbital shaker is important as it increases nutrient exchange throughout the constructs. It should be carried out at least three times a week with each time lasting for a minimum of 12 h.
8. Dissolution of the gelatin microspheres should only occur after gelation of the alginate, upon exposure to 37 °C, and not during the fabrication process. After 24 h, all gelatin microspheres

should be dissolved, creating cavities of corresponding sizes (~200 μm) within the alginate construct.

9. There may be small cell islets seen in the citrate solution, but the overall construct integrity should be intact. If the construct collapses, it means that the construct is not ready for citrate treatment. Longer culture periods may be required but should not exceed a total of 40 days.

Acknowledgement

This work was supported by AcRF Tier 1 RG 36/12 and AcRF Tier 2 ARC 1/13, Ministry of Education, Singapore.

References

1. Gong Y, Su K, Lau TT et al (2010) Microcavitary hydrogel-mediating phase transfer cell culture for cartilage tissue engineering. Tissue Eng 16:3611–3622
2. Su K, Lau TT, Leong W et al (2012) Creating a living hyaline cartilage graft free from non-cartilaginous constituents: an intermediate role of a biomaterial scaffold. Adv Funct Mater 22:972–978
3. Masuda K, Sah RL, Hejna MJ et al (2003) A novel two step method for the formation of tissue engineered cartilage by mature bovine chondrocytes: the alginate recovered chondrocyte (ARC) method. J Orthop Res 21:139–148
4. Leong W, Lau TT, Wang D-A (2013) A temperature-cured dissolvable gelatin microsphere-based cell carrier for chondrocyte delivery in a hydrogel scaffolding system. Acta Biomater 9:6459–6467
5. Lau TT, Peck Y, Huang W et al (2014) Optimization of chondrocyte isolation and phenotype characterization for cartilage tissue engineering. Tissue Eng Part C Method. doi:10.1089/ten.TEC.2014.0159

Chapter 11

Bioprinted Scaffolds for Cartilage Tissue Engineering

Hyun-Wook Kang, James J. Yoo, and Anthony Atala

Abstract

Researchers are focusing on bioprinting technology as a viable option to overcome current difficulties in cartilage tissue engineering. Bioprinting enables a three-dimensional (3-D), free-form, computer-designed structure using biomaterials, biomolecules, and/or cells. The inner and outer shape of a scaffold can be controlled by this technology with great precision. Here, we introduce a hybrid bioprinting technology that is a co-printing process of multiple materials including high-strength synthetic polymer and cell-laden hydrogel. The synthetic polymer provides mechanical support for shape maintenance and load bearing, while the hydrogel provides the biological environment for artificial cartilage regeneration. This chapter introduces the procedures for printing of a 3-D scaffold using our hybrid bioprinting technology and includes the source materials for preparation of 3-D printing.

Key words Cartilage, Scaffold, Bioprinting, Cell printing, Chondrocyte, Fibrin gel, Polycaprolactone

1 Introduction

Bioprinting technology can produce a computer-designed 3-D structure using biomaterials, biomolecules, and/or living cells [1]. Currently, a number of researchers are applying bioprinting technology to cartilage tissue engineering to improve outcomes [2–5]. The inner architecture of a scaffold affects not only its mechanical properties but also the biological environment for cartilage regeneration. To control the inner architecture of a bioprinted scaffold, several researchers have investigated the physical and biological properties for cartilage tissue engineering. Bioprinting technology can produce a hybrid scaffold composed of a soft hydrogel, which provides a biological environment that allows cartilage formation, and a hard plastic material, which gives physical support [6–9]. Additionally, bioprinting technology has applications in osteochondral tissue engineering and allows for the production of an inner architecture and material composition that offers advantages in the study of composite tissue [10–13].

Pauline M. Doran (ed.), *Cartilage Tissue Engineering: Methods and Protocols*, Methods in Molecular Biology, vol. 1340, DOI 10.1007/978-1-4939-2938-2_11, © Springer Science+Business Media New York 2016

Our group has developed a new hybrid printing technology that is a co-printing process of multiple materials, including a high-strength thermoplastic and cell-laden hydrogel [14]. Most cell printing technologies use a hydrogel system for cell delivery; however, the hydrogel cannot be applied to produce a complex 3-D shape and load-bearing site due to the weakness of the hydrogel. Our hybrid technology combines a high-strength material to provide sufficient mechanical support and hydrogel to provide for artificial regeneration of cartilage. In addition, our system can deliver biomolecules with the hydrogel. These features are very useful in the study of cartilage regeneration. Figure 1 shows a schematic diagram and photograph of our bioprinting system. As shown in the figure, the system has a multi-syringe module for delivery of multiple materials. Each syringe can contain different materials for printing. A heating unit can be used to obtain dispensable thermoplastic by melting. Each syringe is connected to a micro-nozzle and pneumatic controller. The nozzle creates micro-patterns with the dispensing volume controlled by adjustment of the pneumatic pressure. The chamber is equipped with an air conditioner, which allows for temperature control during the printing process in which a thermosensitive material is used for cell delivery, and a humidifier, which prevents the hydrogel from drying too quickly. Finally, the printing procedures are managed by the main computer. With this system, we have produced a hybrid scaffold for cartilage tissue engineering that is composed of a polycaprolactone and chondrocyte-laden fibrin gel. In this chapter, we describe the procedures from material preparation to printing.

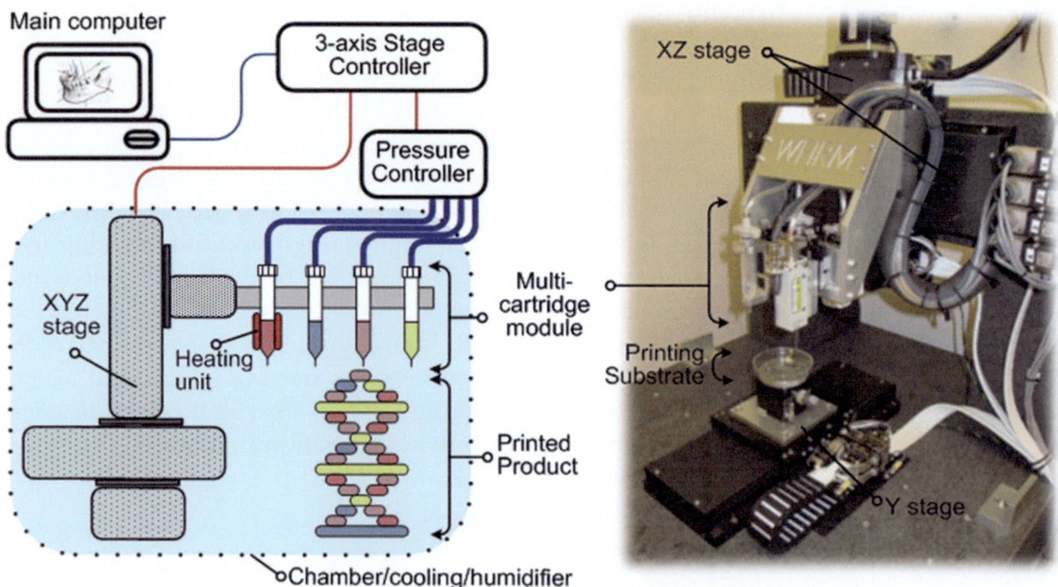

Fig. 1 Schematic diagram (*left*) and photograph (*right*) of a hybrid printing system

2 Materials

All chemicals are used at ambient temperature, unless indicated otherwise, and are stored as recommended by the manufacturer.

2.1 Bioprinting Materials

1. Serum-free DMEM: Dulbecco's modified Eagle medium (DMEM) without serum.
2. Saline solution: 0.90 % w/v NaCl in distilled water.
3. PBS solution: Phosphate-buffered saline solution.
4. Cell carrier material: Weigh 15 mg of hyaluronic acid sodium salt (HA) and transfer it to a 15-mL conical tube containing 4.5 mL of serum-free DMEM (*see* **Note 1**). Shake the material overnight at 37 °C (*see* **Note 2**) and add 0.5 mL of glycerol. After dissolving the glycerol by shaking several times, weigh 225 mg of gelatin (90–110 g bloom) and 150 mg of fibrinogen. Then, add the two materials to the conical tube. Slowly rotate the mixture at 37 °C for 30–45 min to dissolve the materials (*see* **Note 3**). Sterilize with a 0.45-μm syringe filter and store at −20 °C. Application of gelatin with a higher bloom number will reduce the concentration required for patterning (*see* **Note 4**).
5. Sacrificial material: Weigh 9 g of poly(ethylene oxide)-poly(propylene oxide)-poly(ethylene oxide) (PEO-PPO-PEO) triblock copolymer and sterilize the powder with ethylene oxide gas. Add 30 mL of cold PBS solution and slowly rotate the mixture at 4 °C to dissolve the PEO-PPO-PEO triblock copolymer. Store at −20 °C (*see* **Note 5**).
6. Polycaprolactone (molecular weight: 43,000–50,000).
7. Chondrocytes: Chondrocytes for bioprinting can be isolated, for example, from New Zealand White rabbit, by digesting with collagenase type I. After excising cartilage tissue from the ear, digest the tissue in 0.3 % w/v collagenase type 1 solution by shaking for 1 h at 37 °C and 200 rpm. Culture the isolated cells in medium suitable for chondrocytes.

2.2 Other Reagents and Materials

1. $CaCl_2$ solution: 40 mM $CaCl_2$ in saline solution (*see* **Note 6**).
2. Thrombin solution: 5 mL of $CaCl_2$ solution in a bottle of thrombin containing 1000 NIH units. After dissolving the thrombin by gently shaking the bottle several times, make a 10× dilution with saline solution and sterilize using a 0.2-μm syringe filter. Store at −20 °C.
3. Cone-shaped metal nozzle (300 μm diameter): for printing of thermoplastic material (SHN series, Musashi Engineering, Japan) (*see* **Note 7**).

4. Polytetrafluoroethylene (PTFE) nozzle (300 μm diameter): for cell printing (TN-SUSG series, Musashi Engineering, Japan).
5. Needle (250 μm diameter): for sacrificial material printing (DPN series, Musashi Engineering, Japan).
6. Metal syringe and polypropylene syringe, 10 ml: for polycaprolactone and cell-laden hydrogel printing, respectively (Musashi Engineering, Japan).
7. Trypsin solution: 0.05 % w/v trypsin and 0.5 mM EDTA in PBS.

3 Methods

3.1 Bio-ink Preparation

1. Set up the bioprinting apparatus as shown in Fig. 1. Set the chamber temperature of the printing system to 18 °C.
2. Detach the chondrocytes from culture dishes using trypsin solution. Transfer 120×10^6 cells into a 15-mL conical tube and centrifuge for 5 min [15]. Thaw the frozen cell carrier material in a 37 °C water bath and transfer 3 mL of the solution into the conical tube. Gently mix the cells and cell carrier material (*see* **Note 8**). Load the mixture into the polypropylene syringe. After closing the syringe with a stopper, immediately place it into an ice bath for 10 min to induce polymerization of gelatin (*see* **Note 9**). After connection with the 300-μm PTFE nozzle, install the syringe into the bioprinter. Incubate for at least 30 min before printing (*see* **Note 10**).
3. Thaw the sacrificial material at room temperature and place in an ice bath to obtain the liquefied form (*see* **Note 11**). Place the polypropylene syringe in an ice bath and add 5 mL of the liquefied sacrificial material into the syringe (*see* **Note 12**). After connecting the 250-μm needle, install the syringe into the bioprinter. Incubate for at least 30 min to induce hydrogel formation (*see* **Note 13**).
4. Load polycaprolactone into the metal syringe (*see* **Note 14**). After connection with the cone-shaped metal nozzle, install into the printing system. Turn on the syringe heater and increase the temperature to 95 °C (*see* **Note 15**).

3.2 Printer Setup

1. Create a high-humidity environment in the bioprinter chamber before printing by using the humidifier with sterilized deionized water (*see* **Note 16**).
2. Secure a sterile Petri dish on the stage of the bioprinter using double-sided tape.
3. Calibrate the Z-axis position of a nozzle tip by adjusting the manual stage and using a microscope (*see* **Note 17**). Move the

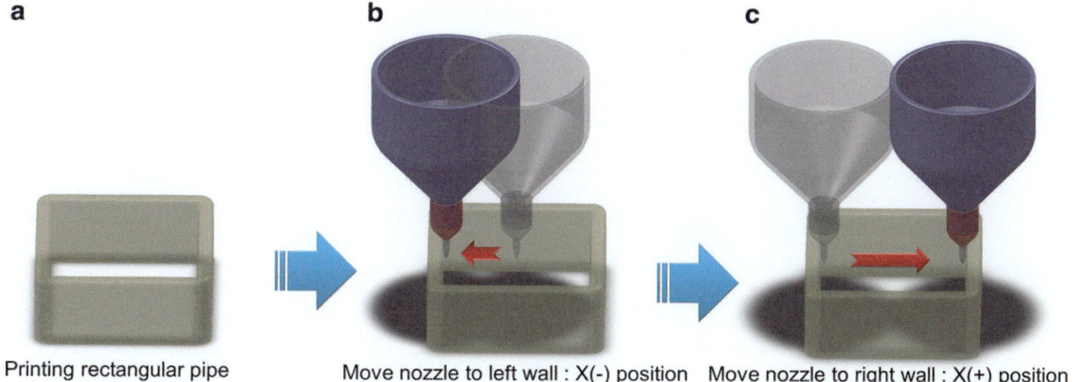

Fig. 2 Calibration of nozzle tip: (**a**) Printing rectangular-shaped pipe. (**b**) Find X(−) reference, and (**c**) find X(+) reference, for calibration of tip position

nozzle into the printing position. Adjust the manual Z-axis stage to contact the nozzle tip with the substrate (Petri dish) (*see* **Note 18**). Repeat this procedure for every syringe.

4. Measure the X- and Y-distance between nozzle tips (*see* **Note 19**). Print a rectangular-shaped pipe using the first syringe (Fig. 2a). Place the nozzle of the first syringe into the center of the pipe and record the position data of the X- and Y-stages. Insert the tip of the second nozzle into the pipe by moving the XYZ stages. Slowly move the nozzle to the left and stop the movement when the nozzle touches the wall (Fig. 2). Record the data of the X-stage position. Repeat for the right direction. Then, average the two positions to calculate the X-position of the second nozzle. Repeat this procedure for the Y-axis. Measurement of the positions of the other nozzles also can be obtained by the same procedure. Calculate the X- and Y-axis distance between nozzles using the recorded data. These data will be used in the generation of a motion program for the exchange of nozzles for printing.

3.3 Motion Program Preparation and Printing

1. Design the scanning paths: Design micro-patterns to construct a desired 3-D structure. Based on the designed patterns, scanning paths can be constructed. The paths should include dispensing paths, speed, exchanging movement of printing syringes, and other factors (Fig. 3).
2. Prepare the motion program (*see* **Note 20**): Record the command list to generate the motion for the designed scanning path (Fig. 3) (*see* **Note 21**).
3. Set the pressure levels for patterning of the biomaterials: 780 kPa for polycaprolactone, 30 kPa for chondrocyte-laden hydrogel, and 200 kPa for sacrificial material (*see* **Note 22**).

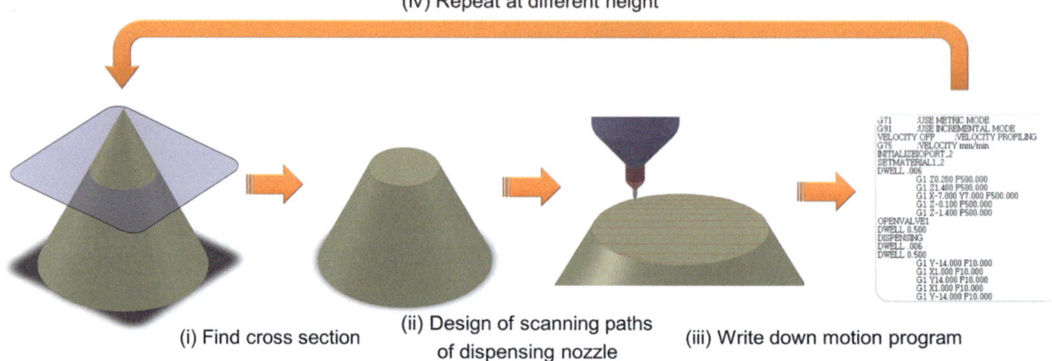

Fig. 3 Generation of motion program to print a desired 3-D shape: The procedure starts from finding the shape of the cross section at a specific height (i). Then, scanning paths of the dispensing nozzle are designed, which can print the shape of the cross section (ii). A motion program composed of a command list can be constructed to generate the motions of the designed paths (iii). These procedures are repeated to construct 3-D patterning at different heights (iv)

4. Bioprint the designed cartilage scaffold: Load the prepared text-based command list (motion program) and execute it to obtain the designed shape (*see* **Note 23**).

3.4 After Printing

1. Selectively remove the sacrificial material from the bioprinted structure by washing out the printed PEO-PPO-PEO triblock copolymer hydrogel using cold PBS solution (*see* **Note 24**).
2. Add thrombin solution on the bioprinted structure and incubate for 30 min to 1 h at room temperature to induce fibrin gel formation.
3. Dissolve the residual sacrificial material by placing the printed structure into cold PBS.
4. Add culture medium suitable for chondrocyte culturing and transfer into a CO_2 incubator at 37 °C (*see* **Note 25**).
5. Exchange the culture medium after 1 day and apply the scaffold in in vitro/in vivo experiments for cartilage tissue engineering.

4 Notes

1. Serum can induce fibrin gel formation.
2. Undissolved HA is transparent. Carefully check the solution to confirm full dissolution.
3. Vigorous shaking of fibrinogen can damage the protein; therefore, it should be shaken gently.
4. Gelatin, glycerol, and HA are used to obtain dispensable materials. Gelatin is the base material to achieve the dispensing

property. A weak hydrogel form of gelatin is used for printing. HA increases the viscosity and is used to improve dispensing properties. Glycerol is used to prevent clogging of the nozzle by reducing the drying speed of the cell carrier material. Finally, fibrinogen is used to provide a suitable biological environment for cartilage tissue regeneration. Different materials, such as alginate and cross-linkable hyaluronic acid, can be used.

5. PEO-PPO-PEO triblock copolymer was used for this experiment and should be slowly dissolved at cold room. Rapid rotation can generate foam, which makes the solution difficult to handle. Due to the very high viscosity, this material cannot be sterilized by filtration.
6. Do not use PBS solution in preparing the $CaCl_2$ solution as it induces some precipitation of $CaCl_2$.
7. Nozzle diameter and length greatly affect printing speed. Shorter nozzle is highly recommended.
8. Molten cell carrier material is very viscous. A piston syringe or positive displacement pipette is recommended for handling the solution.
9. The incorporated chondrocytes can be dropped down in the cell carrier material by gravity. Therefore it should immediately be moved into the ice bath for fast gelation of the material.
10. Gelatin is a thermosensitive material. It is a liquid at high temperature, and gel formation is induced at a low temperature; therefore, the degree of gelation can be affected by temperature.
11. PEO-PPO-PEO triblock copolymer solution is a thermosensitive material. A molten PEO-PPO-PEO triblock copolymer solution of 30 % w/v will be a gel form at high temperature and a liquid at low temperature.
12. The cooling of the polypropylene syringe before adding the sacrificial material will minimize attachment of the residue to the inside wall of the syringe.
13. The hydrogel form of the sacrificial material is used for printing. PEO-PPO-PEO triblock copolymer hydrogel has superior performance for 3-D patterning compared with the other hydrogel.
14. The grain-shaped raw material of polycaprolactone will be less affected by humidity than the powder form during storage.
15. Allow at least 20 min of preheating before printing for stable dispensing of polycaprolactone.
16. High humidity will prevent drying of the printed hydrogel. In most cases, 80 % relative humidity is used for printing. After the experiment, the humidity should be immediately removed to prevent an adverse effect on the printing system.

17. Additional manual Z-axis stages are equipped on each syringe.
18. The contact is monitored by using a microscope.
19. This measurement is critical to achieve micro-patterning with multiple materials.
20. The motion program is a command list to achieve the designed printing motions. Our system uses a text-based and modified G-code for printing.
21. The command list should include stage movement, start and stop commands for dispensing, pressure exchange, and other factors. A simple motion program can be manually prepared, but a complex 3-D shape requires the use of computer-aided manufacturing software.
22. The pressure level greatly affects the printing results. Conditions for use are obtained by conducting pre-patterning experiments for each biomaterial.
23. The dispensing property of polycaprolactone is very reliable in comparison with other materials. The dispensing of hydrogel should be carefully monitored for printing. Drying of material can clog the nozzle and the dispensing rate of hydrogel can be changed in the printing procedure because of its thermosensitive property.
24. Sterilized gauze can be used to wipe off sacrificial material remaining on the outside of the structure. The wiping off can increase the dissolving speed of the sacrificial material.
25. Uncross-linked components such as remnant sacrificial material, gelatin, glycerol, and HA dissolve into the culture medium. A high concentration of chondrocytes enhances the degradation speed of the fibrin gel. Apply aprotinin (20 μg/mL) into the culture medium to inhibit the degradation of fibrin gel caused by enzymes secreted by the chondrocytes.

References

1. Derby B (2012) Printing and prototyping of tissues and scaffolds. Science 338:921–926
2. Cui X, Boland T, D'Lima DD, Lotz MK (2012) Thermal inkjet printing in tissue engineering and regenerative medicine. Recent Pat Drug Deliv Formul 6:149
3. Klein TJ, Malda J, Sah RL, Hutmacher DW (2009) Tissue engineering of articular cartilage with biomimetic zones. Tissue Eng Part B Rev 15:143–157
4. Lee M, Wu BM (2012) Recent advances in 3D printing of tissue engineering scaffolds. Methods Mol Biol 868:257–267
5. Moutos FT, Guilak F (2008) Composite scaffolds for cartilage tissue engineering. Biorheology 45:501–512
6. Kundu J, Shim JH, Jang J, Kim SW, Cho DW (2013) An additive manufacturing-based PCL–alginate–chondrocyte bioprinted scaffold for cartilage tissue engineering. J Tissue Eng Regen Med. doi:10.1002/term.1682
7. Liao E, Yaszemski M, Krebsbach P, Hollister S (2007) Tissue-engineered cartilage constructs using composite hyaluronic acid/collagen I hydrogels and designed poly (propylene fumarate) scaffolds. Tissue Eng 13:537–550

8. Xu T, Binder KW, Albanna MZ, Dice D, Zhao W, Yoo JJ, Atala A (2013) Hybrid printing of mechanically and biologically improved constructs for cartilage tissue engineering applications. Biofabrication 5:015001
9. Yen H-J, Tseng C-S, Hsu S-h, Tsai C-L (2009) Evaluation of chondrocyte growth in the highly porous scaffolds made by fused deposition manufacturing (FDM) filled with type II collagen. Biomed Microdevices 11:615–624
10. Cui X, Breitenkamp K, Finn M, Lotz M, D'Lima DD (2012) Direct human cartilage repair using three-dimensional bioprinting technology. Tissue Eng Part A 18:1304–1312
11. Liu L, Xiong Z, Zhang R, Jin L, Yan Y (2009) A novel osteochondral scaffold fabricated via multi-nozzle low-temperature deposition manufacturing. J Bioact Compat Pol 24:18–30
12. Schek RM, Taboas JM, Segvich SJ, Hollister SJ, Krebsbach PH (2004) Engineered osteochondral grafts using biphasic composite solid free-form fabricated scaffolds. Tissue Eng 10:1376–1385
13. Sherwood JK, Riley SL, Palazzolo R, Brown SC, Monkhouse DC, Coates M, Griffith LG, Landeen LK, Ratcliffe A (2002) A three-dimensional osteochondral composite scaffold for articular cartilage repair. Biomaterials 23:4739–4751
14. Atala A, Kang H-W, Lee SJ, Yoo JJ (2011) Integrated organ and tissue printing methods, system and apparatus. Google Patents
15. Lee SJ, Broda C, Atala A, Yoo JJ (2010) Engineered cartilage covered ear implants for auricular cartilage reconstruction. Biomacromolecules 12:306–313

Chapter 12

Scaffolds for Controlled Release of Cartilage Growth Factors

Marie Morille, Marie-Claire Venier-Julienne, and Claudia N. Montero-Menei

Abstract

In recent years, cell-based therapies using adult stem cells have attracted considerable interest in regenerative medicine. A tissue-engineered construct for cartilage repair should provide a support for the cell and allow sustained in situ delivery of bioactive factors capable of inducing cell differentiation into chondrocytes. Pharmacologically active microcarriers (PAMs), made of biodegradable and biocompatible poly (D,L-lactide-co-glycolide acid) (PLGA), are a unique system which combines these properties in an adaptable and simple microdevice. This device relies on nanoprecipitation of proteins encapsulated in polymeric microspheres with a solid in oil in water emulsion-solvent evaporation process, and their subsequent coating with extracellular matrix protein molecules. Here, we describe their preparation process, and some of their characterization methods for an application in cartilage tissue engineering.

Key words Protein encapsulation, Pharmacologically active microcarriers, Mesenchymal stem cells, Transforming growth factor, Chondrogenic differentiation

1 Introduction

The use of cell therapy to treat degenerative diseases (arthritis, neurodegenerative disease, …) holds great promises and answer to the healthcare requirement of the aging population. Nevertheless, to date, there are still hurdles to cross to obtain an efficient cell therapy. Indeed, after transplantation, the majority of cells die, or, if previously induced toward a differentiated phenotype, do not maintain this induced phenotype. Consequently, due to the small number of surviving cells and generation of non-desired phenotypes, the tissue repair process is not efficient and the cells do not correctly integrate into the host environment. Cell engraftment needs to be ameliorated, particularly the short- but also long-term survival and functional state of the cells after transplantation. Growth factors and morphogens are the main factors orienting

Pauline M. Doran (ed.), *Cartilage Tissue Engineering: Methods and Protocols*, Methods in Molecular Biology, vol. 1340, DOI 10.1007/978-1-4939-2938-2_12, © Springer Science+Business Media New York 2015

stem cell fate and may also affect the immediate environment, thus allowing better graft integration. Various growth factors, cytokines or morphogens, have been widely used for directing the differentiation of mesenchymal stem cells (MSCs). Nevertheless, the administration of these factors still remains a technological challenge, due to their short half-life, pleïotropic actions, and limited passage through biological barriers. Therefore, the use of delivery carriers for these factors, such as nano- or microdevices, is now crucial to both protect and allow a controlled and sustained release of, for example, a protein. In this context, we developed pharmacologically active microcarriers (PAMs), which are polymeric microspheres providing a three-dimensional biomimetic support for transplanted cells and the sustained release of soluble factors. The proof of concept for this unique and simple microdevice delivering cells and proteins was first validated for neuroprotection and tissue repair for the treatment of neurological disorders using a neuronal cell line, neuronal precursors, and adult stem cells. They were combined to these surface-functionalized microspheres with different extracellular matrix/cell adhesion molecules (laminin, fibronectin, poly-D-lysine) and/or growth factors (nerve growth factor (NGF), glial cell line-derived neurotrophic factor (GDNF), neurotrophin 3 (NT-3)) [1–4].

Initially produced with hydrophobic PLGA, these microspheres were recently formulated with a more hydrophilic polymer. These new PAMs presenting a fibronectin- and poly-D-lysine-covered surface permitted enhanced MSC survival and proliferation [5]. They allowed the sustained release of 70 % of the incorporated transforming growth factor (TGF)-β3 over time and exhibited superior chondrogenic differentiation potential compared to the previous formulation, with an increased expression of specific cartilaginous markers such as collagen type II, aggrecan, cartilage oligomeric matrix protein (COMP), and link protein [5]. These systems are currently under investigation to establish their interest in vivo in a mouse osteoarthritis model.

This microdevice represents an efficient easy-to-handle and injectable tool for cartilage repair. This chapter describes the general methods used to produce and characterize such scaffolds for controlled release of TGF-β3 for cartilage repair.

2 Materials

Prepare all solutions using sterile water for injection (WFI) and analytical grade reagents. Prepare and store all reagents at room temperature (unless indicated otherwise). PLGA-P188-PLGA polymer was synthesized following **Note 1**, but PAM formation could also be performed with commercially available PLGA.

2.1 Nanoprecipitation Reagents

1. Tris–hydrochloride buffer: 0.75 M Tris–hydrochloride in WFI. Add 45.5 g of Tris (hydroxymethyl) aminomethane to 400 mL of WFI, adjust the pH to 7.4 with 37 % hydrochloric acid, and dilute to 500.0 mL with WFI.
2. TGF-β3 solution (A1): 10 mg/mL of TGF-β3 in Tris–hydrochloride buffer.
3. Human serum albumin (HSA) solution (A2): 12.5 mg/mL of HSA in 0.3 M NaCl in WFI.
4. Polymer P188 solution (B1): 200 mg/mL of P188 poloxamer in 4 M NaCl in WFI.
5. Polymer P188 solution (B2): 250 mg/mL of P188 poloxamer in 0.3 M NaCl in WFI.

2.2 Microencapsulation and Coating Reagents

1. Organic phase: 50 mg of polymer (PLGA–P188–PLGA: *see* **Note 1**) in 670 μL of a 3:1 methylene chloride:acetone solution. Combine in a 5-mL silanized glass tube.
2. Water phase: 4 % polyvinyl alcohol (PVA) (Mowiol® 4-88) in WFI. Add 4 g of PVA to 80 mL of WFI. Warm the solution under magnetic agitation at 80 °C. After total dissolution, cool the solution at room temperature (RT), and adjust the volume to 100 mL with WFI. The water phase must be prepared at least 24 h before microsphere formulation. Store the solution at 4 °C until use (this solution may be stored for 1 week at 4 °C).
3. Coating solution: 6 μg/mL of fibronectin (FN) and 9 μg/mL of poly-D-lysine (PDL) in Dulbecco's phosphate-buffered saline (DPBS).

2.3 Characterization Reagents

1. ELISA kit for TGF-β3.
2. NaCl solution: 0.3 mg/mL NaCl in WFI.
3. Protein stability medium: 1 % bovine serum albumin (BSA) in PBS.
4. DPBS.
5. BSA/Tween solution: 4 % BSA and 0.2 % Tween®20 in DPBS.
6. Anti-FN antibody solution: 100 μg/mL of mouse monoclonal primary antibody against FN in DPBS.
7. Biotinylated antibody solution: 2.5 μg/mL of biotinylated anti-mouse IgG antibody in DPBS.
8. Streptavidin-fluoprobe solution: 1/500 streptavidin-fluoprobe® 547 solution in DPBS.

2.4 Cell Culture

1. Mesenchymal stem cells (*see* **Note 2**).
2. Adhesion medium: α-Minimum essential medium (α-MEM) supplemented with 3 % fetal bovine serum (FBS), 2 mM glutamine, 100 U/mL penicillin, 100 mg/mL streptomycin.

3. Chondrogenic medium: Dulbecco's modified Eagle's medium (DMEM) supplemented with 0.1 mM dexamethasone, 0.17 mM ascorbic acid, 35 mM proline, and 1 % insulin–transferrin–selenic acid (ITS) supplement.

3 Methods

3.1 Equipment Setup

The whole installation (Fig. 1) has to be set up and at 4 °C before the first formulation steps are carried out.

1. Prepare the glassware as presented in Fig. 1.
2. Switch on the cryostat to allow double-wall beaker 1 (100 mL) and double-wall beaker 2 (500 mL) to be at 4 °C.
3. Place the double-wall beaker 1 under a laboratory motor agitator connected to a polytetrafluoroethylene (PTFE) impeller. Adjust the impeller to optimally mix the final volume of emulsion (about 34 mL) at 550 rpm.
4. Place the thermostated beaker 2 for the solvent extraction step under a laboratory motor agitator with a PTFE impeller operated at 300 rpm.

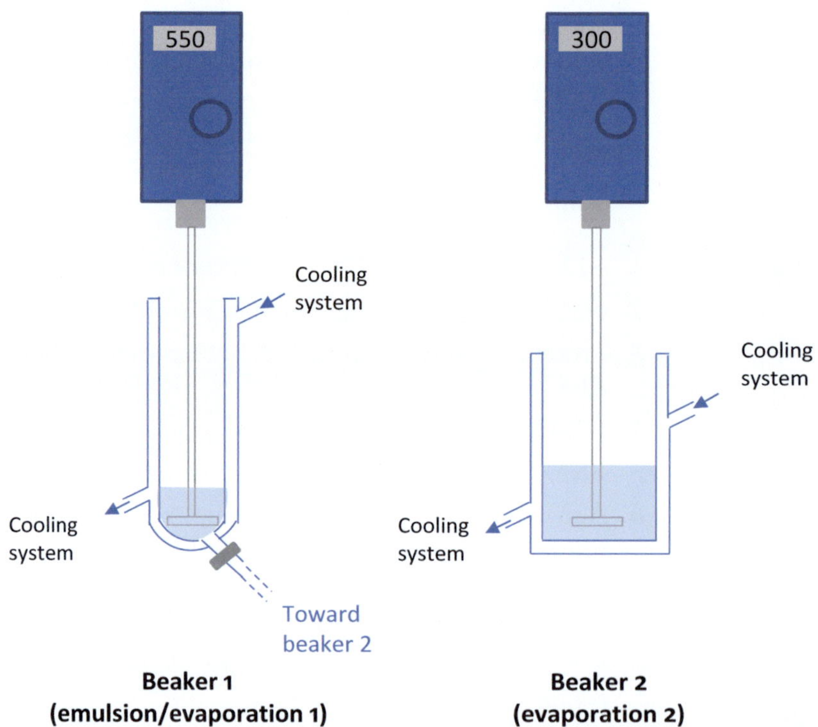

Fig. 1 Apparatus for microsphere preparation by emulsion/solvent extraction. The emulsion takes place in the first beaker containing the water phase into which the organic phase containing the protein nanosuspension is injected with stirring (550 rpm). The first extraction is done in beaker 1 (550 rpm). The second extraction is performed after transfer to the second beaker containing a specified volume (Subheading 3.3, **step 4**) of WFI under agitation (300 rpm)

3.2 Nanoprecipitation of Proteins

Note: Protein loading is 1 μg of TGF-β3 and 5 μg of HSA per mg of microspheres. The following protocol is performed for production of 50 mg of microspheres.

3.2.1 TGF-β3 Nanoprecipitation

1. Mix solution A1 with solution B1 in a PTFE tube to obtain a protein:poloxamer ratio of 1:20 (*see* **Note 3**).
2. Add 1.077 g of glycofurol.
3. Incubate for 30 min at 4 °C.
4. Centrifuge for 30 min at 10,000 × *g*.
5. Remove the supernatant and reserve the precipitate at 4 °C while waiting for Subheading 3.3, **step 1**.

3.2.2 HSA Nanoprecipitation

1. Mix solution A2 with solution B2 in a PTFE tube to obtain a protein:poloxamer ratio of 1:20.
2. Add 1.077 g of glycofurol.
3. Incubate for 30 min at 4 °C.
4. Centrifuge for 30 min at 10,000 × *g*.
5. Remove the supernatant and reserve the precipitate at 4 °C while waiting for Subheading 3.3, **step 1**.

3.3 Microsphere Preparation by Solid/Oil/Water (s/o/w) Emulsion Solvent Evaporation-Extraction

1. Suspend the TGF-β3 and HSA nanoprecipitates in 670 μL of organic phase containing the PLGA-P188-PLGA polymer.
2. With a glass insulin syringe, carefully harvest the total volume of the organic phase containing the HSA and TGF-β3 nanoprecipitates and inject it into the water phase (4 % PVA) in double-wall beaker 1 under agitation (550 rpm) (*see* **Note 4**). Agitate the thus formed emulsion for 1 min.
3. Add 33 mL of WFI to the double-wall beaker 1 (extraction 1). Agitate the solution for another 10 min.
4. Quickly transfer the suspension from double-wall beaker 1 to double-wall beaker 2 already containing 167.5 mL of WFI under agitation (300 rpm). Rinse double-wall beaker 1 with 23 mL of WFI and thereafter add this volume in double-wall beaker 2.
5. Agitate for 20 min at 300 rpm to extract the organic solvent.
6. Filter the microsphere suspension on a hydrophilic 5-μm SVLP (space-variant low-pass) type filter. Carefully rinse double-wall beaker 2 to harvest the totality of microspheres.
7. Harvest the microspheres retained on the SVLP filter surface with a micro-spatula and transfer them to a 5-mL flat-bottomed tube (*see* **Note 5**).
8. Add 500 μL of WFI and keep at −20 °C before freeze-drying.

3.4 Microsphere Characterization

3.4.1 Production Yield

1. After freeze-drying, weigh the flat-bottomed tube with microspheres and subtract the tube weight (as measured in **Note 5**) to obtain the total microsphere weight.
2. Compare the total microsphere weight and the amount of provided polymer (50 mg) to calculate the production yield (*see* **Notes 6** and **8**).

3.4.2 Encapsulation Yield

1. Weigh 5 mg of microspheres in a microtube.
2. Add 200 μL of dimethyl sulfoxide (DMSO) and agitate for 1 h.
3. Centrifuge at 2800 × *g*.
4. Measure the amount of TGF-β3 in the supernatant using an ELISA or bioassay [6].
5. Compare the amount of TGF-β3 protein measured to the amount of TGF-β3 theoretically present in 5 mg of microspheres (*see* **Note** 7 and **8**).

3.4.3 Size Distribution and Zeta Potential

1. Size distribution: Weigh 1 mg of microspheres and disperse it in 1 mL of ionic solution (such as Isoton II diluent, Beckman Coulter). Measure the particle size distribution using, for example, a Beckman Coulter Multisizer® Coulter Counter.
2. Zeta potential: Weigh 1 mg of microspheres and disperse it in 1 mL of NaCl solution. Use 1 mL of the suspension to measure the zeta potential using, for example, a Malvern Zetasizer®.

3.4.4 Release Kinetics

1. Weigh 5 mg of microspheres in a microtube.
2. Add 500 μL of protein stability medium.
3. Place the tube on a water bath at 37 °C under agitation (300 rpm).
4. At the desired time, centrifuge the microtube (2800 × *g*), collect 100 μL of the supernatant, and place it at −20 °C for measurement of TGF-β3 concentration by ELISA or bioassay [6] (*see* **Note 8**).
5. Add 100 μL of new medium to the same tube and place the tube in the water bath at 37 °C under agitation (300 rpm) until the next kinetic point.

3.5 Coating of Microspheres: Preparation of PAMs

NB: The following steps are carried out under sterile conditions.

1. Weigh 5 mg of microspheres and disperse them in 5 mL of DPBS in a 15-mL silanized tube (*see* **Note 9**).
2. Add 5 mL of coating solution to the tube.
3. Place the tube on a rotator at 15 rpm for 1 h and 30 min at 37 °C.
4. Mix the tube (*see* **Note 9**).
5. Centrifuge at 2800 × *g* and remove the supernatant.

6. Add 5 mL of WFI with 2 % antibiotics (final concentration: 200 U/mL penicillin, 200 U/mL streptavidin), mix, centrifuge, and remove the supernatant. Repeat this step two times.
7. Freeze the microspheres at −80 °C for further freeze-drying.

3.6 Coating Characterization (FN Immunostaining)

1. In a 1.5-mL microtube containing 1 mg of PAMs, add 1 mL of BSA/Tween solution.
2. Incubate for 30 min at room temperature (RT) with stirring (15 rpm).
3. Centrifuge the tube and discard the supernatant.
4. Add DPBS and centrifuge at 9000 × *g*. Repeat this step two times.
5. Add 1 mL of anti-FN antibody solution.
6. Incubate for 1 h and 30 min at 37 °C still under rotation.
7. Centrifuge the tube and discard the supernatant. Add 1 mL of DPBS and centrifuge at 9000 × *g*. Repeat this step three times.
8. Add 1 mL of biotinylated antibody solution.
9. Incubate for 1 h at RT still under rotation.
10. Centrifuge the tube and discard the supernatant. Add 1 mL of DPBS and centrifuge at 9000 × *g*. Repeat this step two times.
11. Add 1 mL of streptavidin-fluoprobe solution.
12. Incubate for 40 min at RT still under rotation.
13. Centrifuge the tube and discard the supernatant. Add 1 mL of DPBS and centrifuge at 9000 × *g*. Repeat this step three times.
14. Place a small volume of PAM suspension on appropriate microscope slides for confocal microscopy (examples of obtained images are shown in Fig. 2).

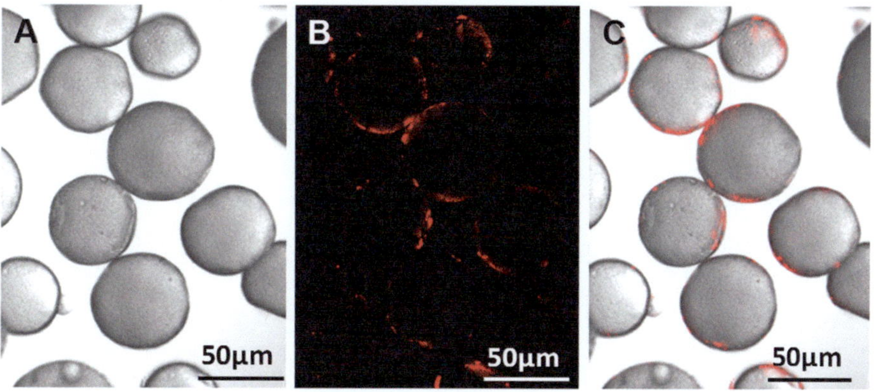

Fig. 2 Example of observation after coating of PLGA-P188-PLGA microspheres with FN and PDL. Differential interference contrast (DIC) microscopy image (*a*), confocal microscopy image (*b*), and superposition of DIC and confocal microscopy images showing anti-FN immunofluorescence of PAMs (*c*)

3.7 Association of PAMs with MSCs

1. Add adhesion medium to PAMs to give a PAM concentration of 1 mg/mL (*see* **Note 10**).
2. Harvest MSCs and adjust the concentration to obtain 0.5×10^6 cells in 1 mL.
3. Add 500 μL of the thus prepared cell suspension to a well of an ultralow adhesion (ULA) 24-well culture plate (i.e., 0.25×10^6 cells per well).
4. Just before adding PAMs to the cells, vortex the PAM suspension and use ultrasound to achieve total dispersion.
5. Add 500 μL of PAM suspension to the ULA plate well already containing cells (final volume = 1 mL).
6. Homogenize the well contents by gently aspirating and refluxing using a 1-mL micropipette.
7. To allow cell-PAM complex formation (as observed in Fig. 3), incubate for a minimum of 4 h at 37 °C and 5 % CO_2, before conducting a cell survival assay or animal injection.

3.8 MSC Chondrogenesis with PAMs

1. 15 min before cell association, add chondrogenic medium to PAMs to give a PAM concentration of 1 mg/mL.
2. Harvest MSCs and adjust the concentration to obtain 0.5×10^6 cells in 1 mL.

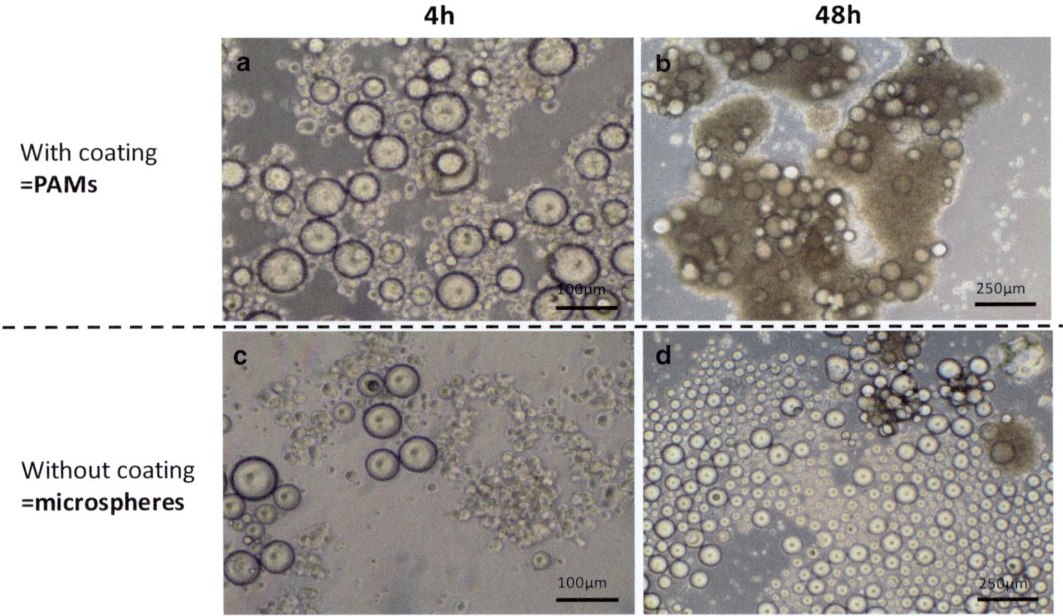

Fig. 3 Example of association between human MSCs and PLGA-P188-PLGA microspheres (*c* and *d*) or coated PLGA-P188-PLGA microspheres (= PAMs) (*a* and *b*). As observed in (*a*) and (*b*) versus (*c*) and (*d*), the coating of microspheres with FN and PDL allows better association of MSCs at the PAM surfaces

3. Add 500 μL of the thus prepared cell suspension to a 15-mL tube (i.e., 0.25×10^6 cells per tube).
4. Just before adding PAMs to the cells, vortex the PAM suspension and use ultrasound to achieve total dispersion.
5. Add 500 μL of PAM suspension to the 15-mL tube already containing cells.
6. Homogenize the tube contents by gently aspirating and refluxing using a 1-mL micropipette.
7. Incubate for 4 h at 37 °C and 5 % CO_2.
8. Centrifuge for 5 min at $300 \times g$.
9. Incubate for 24 h at 37 °C and 5 % CO_2.
10. If the formed micropellet seems adhered to the tube, agitate the tube to suspend it.
11. Incubate for 21 days at 37 °C and 5 % CO_2 with a medium change every 3 days (*see* **Note 8**).

4 Notes

1. The triblock copolymer PLGA–P188–PLGA (ABA copolymer) was prepared by ring-opening polymerization (ROP) of DL-lactide and glycolide using P188 as an initiator, and stannous octoate [Sn(Oct) 2] as catalyst. A mixture of P188, DL-lactide, and glycolide was introduced into 100 mL round-bottom flasks with the catalyst. The mixture was heated to 140 °C and degassed by 15 vacuum-nitrogen purge cycles in order to remove the moisture and the oxygen, inhibitors of this polymerization. Flasks were then frozen at 0 °C and sealed under dynamic vacuum at 10^{-3} mbar. Polymerization was allowed to proceed at 140 °C under constant agitation. After 5 days, the products were recovered by dissolution in dichloromethane and then precipitated by adding an equal volume of ethanol. Finally, the polymer was filtered, washed with cold ethanol, and dried overnight at 45 °C under reduced pressure, to constant weight.
2. Various cells could be added to the PAMs depending on the required application, such as fetal dopaminergic neurons [1], PC 12 cells [2], marrow-isolated adult multilineage-inducible (MIAMI) cells [4], and more recently endothelial progenitor cell (EPC) [7].
3. Homogenize the 10 mg/mL TGF-β3 solution by reflux aspiration (avoid strong mixing with vortex).
4. The injection should occur at the surface of the vortex.
5. Before adding microspheres, weigh the empty tube and note this weight to allow the production yield measurement.

6.

$$\text{Production yield} = \frac{\text{Amount of obtained microspheres (mg)}}{\text{Amount of drymatter afforded to the process (PLGA} - \text{P188} - \text{PLGA) (mg)}} \times 100$$

7.

$$\text{Experimental encapsulation yield} = \frac{\text{Amount of TGF} - \dagger 3 \text{ measured inside the microspheres}}{\text{Amount of TGF} - \dagger 3 \text{ afforded to the process}} \times 100$$

8. Results expected from each of these analyses may be found in Morille et al. [5].
9. If necessary, quickly sonicate (5–15 s) to achieve full dispersion of the microspheres.
10. Before associating to cells, incubate the microspheres for 15 min with culture medium (adhesion or chondrogenic, depending on the experiment performed) to fully hydrate the microspheres.

References

1. Tatard VM, Sindji L, Branton J, Aubert-Poueÿssel A, Colleau J, Benoit JP, Montero-Menei CN (2007) Pharmacologically active microcarriers releasing glial cell line – derived neurotrophic factor: survival and differentiation of embryonic dopaminergic neurons after grafting in hemiparkinsonian rats. Biomaterials 28:1978–1988
2. Tatard VM, Venier-Julienne MC, Benoit JP, Menei P, Montero-Menei CN (2004) In vivo evaluation of pharmacologically active microcarriers releasing nerve growth factor and conveying PC12 cells. Cell Transplant 13:573–583, Published online Epub 2004 (10.3727/000000004783983675)
3. Delcroix GJR, Garbayo E, Sindji L, Thomas O, Vanpouille-Box C, Schiller PC, Montero-Menei CN (2011) The therapeutic potential of human multipotent mesenchymal stromal cells combined with pharmacologically active microcarriers transplanted in hemi-parkinsonian rats. Biomaterials 32:1560–1573, Published online EpubFeb (10.1016/j.biomaterials.2010.10.041
4. Garbayo E, Curtis K, Raval A, DellaMorte D, Gomez LA, D'Ippolito G, Reiner T, Perez-Stable C, Howard G, Perez-Pinzon M, Montero-Menei C, Schiller P (2011) Neuroprotective properties of Marrow-Isolated Adult Multilineage-Inducible (MIAMI) cells in rat hippocampus following global cerebral ischemia are enhanced when complexed to biomimetic microcarriers. Xenotransplantation 18: 286–286, Published online EpubSep-Oct
5. Morille M, Tran V-T, Garric X, Cayon J, Coudane J, Noel D, Venier-Julienne M-C, Montero-Menei CN (2013) New PLGA-P188-PLGA matrix enhances TGF-beta 3 release from pharmacologically active microcarriers and promotes chondrogenesis of mesenchymal stem cells. J Control Release 170:99–110, Published, online EpubAug 28
6. Tesseur I, Zou K, Berber E, Zhang H, Wyss-Coray T (2006) Highly sensitive and specific bioassay for measuring bioactive TGF-beta. BMC Cell Biol 7:15
7. Penna C et al (2013) Pharmacologically active microcarriers influence VEGF-A effects on mesenchymal stem cell survival. J Cell Mol Med 17:192–204

Chapter 13

Nanostructured Capsules for Cartilage Tissue Engineering

Clara R. Correia, Rui L. Reis, and João F. Mano

Abstract

Polymeric multilayered capsules (PMCs) have found great applicability in bioencapsulation, an evolving branch of tissue engineering and regenerative medicine. Here, we describe the production of hierarchical PMCs composed by an external multilayered membrane by layer-by-layer assembly of poly(L-lysine), alginate, and chitosan. The core of the PMCs is liquified and encapsulates human adipose stem cells and surface-functionalized collagen II-TGF-β3 poly(L-lactic acid) microparticles for cartilage tissue engineering.

Key words Bioencapsulation, Capsules, Cartilage regeneration, Collagen II, Layer-by-layer, Microparticles, Stem cells, Tissue engineering, TGF-β3

1 Introduction

Since the pioneering study of Decher [1], the layer-by-layer technique (LbL) has been used in a wide range of biomedical applications, including the production of polymeric multilayered capsules (PMCs). Particularly, PMCs have found great applicability in bioencapsulation systems for tissue engineering and regenerative medicine purposes [2–8]. The production of PMCs is based on the LbL setwise adsorption of polyelectrolytes with oppositely charged macromolecules. Polyelectrolytes are assembly on the surface of spherical particles used as sacrificial templates, and, ultimately, the core is dissolved or eliminated, originating PMCs [9–13]. Due to the high versatility of the LbL technique, different properties of the capsules can be easily tailored; for example, the permeability of the shell can be controlled by varying the number of layers deposited and the surface of the capsules can be customized by endowing nanoparticles, lipids, viruses, among others [14]. While the liquified core ensures the viability of the encapsulated cells by allowing a rapid diffusion of nutrients, oxygen, waste

Pauline M. Doran (ed.), *Cartilage Tissue Engineering: Methods and Protocols*, Methods in Molecular Biology, vol. 1340, DOI 10.1007/978-1-4939-2938-2_13, © Springer Science+Business Media New York 2015

products, and metabolites, the LbL membrane provides an immune privilege environment by blocking the entrance of high-molecular-weight immune system compounds, such as immunoglobulins and immune cells [15]. The main drawback of capsules in cell encapsulation approaches is related with the fact that most cells are anchorage dependent and, thus, deprived of a physical support cells are not able to adhere and proliferate [16]. Therefore, the liquefaction of the core to achieve an excellent diffusion of essential molecules will, on the other hand, compromise the biological outcome of the encapsulated cells. To ensure an efficient diffusion provided by the liquified environment of PMCs, and simultaneously provide cell adhesion sites to the encapsulated cells, we developed liquified nanostructured capsules encapsulating microparticles as cell supports [17, 18]. Here, we describe the combination of the well-studied *proof-of-concept* system with surface-functionalized poly(L-lactic acid) (PLLA) microparticles for cartilage tissue engineering. Therefore, microparticles have a dual functionality by providing physical support for cellular adhesion as well as triggering encapsulated stem cells to differentiate into the chondrogenic lineage. The modification of PLLA microparticles is achieved by binding to its surface (1) collagen II, a main component of the extracellular matrix of the hyaline cartilage tissue [19], and (2) the transforming growth factor-β3 (TGF-β3), a required bioactive agent to induce chondrogenesis of stem cells [20]. Transglutaminase-2 enzyme, a potent cross-linking mediator of several cartilage components [21, 22], is used as the cross-linking agent.

2 Materials

2.1 Microparticles

1. PLLA/CH_2Cl_2 solution (5 % w/v, transparent): 1 g of PLLA (molecular weight (Mw) ~1600–2400, 70 % crystallinity) in 20 mL of dichloromethane (CH_2Cl_2).
2. PVA solution (0.5 % w/v): 0.5 g of polyvinyl alcohol (PVA, 87–90 % hydrolyzed, Mw ~30,000–70,000) in 100 mL of distilled water.
3. Collagen II solution: 0.1 mg/mL collagen II in 0.02 M acetic acid.
4. TG-2/TGF-β3 solution: 0.01 U/mL of transglutaminase-2 enzyme (TG-2) and 10 ng/mL of human TGF-β3.
5. Phosphate-buffered saline (PBS).

2.2 Core

1. Human subcutaneous adipose tissue from liposuction procedures.
2. Sterile PBS.

3. Collagenase type II solution: 0.05 % collagenase type II in PBS solution. Sterilize by 0.22-μm filtration.
4. Red blood cell lysis buffer: 155 mM of ammonium chloride, 12 mM of potassium bicarbonate, and 0.1 M of EDTA. Sterilize by 0.22-μm filtration.
5. Supplemented α-MEM cell culture medium: α-MEM medium supplemented with 10 % fetal bovine serum (FBS) and 1 % antibiotic-antimycotic solution (Life Technologies).
6. Recombinant enzyme TrypLE™ Express reagent solution (Life Technologies).
7. Supplemented DMEM cell culture medium: DMEM medium supplemented with 10 % FBS, 1 % antibiotic-antimycotic solution, 0.4 mM proline, 0.2 mM ascorbic acid, sodium pyruvate (1:100), ITS+premix (1:100), and 0.2 mM dexamethasone. The DMEM medium used is commercially modified with high glucose and L-glutamine, and without sodium pyruvate and HEPES (Sigma-Aldrich).
8. 1.5 % alginate solution: Dissolve 3 % w/v low-viscosity sodium alginate in 0.15 M of sodium chloride containing 25 mM MES buffer. Set the pH to 7. Sterilize by 0.22-μm filtration. Inside the sterilized environment of a flow chamber, add supplemented DMEM cell culture medium (*see* **Note 1**) to dilute the alginate concentration to 1.5 % w/v.
9. Calcium chloride solution: 0.1 M calcium chloride in distilled water containing 25 mM MES buffer. Set the pH to 7 and sterilize by 0.22-μm filtration.
10. Washing solution: 0.15 M of sodium chloride and 25 mM of MES buffer at pH 7. Sterilize by 0.22-μm filtration.

2.3 Membrane

1. PLL solution: 0.5 mg/mL of poly(L-lysine) hydrobromide (PLL, Mw ~30,000–70,000) in 0.15 M of sodium chloride containing 25 mM of MES buffer. Set the pH to 7 (*see* **Note 2**).
2. ALG solution: 0.5 mg/mL of low-viscosity sodium alginate (ALG) in 0.15 M of sodium chloride containing 25 mM of MES buffer. Set the pH to 7 (*see* **Note 2**).
3. CHT solution: 0.5 mg/mL of chitosan (CHT, water-soluble highly purified chitosan, Protasan UP CL 213, NovaMatrix, *see* **Note 3**) in 0.15 M of sodium chloride containing 25 mM of MES buffer. Set the pH to ~6.3–6.4 (*see* **Note 2**).
4. EDTA solution: 0.02 M of ethylenediaminetetraacetic acid (EDTA, anhydrous, crystalline, suitable for cell culture) in distilled water. Set the pH to 7 (*see* **Note 4**).

3 Methods

3.1 Microparticles

The scheme for production and surface functionalization of the PLLA microparticles is shown in Fig. 1.

1. For microparticle production, under agitation with a magnetic stirrer, slowly add the PLLA/CH_2Cl_2 solution to the PVA solution (*see* **Note 5**). Let the resulting solution to stir for 48 h at room temperature (RT) inside a fume hood, in order to evaporate the organic solvent.
2. To wash and collect the microparticles, transfer the solution prepared in **step 1** to a centrifuge tube with a 100-μm filter to discard aggregates. Centrifuge at 300 ×*g* for 5 min (*see* **Note 6**).

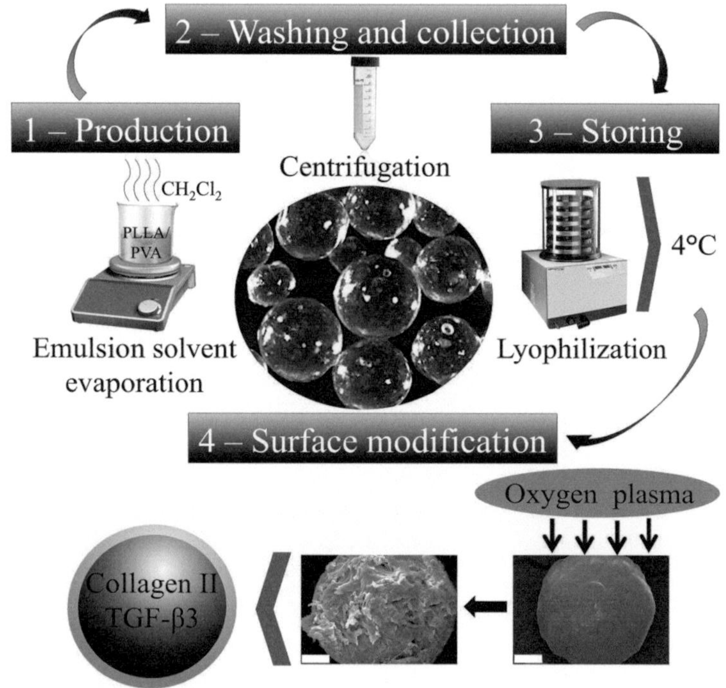

Fig. 1 Schematic representation of the production and surface functionalization steps of the poly(L-lactic acid) (PLLA) microparticles. (*1*) First, PLLA microparticles are obtained by solvent evaporation technique. (*2*) Then, microparticles are collected and washed by centrifugation steps. (*3*) To obtain a lyophilized powder, the washed microparticles are freeze-dried for 3 days and stored at 4 °C until further use. At the center of the scheme an image of the obtained microparticles is presented. (*4*) Ultimately, PLLA microparticles are surface modified by plasma treatment to increase the roughness of the surface, as evidenced by the scanning electron microscopy images of one particle before and after treatment (scale bar is 200 μm). Subsequently, microparticles are immersed in a transglutaminase/collagen-II/TGF-β3 solution for protein immobilization

Discard the supernatant. Wash the microparticles with distilled water. Repeat the centrifugation and washing steps until the supernatant becomes transparent.

3. Freeze the microparticles at −80 °C and then lyophilize for 3 days. Store at 4 °C until use.
4. For plasma surface modification, first place the PLLA microparticles inside a plasma reactor chamber fitted with a radio frequency generator. Use air as the working atmosphere. Let the pressure of the chamber stabilize to ~0.2 mbar. A glow discharge plasma is created by controlling the electrical power at 30 V of electrical potential difference. Treat the microparticles for 5 min. Remove the microparticles from the chamber and employ a gentle mix in order to maximize the PLLA surface exposure to plasma treatment. Repeat this procedure three times to apply a total plasma reaction time of 15 min (*see* **Note 7**).
5. For protein surface modification, sterilize the microparticles by 1 h of UV radiation, with a gentle mixing after 30 min. Immerse the PLLA microparticles in the collagen II solution for 8 h at 4 °C in agitation. Repeat the washing and collection steps described in **step 2**. To cross-link TGF-β3 to collagen II-PLLA microparticles, immerse the collagen II-PLLA microparticles in the TG-2/TGF-β3 solution for 8 h at 4 °C in agitation. Wash the surface-modified microparticles with PBS to remove non-bound TGF-β3.

3.2 Multilayered Capsules

The scheme for production of liquified multilayered capsules encapsulating surface-modified microparticles and adipose-derived stem cells is shown in Fig. 2. Perform all the procedures under the sterilized environment of a cell culture laminar flow chamber.

1. To isolate human adipose-derived stem cells (hASCs), wash the subcutaneous adipose tissues with sterile PBS, in order to remove the majority of blood. Under agitation, incubate the tissue sample with collagenase type II solution for 45 min at 37 °C. Filter the digested sample with a 200-μm pore size strainer and centrifuge at 800 ×*g* for 10 min in order to pellet the stromal vascular fraction (SVF). Resuspend the SVF in red blood cell lysis buffer. After 10 min of incubation at RT, centrifuge the obtained mixture at 300 ×*g* for 5 min. Resuspend the red blood cell-free SFV in supplemented α-MEM cell culture medium and transfer to cell culture flasks. Incubate at 37 °C in a humidified air atmosphere of 5 % CO_2. Change the supplemented α-MEM cell culture medium every 2 days. hASCs are selected by plastic adherence.
2. For core template loading, at 90 % confluence, wash the hASCs grown in tissue culture flasks with sterile PBS. Detach the hASCs by a chemical procedure with TrypLE™ Express

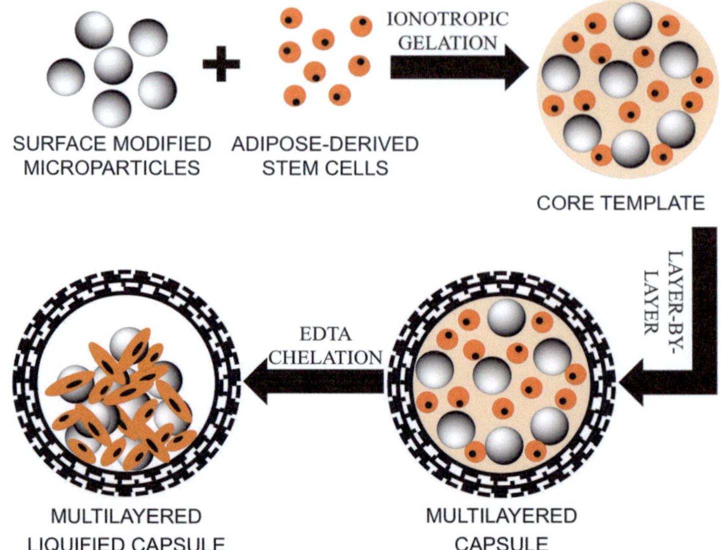

Fig. 2 Schematic representation of the production steps of liquefied multilayered capsules encapsulating surface-modified microparticles and adipose-derived stem cells

reagent solution at 37 °C in a humidified air atmosphere of 5 % CO_2. After 5 min of incubation, add PBS and centrifuge at 300×*g* for 5 min. Discard the supernatant. Add 50 mg of surface-modified PLLA microparticles and 5×10^6 human adipose-derived stem cells per mL of 1.5 % alginate solution. Gently mix the different contents by pipetting up and down.

3. For core template production, under agitation, add the alginate solution prepared in **step 2** dropwise using a 21 G needle to the calcium chloride solution. Let the hydrogel-loaded beads immediately formed to stir for 20 min at RT. Wash the hydrogel-loaded beads in the washing solution.
4. To produce the multilayered membrane, immerse the hydrogel-loaded beads first in PLL solution and, subsequently, in ALG, CHT, and ALG solutions. Set the time of immersion for 10 min at RT. Between each polyelectrolyte immersion, wash the hydrogel-loaded beads with the washing solution for 5 min to remove excess non-absorbed macromolecules. Repeat this process three times to obtain a 12-layered membrane surrounding the hydrogel-loaded beads. All the procedure is performed with the aid of standard sieves.
5. To liquefy the core, immerse the multilayered hydrogel-loaded beads in the EDTA solution for 3 min at RT. With a spatula, transfer the liquified multilayered capsules to non-treated cell culture plates already filled with supplemented DMEM cell culture medium. Incubate at 37 °C in a humidified air atmosphere of 5 % CO_2.

4 Notes

1. It is important to add to the core solution culture medium supplemented with 10 % fetal bovine serum. This will ensure cell survival during the production of the PMCs at RT, mainly during the most time-consuming step of the production of the layer-by-layer membrane.
2. Since layer-by-layer development is based on the adsorption of polyelectrolytes with oppositely charged macromolecules, it is important to ensure that the selected polymers retain their positive or negative charge. At pH 7, both PLL and the carboxylate groups of alginate remain deprotonated due to the dissociation constants (pKa) of ~10 for PLL and 3.38 for mannuronic (M) and 3.65 for guluronic (G) acid alginate monomers [23, 24]. However, at pH 7 the amine groups on chitosan glucosamine monomers deprotonate since its pKa is ~6.5 [25, 26], preventing the interaction of chitosan with anionic components. Therefore, the working pH of ~6.3–6.4 was selected, since lower pH ranges could jeopardize the viability of the encapsulated cells.
3. In cell encapsulation it is highly recommended that all the solutions used are non-cytotoxic to ensure cell viability. Since chitosan is soluble in diluted acid solutions, we used a commercially available water-soluble chitosan.
4. EDTA will only start to dissolve when neutralized with sodium hydroxide to a pH of 8. After a complete dissolution, change the pH to 7.
5. Use a syringe to slowly add the PLLA/CH_2Cl_2 to the PVA solution. This will diminish the waste aggregates of PLLA formed when large amounts of the solution reach the PVA bath, and thus, the efficacy of the production of microparticles is increased.
6. Centrifuge tubes with the same volume of microparticle suspension have variable weights according to the quantity of microparticles per tube. Therefore, before centrifugation, calibrate the centrifuge tubes by weight.
7. The induced reactivity by plasma treatment on the surface of PLLA microparticles is very unstable. Therefore, the protein surface modification step should be performed immediately after plasma treatment.

Acknowledgements

This work was financially supported by the Portuguese Foundation for Science and Technology (FCT) through the project PTDC/CTM-Bio/1814/2012 and Ph.D. grant with the reference no.

SFRH/BD/69529/2010, co-financed by the Operational Human Potential Program (POPH), developed under the scope of the National Strategic Reference Framework (QREN) from the European Social Fund (FSE). The authors also acknowledge the financial support from Seventh Framework Programme of European Union (FP7/2007–2013) under grant agreement number REGPOT-CT2012-316331-POLARIS.

References

1. Decher G, Hong JD (1991) Buildup of ultrathin multilayer films by a self-assembly process, 1 consecutive adsorption of anionic and cationic bipolar amphiphiles on charged surfaces. Makromol Chem Macromol Symp 46: 321–327
2. Facca S, Cortez C, Mendoza-Palomares C et al (2010) Active multilayered capsules for in vivo bone formation. PNAS 107: 3406–3411
3. Mendes AC, Baran ET, Pereira RC et al (2012) Encapsulation and survival of a chondrocyte cell line within xanthan gum derivative. Macromol Biosci 12:350–359
4. Ying Ma Y, Zhang Y, Liu Y (2013) Investigation of alginate-ε-poly-L-lysine microcapsules for cell microencapsulation. J Biomed Mater Res A 101:1265–1273
5. Tan W-H, Takeuchi S (2007) Monodisperse alginate hydrogel microbeads for cell encapsulation. Adv Mater 19:2696–2701
6. Kim J, Arifin DR, Muja N (2011) Multifunctional capsule-in-capsules for immunoprotection and trimodal imaging. Angew Chem 123:2365–2369
7. Hillberg AL, Kathirgamanathan K, Lam JBB (2013) Improving alginate-poly-L-ornithine-alginate capsule biocompatibility through genipin crosslinking. J Biomed Mater Res B Appl Biomater 101B:258–268
8. Costa RR, Mano JF (2014) Polyelectrolyte multilayered assemblies in biomedical technologies. Chem Soc Rev. doi:10.1039/c3cs60393h
9. Costa RR, Castro E, Arias FJ et al (2013) Multifunctional compartmentalized capsules with a hierarchical organization from the nano to the macro scales. Biomacromolecules 14: 2403–2410
10. Costa RR, Custódio CA, Arias FJ et al (2013) Nanostructured and thermoresponsive recombinant biopolymer-based microcapsules for the delivery of active molecules. Nanomedicine 9:895–902
11. Caruso F, Caruso RA, Mohwald H (1998) Nanoengineering of inorganic and hybrid hollow spheres by colloidal templating. Science 282:1111–1114
12. Donath E, Sukhorukov GB, Caruso F et al (1998) Novel hollow polymer shells by colloid-templated assembly of polyelectrolytes. Angew Chem Int Ed 37:2201–2205
13. Costa NL, Sher P, Mano JF (2011) Liquefied capsules coated with multilayered polyelectrolyte films for cell immobilization. Adv Eng Mater 13:B218–B224
14. Shchukin D, Sukhorukov G, Mohwald H (2003) Smart inorganic/organic nanocomposite hollow microcapsules. Angew Chem Int Ed 42:4472–4475
15. Hernández RM, Orive G, Murua A et al (2010) Microcapsules and microcarriers for in situ cell delivery. Adv Drug Deliv Rev 62: 711–730
16. Brun-Graeppi AKAS, Richard C, Bessodes M et al (2011) Cell microcarriers and microcapsules of stimuli-responsive polymers. J Control Release 149:209–224
17. Correia CR, Sher P, Reis RL et al (2013) Liquified chitosan–alginate multilayer capsules incorporating poly(L-lactic acid) microparticles as cell carriers. Soft Matter 9:2125–2130
18. Correia CR, Reis RL, Mano JF (2013) Multilayered hierarchical capsules providing cell adhesion sites. Biomacromolecules 14: 743–751
19. Mahmoudifar N, Doran PM (2012) Chondrogenesis and cartilage tissue engineering: the longer road to technology development. Trends Biotechnol 30:166–176
20. Madry H, Rey-Rico A, Venkatesan JK et al (2013) Transforming growth factor beta-releasing scaffolds for cartilage tissue engineering. Tissue Eng B Rev. doi:10.1089/ten.teb.2013.0271
21. Jones ME, Messersmith PB (2007) Facile coupling of synthetic peptides and peptide-polymer conjugates to cartilage via transglutaminase enzyme. Biomaterials 28:5215–5224
22. Niger C, Beazley KE, Nurminskaya M (2013) Induction of chondrogenic differentiation in mesenchymal stem cells by TGF-beta cross-linked to collagen-PLLA [poly(L-lactic acid)] scaffold by transglutaminase 2. Biotechnol Lett 35:2193–2199

23. Amali AJ, Awwad NH, Rana RK et al (2011) Nanoparticle assembled microcapsules for application as pH and ammonia sensor. Anal Chim Acta 708:75–83
24. Kurayama F, Suzuki S, Oyamada T et al (2010) Facile method for preparing organic/inorganic hybrid capsules using amino-functional silane coupling agent in aqueous media. J Colloid Interface Sci 349:70–76
25. Jean M, Alameh M, De Jesus D et al (2012) Chitosan-based therapeutic nanoparticles for combination gene therapy and gene silencing of in vitro cell lines relevant to type 2 diabetes. Eur J Pharm Sci 45:138–149
26. Pujana MA, Pérez-Álvarez L, Iturbe LCC et al (2012) Water dispersible pH-responsive chitosan nanogels modified with biocompatible crosslinking-agents. Polymer 53:3107–3116

Chapter 14

Stratified Scaffolds for Osteochondral Tissue Engineering

Patcharakamon Nooeaid, Gundula Schulze-Tanzil, and Aldo R. Boccaccini

Abstract

Stratified scaffolds are promising devices finding application in the field of osteochondral tissue engineering. In this scaffold type, different biomaterials are chosen to fulfill specific features required to mimic the complex osteochondral tissue interface, including cartilage, interlayer tissue, and subchondral bone. Here, the biomaterials and fabrication methods currently used to manufacture stratified multilayered scaffolds as well as cell seeding techniques for their characterization are presented.

Key words Stratified scaffold, Osteochondral tissue engineering, Biomaterial, Alginate, Mesenchymal stem cells

1 Introduction

Due to the fact that cartilage lacks intrinsic repairing capacity and considering that limitations of clinical treatments still exist to heal defects at the cartilage-subchondral bone interface [1], tissue engineering becomes a promising approach for the repair of such osteochondral defects [2, 3]. To develop biomaterial scaffolds suitable for osteochondral tissue engineering, stratified scaffolds, including biphasic scaffolds and multilayered scaffolds, are being intensively studied and investigated [4–6].

Suitable scaffold materials and manufacturing techniques are required to develop such complex scaffolds with improved physical, biochemical, and biomechanical properties which are able to mimic the intrinsic properties of native osteochondral tissues [5].

Stratified composite scaffolds combining inorganic and organic biomaterials, which exhibit high biocompatibility, controlled biodegradability, and sufficient mechanical properties, are emerging to provide a promising 3D environment for cell proliferation and differentiation [7]. Moreover, such complex constructs are necessary to induce the regeneration of the different involved tissues at the osteochondral interface (i.e., cartilage and bone) simultaneously [7].

Pauline M. Doran (ed.), *Cartilage Tissue Engineering: Methods and Protocols*, Methods in Molecular Biology, vol. 1340, DOI 10.1007/978-1-4939-2938-2_14, © Springer Science+Business Media New York 2015

Single structures incorporating two different phases to mimic the cartilage and bone tissues, namely biphasic scaffolds, have been fabricated by fusing two phases together without an artificial connection such as by using glue or sutures [2, 7, 8]. This approach seems to avoid the development of shear stresses between cartilage and subchondral bone, which may lead to failure after implantation [2, 7, 8]. Alternative convenient structures incorporating a functional interface between the cartilage and bone phases can be achieved by developing bi- or multilayered scaffolds [2, 7, 9].

The combination of material phases and structures resembling the properties and structures of cartilage, interface, and bone phases is a promising approach for osteochondral defect regeneration [10, 11]. The proper selection of suitable scaffold materials is crucial for the success of the approach. Biodegradable polymers, including synthetic polymers (e.g., polylacticglycolic acid [PLGA], polylactic acid [PLA], polycaprolactone [PCL], and polyhydroxybutyrate [PHB]) and natural polymers (e.g., collagen, gelatin, hyaluronic acid, alginate, silk, and chitosan), are widely used to construct cartilage scaffolds due to their biocompatibility, degradability, and elasticity [1]. In contrast, the scaffolds for mineralized bone require bioactivity, osteoconductivity, osteoinductivity, and sufficient mechanical integrity and rigidity. Bioceramics (hydroxyapatite and calcium phosphate), bioactive glasses (45S5 Bioglass®), and their composites are extensively used as a scaffold for bone regeneration [1, 12]. According to these different requirements, a successful osteochondral tissue engineering strategy should involve the suitable combination of two or more different materials together in a composite structure, forming either biphasic or multilayered scaffolds.

The second essential component in a tissue engineering approach is the cells. Mesenchymal stroma cells (MSCs) represent a cell source capable of undergoing chondrogenic and osteogenic differentiation [13–15]. Therefore, they are ideal candidates for osteochondral tissue engineering [13]. For example, implantation of scaffolds seeded with chondrogenically predifferentiated MSCs into osteochondral defects in a sheep model led to superior histological results compared with the implantation of articular chondrocytes [16]. Additional advantages of MSCs compared with chondrocytes and osteoblasts are that harvesting them is associated with only low donor site morbidity, that they possess high proliferative capacity, low immunogenicity, high migratory capacity, as well as trophic and stimulatory effects on other cell types [17–19]. Compared with induced embryonic or pluripotent stem cells, MSCs bear no major risk of tumorigenesis [20]. This is particularly important in view of achieving blood vessel in-growth in the bone phase after implantation of a tissue-engineered osteochondral cylinder. Overall, it is known that under physiological conditions bone healing is not as challenging as cartilage repair, and bone restoration can be achieved

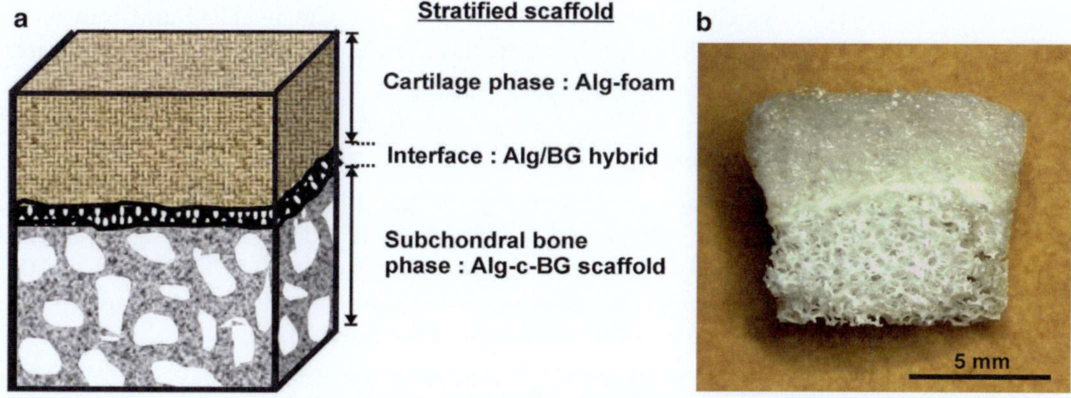

Fig. 1 (**a**) Schematic diagram showing the stratified scaffold composed of alginate freeze-dried foam (Alg-foam) acting as a scaffold for cartilage, alginate/Bioglass® (Alg/BG) hybrid adhesive acting as the interface layer, and alginate-coated Bioglass®-based scaffold (Alg-c-BG) acting as a scaffold for subchondral bone, (**b**) digital photograph of the obtained stratified scaffold, according to the scheme in figure (**a**). The thickness of both the cartilage and bone phases was 5 mm

by migration of activated resident MSCs into the osteochondral defect and subsequent osteogenic differentiation [1]. Hence, the preparation of the cartilage phase of a biphasic scaffold is more challenging and is the focus of this chapter.

In this chapter, we describe one type of suitable stratified scaffold for osteochondral regeneration, namely a multilayered scaffold, which is composed of an alginate foam, an alginate/bioactive glass hybrid interlayer, and an alginate-coated bioactive glass-based scaffold, serving as a complex scaffold for cartilage, interface, and subchondral bone, respectively, as shown in Fig. 1. The characterization of the scaffolds in cell culture studies using MSCs is also described.

2 Materials

2.1 Scaffold Fabrication

1. Alginate solution: Sodium alginate is dissolved in deionized (DI) water at a concentration of 3 % w/v. Weigh 1.2 g of sodium alginate and transfer to a glass beaker. Add water to a volume of 38.8 mL. Stir on a magnetic stirrer at room temperature for about 2 h.
2. Alginate/bioactive glass (type 45S5) adhesive paste: Mix 0.15 g of 45S5 bioactive glass (particle size ~2 μm) with 10 mL of alginate solution from **step 1**. Stir for around 1 h.
3. Alginate coating solution: Weigh 0.3 g of sodium alginate and transfer to a glass beaker. Add water to a volume of 19.7 mL. Stir on a magnetic stirrer at room temperature for 2 h.

4. Gelation agent: Weigh 14.7 g of $CaCl_2{\cdot}2H_2O$ and transfer to a 100-mL volumetric flask. Make up to 100 mL with DI water.
5. Cross-linking agent: Weigh 36.75 g of $CaCl_2{\cdot}2H_2O$ and transfer to a 500-mL volumetric flask. Make up to 500 mL with DI water.
6. Bioactive glass slurry: Weigh 0.875 g of poly(vinyl alcohol) and transfer to a glass beaker. Add water to a volume of 19.1 mL. Stir on a magnetic stirrer at 80 °C for 1 h. Add 10 g of bioactive glass powder (45S5 bioactive glass) and stir further at room temperature for around 2 h.
7. Polyurethane (PU) templates for bioactive glass scaffold fabrication: Cut PU foam into 10-mm^3 cubic shapes. Clean PU cubic foams with DI water and dry at room temperature for 24 h.

2.2 Cell Culture

1. Phosphate-buffered saline (PBS).
2. Collagenase solution: 0.075 % collagenase type I in Dulbecco's modified Eagle's medium (DMEM, D-glucose content: 1 g/L).
3. Fetal calf serum (FCS).
4. Biocoll separating solution.
5. Stem cell growth medium (EM6F): 5 ng/mL selenium, 5 μg/mL transferrin, 4.7 μg/mL linoleic acid, 5 μg/mL insulin, 1 μg/mL ascorbic acid, 1 μg/mL dexamethasone, 34 mL/100 mL MCDB 201 with L-glutamine solution, 51 mL/100 mL DMEM (1 g/L D-glucose), 15 mL/100 mL FCS, 50 IU/mL streptomycin, 50 IU/mL penicillin.
6. Trypsin–ethylendiamine (EDTA) solution: 0.05 % trypsin–0.02 % EDTA in PBS.
7. Insulin-Transferrin-Selenium (ITS+1) stock solution: 10 mg/L insulin, 5.5 mg/L transferrin, 5 μg/L selenium, 0.5 mg/mL bovine serum albumine (BSA), 4.7 μg/mL linoleic acid.
8. Serum-free chondrogenic medium: 4.5 g/L high-glucose DMEM, 1 mM sodium pyruvate, 0.1 mM ascorbic acid-2-phosphate, 0.1 mM dexamethasone, 1 % ITS+1 stock solution, 10 ng/mL recombinant human transforming growth factor-β1 (TGF-β1).
9. Fluorescein diacetate/ethidium bromide (FDA/EtBr) solution: 9 μg/mL FDA and 10 μg/mL EtBr in PBS.

3 Methods

3.1 Scaffold Fabrication

1. 3D porous alginate foams: Add 1 mL of alginate solution into the wells of a 48-well plate and store at room temperature for 30 min to remove air bubbles. Add 100 μL of $CaCl_2{\cdot}2H_2O$

gelation agent into each well. Store at room temperature for gelation for 30 min. Freeze the well plate containing gels at −20 °C for 24 h and transfer to a freeze-dryer for lyophilizing at ~50 °C under vacuum for 24 h.

2. Alginate-coated 45S5 bioactive glass-based scaffolds: 45S5 bioactive glass-based scaffolds are fabricated by the foam replica method first reported in 2006 [21].
   (a) Dip PU cubes into the bioactive glass slurry for 1 min and squeeze out excess slurry by hand. Dry the coated PU cubes in an oven at 60 °C for 12 h. Dip the dried cubes into the slurry and dry again two more times. Transfer the cubes to a furnace, burn out the PU template at 450 °C for 1 h, and sinter at 1100 °C for 4 h.
   (b) Dip sintered bioactive glass-based scaffolds (around 8 mm^3 in dimensions) into the alginate coating solution for 1 min. Dip two more times and dry at room temperature for 24 h.
3. Multilayered scaffolds: Brush the alginate/45S5 bioactive glass adhesive paste onto one side of the alginate-coated bioactive glass-based scaffold. Place rapidly the 3D porous alginate foam (*see* **Note 1**) onto the glued side of the bioactive glass-based scaffold and slightly (manually) press. Immerse the multilayered scaffold into the cross-linking agent for 4 h. Remove and rinse twice with DI water. Dry at room temperature for 24 h.

3.2 Sterilization

Scaffolds can be sterilized by low-temperature (<55 °C) sterilization procedures such as those using plasma.

3.3 Cell Culture

3.3.1 Isolation of Human Bone Marrow-Derived Mesenchymal Stem Cells (hMSCs) from Femoral Heads

1. Curette whole femoral head spongiosa from the interior of the femoral neck and head, transfer to a course mesh, and press the tissue through the pores of the mesh using a stamp. Rinse the spongiosa fragments with iced PBS. Using a syringe, press the liquid cell suspension through a 140-μm pore diameter sterilized filter membrane. Perform every step under sterile conditions.
2. To increase the cell yield, digest the tissue fragments remaining in the sieve for 25 min at 37 °C under gentle rotation using collagenase solution. Stop the digestion by adding 5 % FCS. Filter the cell suspension through the 140-μm pore diameter filter and add it to the aforementioned cell suspension from **step 1**.
3. Wash the isolated cell suspension with PBS and centrifuge it at 200×*g* and 4 °C. Resuspend the purified cell pellet and overlay it cautiously on the biocoll separating solution before centrifuging it (without brake) at 250×*g* and 4 °C. After 20 min, remove the supernatant and extract very carefully the

interphase containing the MSCs using a 2-mL plastic pipette. Add 15 mL of iced PBS to the MSC layer and centrifuge for 5 min at 200 × *g* and 4 °C.

4. Subsequently, resuspend the MSCs in 20 mL of EM6F medium and seed them into T175 culture flasks. Further MSC culturing and expansion should be performed in EM6F at 37 °C, 90 % air humidity, and 5 % CO_2. After 3 days, discard non-adherent cells and culture the adherent MSCs to 80 % confluence, changing the medium every 2–3 days.
5. The cells should be expanded to sufficient numbers for scaffold seeding by regular passaging using standard protocols. Harvest the cells using trypsin–EDTA solution.

3.3.2 Scaffold Seeding

1. Place dried and sterilized 3D porous alginate foam scaffolds (8 mm diameter and 3 mm height (*see* **Note 2**), mean pore size 237 ± 48 μm, and porosity around 92 %) in a 12-well tissue culture plate. Wash and soak the scaffolds with EM6F medium and then incubate them in a 5 % CO_2 incubator at 37 °C overnight. Aspirate carefully the medium from the scaffolds before cell seeding.
2. Upon 80 % confluence, trypsinize hMSCs at passage 3 (*see* **Note 3**) and count them. Calculate the scaffold void volume by multiplying the total volume of the scaffold by the porosity (*see* **Note 4**). Resuspend 25×10^6 cells/mL in a volume of EM6F medium equal to the scaffold void volume. Let the scaffolds directly absorb the prepared cell suspension. Incubate for 1 h to allow cell adherence in the scaffold (*see* **Note 5**) before adding 3 mL of fresh EM6F medium to each well. Incubate the cultures at 37 °C in 5 % CO_2.
3. Subject the cells seeded on the scaffold to chondrogenic differentiation for 14 days (as detailed below). Culture undifferentiated controls for comparison. At specified time points, scaffolds and supernatant can be harvested and analyzed.

3.3.3 Chondrogenic Induction

1. After 24 h of cultivation on the scaffold, add 20 μmol/L of azacytidine (*see* **Note 6**) to the growth medium (EM6F) for 24 h to prepare MSCs for chondrogenic induction.
2. Subsequently, culture MSCs for 14 days in serum-free chondrogenic medium. According to this protocol, TGF-β1 is added 48 h after scaffold seeding and chondrogenic induction medium is used until the end of culture.
3. Collect the scaffolds and supernatants and analyze them after 14 days. Determine the viability by FDA/EtBr staining of the cells (as detailed below).

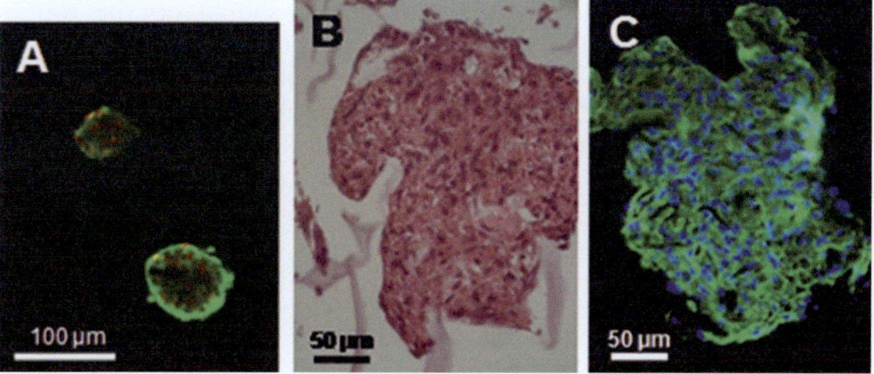

Fig. 2 MSCs cultured for 7 days in alginate scaffolds. The image on the *left side* (**a**) shows cell clusters within an alginate scaffold (cartilage phase) which has been seeded for 7 days with hMSCs (not differentiated). Viable cells are *green* and dead cells stain *red* (magnification 200×, scale bar 100 μm). In the *middle* (**b**), a cell cluster is depicted by Hematoxylin and Eosin staining which reveals deposition of cartilaginous ECM (magnification 400×, scale bar 50 μm). On the *right side* (**c**), the type II collagen deposition is shown in *green* color, surrounding the *blue* stained cell nuclei (magnification 400×, scale bar 50 μm)

3.3.4 Cytocompatibility Testing and Histology

1. To rapidly estimate the cell viability in seeded scaffolds, FDA/EtBr staining can be performed (Fig. 2a). Incubate the scaffolds in FDA/EtBr solution for 2 min in the dark. Rinse the scaffolds gently with PBS (*see* **Note 7**).
2. Put the wet scaffolds on a slide and monitor the green (living cells, FDA) or red (dead cells, EtBr) fluorescence by fluorescence or by confocal laser scanning microscopy.
3. Perform histological staining (Fig. 2b) or immunolabeling for cartilage markers such as type II collagen (Fig. 2c) of paraffin sections after embedding scaffolds in paraffin and following standard protocols.

4 Notes

1. The cylindrical 3D porous alginate foam (10 mm diameter and 8 mm height) produced in the well plates is cut into cube-like shapes (8 mm wide, 8 mm long, and 5 mm high) using a razor blade, in order to match the geometry of the alginate foam to the geometry of the bioactive glass-based scaffold.
2. For the purpose of MSC culture, the 3D porous alginate foam is cut into pieces 8 mm diameter and 3 mm high using a razor blade.

3. MSCs for scaffold seeding should not be too extensively expanded before seeding on the scaffolds since the differentiation capability decreases during culturing. They should be used in rather early passages (< passage 5) and they should not derive from very old donors.

4. The total volume of the scaffold is calculated as follows:

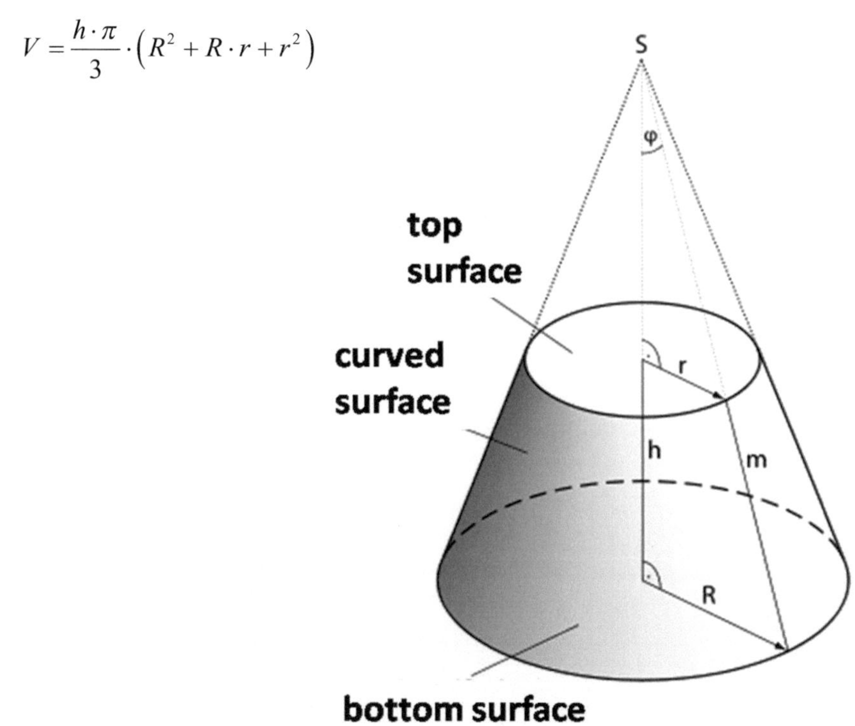

The porosity of alginate foams can be calculated as follows:

$$\text{Porosity}\,(\%) = \left[1 - \left(\frac{\text{Wfoam}}{\text{Vfoam} \times ` \text{alginate}}\right)\right] \times 100$$

where Wfoam is the weight of alginate foam, Vfoam is the total volume of foam according to the dimensions of the foam, and ρalginate is the density of alginate (1.02 g/cm^3).

5. Cell migration from the scaffold to the culture dishes can be avoided by coating the culture dishes with agarose or using low-attachment plates.

6. Azacytidine is used to initiate demethylation of DNA (it is a cytidine analogon, and acts as a DNA methyltransferase inhibitor), so MSCs can subsequently more easily undergo chondrogenic differentiation. Azacytidine can lead to cell cycle arrest and differentiation [22].

7. Gentle handling of the constructs seeded with cells is required to avoid cell loss.

Acknowledgement

The authors acknowledge the financial support from Thai Government Science and Technology Scholarship, for a fellowship for P.N. granted by the Office of the Civil Service Commission (OCSC), Bangkok, Thailand. We would like to thank Benjamin Kohl, Huang Zhao, and Carola Meier for their support.

References

1. Huey DJ, Hu JC, Athanasiou KA (2012) Unlike bone, cartilage regeneration remains elusive. Science 338:917–921. doi:10.1126/science.1222454
2. Nooeaid P, Salih V, Beier JP, Boccaccini AR (2012) Osteochondral tissue engineering: scaffolds, stem cells and applications. J Cell Mol Med 16:2247–2270. doi:10.1111/j.1582-4934.2012.01571.x
3. Hutmacher DW (2000) Scaffolds in tissue engineering bone and cartilage. Biomaterials 21:2529–2543
4. Kon E, Delcogliano M, Filardo G et al (2010) Orderly osteochondral regeneration in a sheep model using a novel nano-composite multilayered biomaterial. J Orthop Res 28:116–124. doi:10.1002/jor.20958
5. Nooeaid P, Roether JA, Weber E et al (2013) Technologies for multilayered scaffolds suitable for interface tissue engineering. Adv Eng Mater. doi:10.1002/adem.201300072
6. Shimomura K, Moriguchi Y, Murawski CD, et al. (2014) Osteochondral tissue engineering with biphasic scaffold: current strategies and techniques. Tissue Eng Part B Rev. 20(5):468–76. doi: 10.1089/ten.teb.2013.0543
7. Liu M, Yu X, Huang F et al (2013) Tissue engineering stratified scaffolds for articular cartilage and subchondral bone defects repair. Orthopedics 36:868–873. doi:10.3928/01477447-20131021-10
8. Martin I, Miot S, Barbero A et al (2007) Osteochondral tissue engineering. J Biomech 40:750–765. doi:10.1016/j.jbiomech.2006.03.008
9. Jeon JE, Vaquette C, Klein TJ, Hutmacher DW (2014) Perspectives in multiphasic osteochondral tissue engineering. Anat Rec (Hoboken) 297:26–35. doi:10.1002/ar.22795
10. Tampieri A, Sandri M, Landi E et al (2008) Design of graded biomimetic osteochondral composite scaffolds. Biomaterials 29:3539–3546. doi:10.1016/j.biomaterials.2008.05.008
11. Naveena N, Venugopal J, Rajeswari R et al (2012) Biomimetic composites and stem cells interaction for bone and cartilage tissue regeneration. J Mater Chem 22:5239. doi:10.1039/c1jm14401d
12. Gerhardt LC, Boccaccini AR (2010) Bioactive glass and glass-ceramic scaffolds for bone tissue engineering. Materials 3:3867–3910. doi:10.3390/ma3073867
13. Berninger MT, Wexel G, Rummeny EJ et al (2013) Treatment of osteochondral defects in the rabbit's knee joint by implantation of allogeneic mesenchymal stem cells in fibrin clots. J Vis Exp 21(75):e4423. doi:10.3791/4423
14. Oldershaw R (2012) Cell sources for the regeneration of articular cartilage: the past, the horizon and the future. Int J Exp Pathol 93:389–400. doi:10.1111/j.1365-2613.2012.00837.x
15. Gardner OF, Archer CW, Alini M et al (2013) Chondrogenesis of mesenchymal stem cells for cartilage tissue engineering. Histol Histopathol 28:23–42
16. Marquass B, Schulz R, Hepp P et al (2011) Matrix-associated implantation of predifferentiated mesenchymal stem cells versus articular chondrocytes: in vivo results of cartilage repair after 1 year. Am J Sports Med 39:1401–1412. doi:10.1177/0363546511398646
17. Pelttari K, Steck E, Richter W (2008) The use of mesenchymal stem cells for chondrogenesis. Injury 39(Suppl 1):S58–S65. doi:10.1016/j.injury.2008.01.038
18. Richter W (2009) Mesenchymal stem cells and cartilage in situ regeneration. J Intern Med 266:390–405. doi:10.1111/j.1365-2796.2009.02153.x
19. Bobis S, Jarocha D, Majka M (2006) Mesenchymal stem cells: characteristics and clinical applications. Folia Histochem Cytobiol 44:215–230
20. Ding DC, Shyu WC, Lin SZ (2011) Mesenchymal stem cells. Cell Transplant 20:5–14. doi:10.3727/096368910X

21. Chen QZ, Thompson ID, Boccaccini AR (2006) 45S5 Bioglass-derived glass-ceramic scaffolds for bone tissue engineering. Biomaterials 27:2414–2425. doi:10.1016/j.biomaterials.2005.11.025

22. Seeliger C, Culmes M, Schyschka L et al (2013) Decrease of global methylation improves significantly hepatic differentiation of Ad-MSCs: possible future application for urea detoxification. Cell Transplant 22:119–131

Part IV

Bioreactors

Chapter 15

Mechanobioreactors for Cartilage Tissue Engineering

Joanna F. Weber, Roman Perez, and Stephen D. Waldman

Abstract

Mechanical stimulation is an effective method to increase extracellular matrix synthesis and to improve the mechanical properties of tissue-engineered cartilage constructs. In this chapter, we describe valuable methods of imposing direct mechanical stimuli (compression or shear) to tissue-engineered cartilage constructs as well as some common analytical methods used to quantify the effects of mechanical stimuli after short-term or long-term loading.

Key words Cartilage, Tissue engineering, Mechanical stimulation, Collagen, Proteoglycans, Chondrocytes, Mechanotransduction

1 Introduction

Articular cartilage of synovial joints is a dense connective tissue mainly comprised of collagen, proteoglycans, and a specialized type of cells called chondrocytes. The primary function of this tissue is to create a near-frictionless surface between the articulating bones while also providing protection against impact [1]. Articular cartilage, however, has a limited capacity for repair leaving it susceptible to damage from trauma or disease such as osteoarthritis (OA) [2]. The progressive loss of articular cartilage associated with OA begins at focal regions with the first pathological changes observed at the articulating surface. Left untreated, these degenerative changes extend to deeper regions in the tissue leading to the full-thickness loss of cartilage at the defect site. While there are surgical techniques available to repair damaged cartilage, few procedures are suitable for the repair of large defects.

The formation of cartilaginous tissue using tissue engineering methods is a promising alternative approach for the repair of damaged cartilage; however, it has been challenging to engineer articular cartilage that possesses similar properties to native tissue [3, 4]. While the engineered tissue constructs can accumulate substantial amounts of cartilaginous extracellular matrix (ECM), it has

Pauline M. Doran (ed.), *Cartilage Tissue Engineering: Methods and Protocols*, Methods in Molecular Biology, vol. 1340, DOI 10.1007/978-1-4939-2938-2_15, © Springer Science+Business Media New York 2015

been difficult to synthesize tissue with levels of ECM constituents comparable with native cartilage and, as a result, the developed constructs display inferior mechanical performance. As the mechanical environment is involved in the development and maintenance of articular cartilage in vivo [5], much attention has focused on the use of mechanical stimuli as a means to upregulate matrix synthesis and to improve tissue properties [6–8]. Previous studies have shown that mechanical stimuli, in the form of compressive or shear loading, can have profound effects on the biosynthetic response of chondrocytes and can serve to impart near-functional properties of the developed tissue constructs. The main objective of this chapter is to describe the methodology of applying different types of mechanical stimuli to tissue-engineered cartilage constructs. These techniques will be described using a simple chondrocyte-seeded hydrogel model which can be readily adapted to other cell-seeded biomaterial scaffolds and will focus on both the application of different loading regimes and some common analytical methods to quantify the effects of mechanical stimuli after short-term or long-term loading.

2 Materials

2.1 Chondrocyte Isolation

1. Articular cartilage tissue: Obtain articular cartilage tissue from bovine metacarpal-phalangeal joints.
2. HEPES stock solution: 1 M 4-(2-hydroxyethyl)-1-piperazineethanesulfonic acid (HEPES). Weigh out 29.79 g of HEPES powder and add to a 200-mL beaker with 100 mL of deionized water. Adjust the pH to 7.4 with 1 N sodium hydroxide. Increase the volume to 125 mL with deionized water (*see* **Note 1**).
3. Culture medium: Ham's F-12 with 25 mM HEPES, pH 7.4. In a 5-L container, add 4 L of deionized water. Add the entire bottle of Ham's F-12 powder (10.7 g/L; Thermo Fisher Scientific) and 125 mL of HEPES stock solution. Mix until the powder has dissolved, and then adjust the pH to 7.4 using 1 N hydrochloric acid and/or 1 N sodium hydroxide. Filter the solution through a 0.2 μm membrane into sterile bottles (*see* **Note 2**) and store at 4 °C. Pre-warm the medium to 37 °C before use.
4. Antibiotic/antimycotic solution: 100× antibiotic/antimycotic solution (containing 10,000 U penicillin, 10 mg streptomycin, and 25 μg amphotericin B per mL).
5. Protease solution: 0.5 % (w/v) protease in culture medium. Weigh 200 mg of microbial protease powder (from *Streptomyces griseus*; activity 3.5 U/mg) into a 50-mL conical tube. Add 20 mL of culture medium. Agitate until dissolved (*see* **Note 3**).

Syringe-filter (0.2 μm) into a sterile 50-mL conical tube. Do not store; protease solution should be used on the same day.

6. Collagenase solution: 0.15 % (w/v) collagenase in culture medium. Weigh 30 mg of microbial collagenase (from *Clostridium histolyticum*; activity >0.15 U/mg) powder into a 50-mL conical tube. Add 20 mL of culture medium. Agitate until dissolved (*see* **Note 3**). Syringe-filter (0.2 μm) into a sterile 50-mL conical tube. Do not store; collagenase solution should be used on the same day.
7. Trypan blue solution: 0.4 % Trypan blue solution.

2.2 Agarose Hydrogel

1. PBS solution: 1 M phosphate-buffered saline (PBS), pH 7.4. Add 800 mL of deionized water to a 1-L glass beaker. Weigh 8 g of sodium chloride, 0.2 g of potassium chloride, 1.44 g of sodium phosphate dibasic, and 0.24 g of potassium phosphate dibasic and transfer to the beaker. Adjust the pH to 7.4 with 1 N hydrochloric acid. Top up the volume to 1 L with deionized water. Autoclave (liquid cycle). Store at room temperature.
2. Agarose powder: Agarose type VII (low-melting-point agarose). Weigh 0.8 g of agarose powder into a 50-mL glass beaker. Add a stirrer bar and cover tightly with aluminum foil. Autoclave and store until use.

2.3 Complete Culture Medium

1. Ascorbic acid stock solution: 50 mg/mL ascorbic acid. Weigh out 1 g of ascorbic acid powder and add to a foil-wrapped 50-mL tube. Add 20 mL of PBS solution and agitate until dissolved. Syringe-filter (0.2 μm) into a clean foil-wrapped tube. Aliquot the solution into sterile black 1.5-mL tubes. Store at −20 °C. Thaw immediately prior to use. Do not refreeze thawed aliquots.
2. Culture medium containing 20 % FBS: Ham's F-12 medium with 20 mM HEPES and 20 % fetal bovine serum (FBS). In a 250-mL filter bottle, add 50 mL of FBS and 200 mL of culture medium, and then filter (*see* **Note 4**). Store at 4 °C. Warm to 37 °C immediately before use.
3. Complete culture medium: Culture medium containing 20 % FBS with 1× antibiotics/antimycotics and 100 μg/mL ascorbic acid. Immediately prior to feeding, add 10 μL/mL of antibiotic/antimycotic solution and 2 μL/mL of ascorbic acid stock solution directly to the prepared culture medium containing 20 % FBS.

2.4 Mechanical Stimulation Equipment

1. Two-axis micromechanical testing machine: e.g., Mach-1 Micromechanical Tester equipped with a 1-kg load cell (V500cs, Biomomentum Inc). The user interface is depicted in Fig. 1 (*see* **Note 5**).

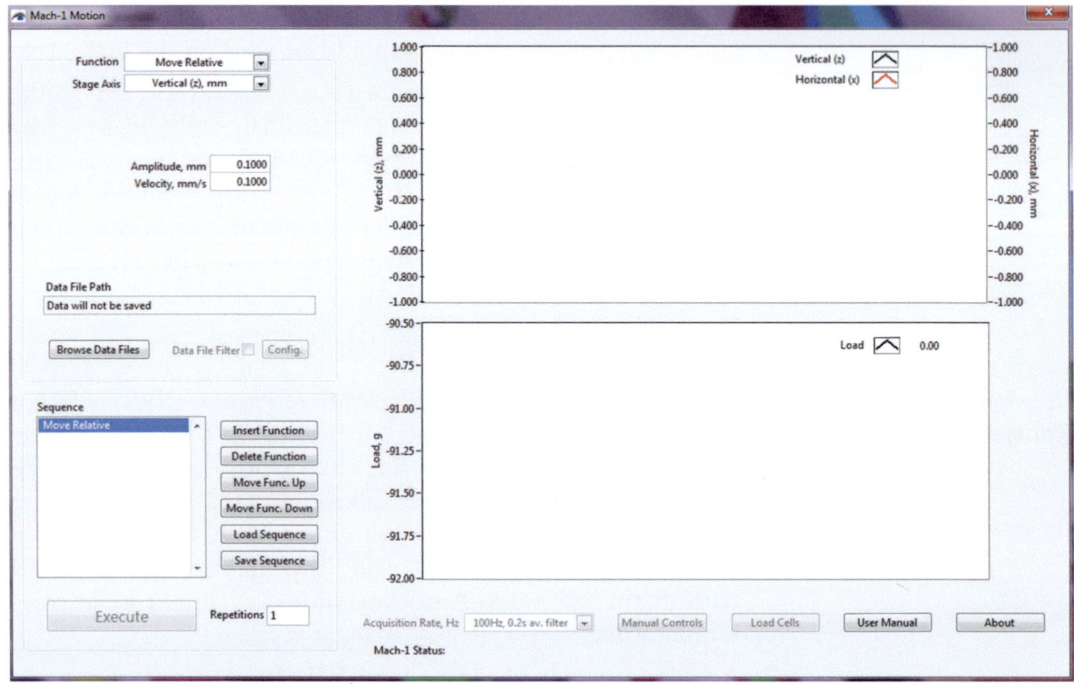

Fig. 1 Screen-shot of the Mach-1 software. The *left side* allows the user to create a program from a series of functions and user-defined parameters. The *right side* tracks real plots of the actuator position (*top*) and load (*bottom*)

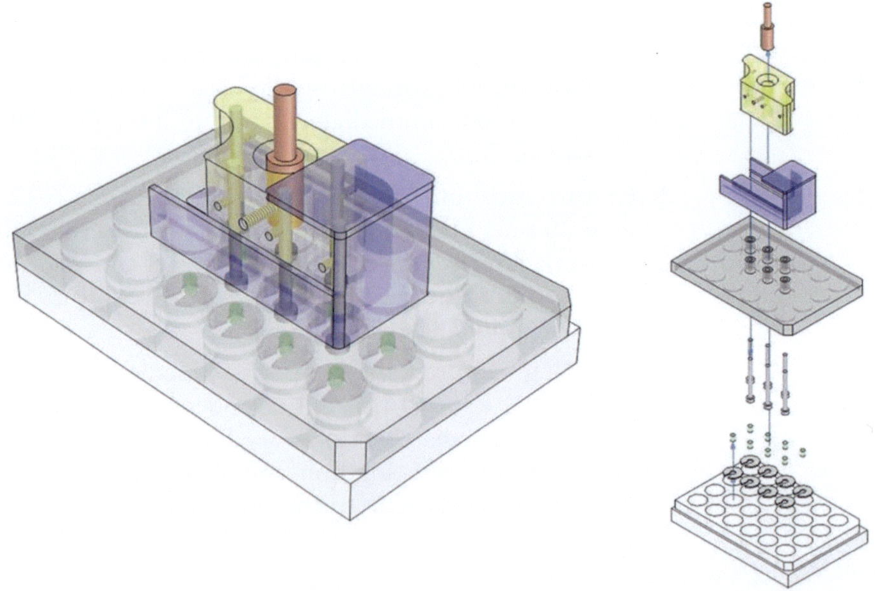

Fig. 2 Schematic of custom-designed loading lid and spacer. Assembled view (*left*), exploded view (*right*). The loading lid fits tightly on a 24-well plate to maintain sterility. The spacer (shown in *blue*), which is removed prior to stimulation, is to ensure that the loading platens do not make contact with the culture plate lid during actuator motion

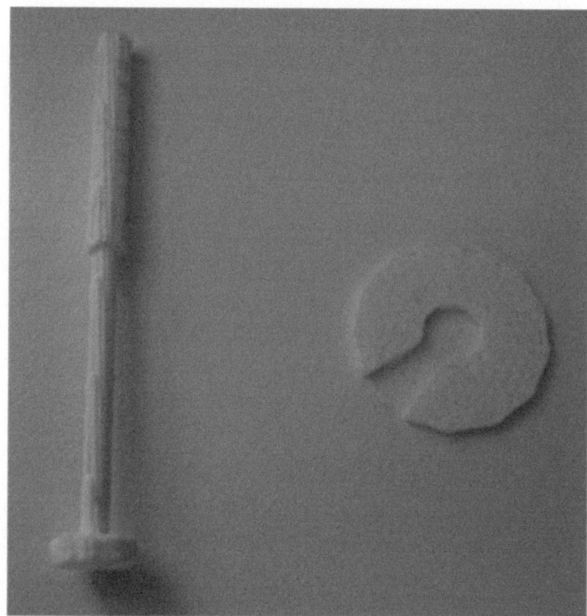

Fig. 3 Rapid prototyped loading pin (*left*) and retaining ring (*right*)

2. Loading lid with removable spacer (Fig. 2).
3. Loading platens (Fig. 3): Semi-porous ended rapid prototyped plastic (ABS) or custom-designed titanium alloy (Ti 6Al 4 V) (*see* **Note 6**).
4. Retaining rings (Fig. 3): Rapid prototyped plastic (ABS) (*see* **Notes** 7 and **8**).

2.5 Digestion Solutions

1. Digestion buffer: 20 mM ammonium acetate, 1 mM ethylenediaminetetraacetic acid, 2 mM dithiothreitol. In a 200-mL beaker on a stirrer plate add 80 mL of deionized water. Weigh out 272 mg of ammonium acetate, 38 mg of ethylenediaminetetraacetic acid, and 31 mg of DL-dithiothreitol and add to the beaker. Stir until dissolved. Adjust the pH to 6.2 using acetic acid and/or sodium hydroxide. Bring up the volume to 100 mL with deionized water. Store at 4 °C.
2. Papain digestion solution: 40 μg/mL papain (from ~25 mg/mL stock; activity 16–40 U/mg protein) (*see* **Note 9**), 20 mM ammonium acetate, 1 mM ethylenediaminetetraacetic acid, 2 mM dithiothreitol. Add 1.6 μL of papain stock solution (from ~25 mg/mL stock) for every 1 mL of digestion buffer.

2.6 Solutions for PicoGreen DNA Assay

1. 1 M Tris–HCl solution: In a 200-mL beaker on a stirrer plate add 80 mL of deionized water. Weigh out 12.14 g of Tris–hydrochloride and add to the beaker. Stir until dissolved. Adjust the pH to 8.0 using 1 N hydrochloric acid. Bring up the volume to 100 mL with deionized water. Store at room temperature.
2. 0.5 M EDTA solution: In a 200-mL beaker on a stirrer plate add 80 mL of deionized water. Weigh out 18.61 g of ethylene-diaminetetraacetic acid and add to the beaker. Stir until dissolved. Adjust the pH to 8.0 using 1 N sodium hydroxide. Bring up the volume to 100 mL with deionized water. Store at room temperature.
3. Tris–EDTA buffer: 10 mM Tris–HCl, 1 mM EDTA, pH 8. In a 200-mL beaker, add 1 mL of 1 M Tris–HCl solution and 200 μL of 0.5 M EDTA solution and bring up the volume to 100 mL with deionized water. Store at room temperature.
4. DNA standard solutions (0–2000 ng/mL): Series of standard solutions containing 0–2000 ng/mL of calf thymus DNA in Tris–EDTA buffer.

2.7 RNA Extraction (See Note 10)

1. EconoSpin silica mini spin column (Epoch Life Science).
2. TRI Reagent: RNA isolation reagent.
3. Guanidine hydrochloride solution: 5 M guanidine hydrochloride, 20 mM Tris–HCl, in 38 % ethanol, pH 5. Weigh out 4.78 g of guanidine hydrochloride into a 15-mL sterile tube. Add 200 μL of nuclease-free 1 M Tris–HCl solution and 3.8 mL of 100 % ethanol. Adjust the pH to 5 using 2 M sodium acetate. Bring up the volume to 10 mL using nuclease-free water.
4. DNase buffer: 1 M NaCl, 20 mM Tris–HCl, 10 mM $MnCl_2$. In a 50-mL tube weigh out 25 mg of manganese(II) chloride and 1.16 g of sodium chloride. Add 400 μL of nuclease-free 1 M Tris–HCl solution. Bring up the volume to 20 mL with nuclease-free water.
5. Column wash buffer: 20 mM NaCl, 2 mM Tris–HCl in 80 % ethanol, pH 7.5. In a 50-mL tube, weigh out 47 mg of sodium chloride. Add 32 mL of 100 % ethanol and 80 μL of nuclease-free 1 M Tris–HCl solution. Adjust the pH to 7.5 using 1 N hydrochloric acid. Bring up the volume to 40 mL with nuclease-free water.

2.8 Solutions for Construct Harvesting and Analyses

1. DNA standard solutions (0–6 μg/mL): Series of standard solutions containing 0–6 μg/mL of calf thymus DNA in PBS solution.
2. 0.2 % Formic acid solution: 0.2 % formic acid in water. In a 1-L beaker, add 1.135 mL of 88 % formic acid and 400 mL of

deionized water. Adjust the pH to 5.3 using formic acid or NaOH. Bring up the volume to 500 mL.

3. DMMB dye solution: In a 15-mL tube protected from light, dissolve 8 mg of dimethylmethylene blue (DMMB) in 5 mL of 100 % ethanol. In a 1-L container protected from light, add the dissolved DMMB and 495 mL of 0.2 % formic acid solution. Store at room temperature and protect from light.
4. Glycosaminoglycan (GAG) stock solution: 10 mg of chondroitin sulfate sodium salt in 1 mL of PBS solution.
5. 1 % BSA solution: 1 % bovine serum albumin (BSA) in PBS. In a 50-mL tube, dissolve 0.3 g of BSA in 30 mL of PBS solution. Do not spin or vortex; use a rocker table only. Prepare fresh.
6. GAG standard solutions: In a 15-mL tube, dilute 60 μL of GAG stock solution in 2940 μL of 1 % BSA solution (final chondroitin sulfate concentration 200 μg/mL). Using this solution, create a series of standard solutions containing 0–120 μg/mL chondroitin sulfate in 1 % BSA.
7. 6 N HCl solution: 6 N HCl in deionized water.
8. 5.7 N NaOH solution: In a 200-mL beaker on a stirrer plate add 80 mL of deionized water. Weigh out 22.80 g of NaOH and add to the beaker. Stir until dissolved. Bring up the volume to 100 mL with deionized water. Store at room temperature.
9. Hydroxyproline standard solutions: Series of standard solutions containing 0–5 μg/mL of L-hydroxyproline (*cis* or *trans*) in deionized water.
10. Hydroxyproline assay buffer: In a 200-mL beaker on a stir plate add 80 mL of deionized water. Weigh out 5 g of citric acid, 7.23 g of sodium acetate, and 3.4 g of NaOH and add to the beaker. Stir until dissolved. To the beaker, add 1.2 mL of glacial acetic acid. Adjust the pH to 6.0 using acetic acid. Bring up the volume to 100 mL with deionized water. Store at room temperature.
11. 0.05 N Chloramine-T solution: In a 50-mL beaker add 4 mL of deionized water. Weigh out 0.282 g of chloramine-T and add to the beaker. Stir until dissolved. To the beaker, add 6 mL of methyl-cellosolve and 10 mL of hydroxyproline assay buffer. Prepare fresh.
12. 3.15 N Perchloric acid solution: 4.6 mL of 70 % perchloric acid in 14.6 mL of deionized water. Preparation should be done in a fume hood that can be used with perchloric acid. Store in a glass container at room temperature.
13. Ehrlich's reagent solution: In a 50-mL beaker, add 20 mL of methyl-cellosolve. Weigh out 4 g of Ehrlich's reagent

(*p*-dimethylaminobenzaldehyde) and add to the beaker. Stir until dissolved. This can be facilitated by heating the mixture to no greater than 60 °C. Prepare fresh.

14. 70 % ethanol solution: 70 % ethanol in deionized water.
15. Paraformaldehyde solution: 4 % paraformaldehyde in PBS solution, pH 7.4. In a fume hood, add 60 mL of warm PBS solution (no more than 60 °C) to a 200-mL beaker on a stirrer/hot plate. Weigh out 4 g of paraformaldehyde powder and add to the beaker. Add a pellet of sodium hydroxide. Once the solution becomes clear, remove from the heat. Once cool, add 10 mL of PBS solution and mix. Adjust the pH to 7.4 using 1 N sodium hydroxide and/or 1 N hydrochloric acid. Top up the volume to 100 mL with PBS. Store at 4 °C for up to 1 month. Warm to 37 °C before use (*see* **Note 11**).

3 Methods

3.1 Isolation of Primary Articular Chondrocytes

1. Harvest cartilage slices into a 120-mm Petri dish with 20 mL of culture medium and 500 μL of antibiotic/antimycotic solution. Rinse the slices with culture medium, and then incubate in 20 mL of protease solution at 37 °C for 1–2 h. Rinse again three times with culture medium, and then incubate with 20 mL of collagenase solution overnight (about 18 h) at 37 °C.
2. Agitate the cell suspension by pipetting (*see* **Note 12**). Filter the suspension through a 200-mesh screen filter into a clean 120-mm Petri dish. Transfer the filtered suspension to a 50-mL conical tube. Wash the Petri dishes and filter with culture medium and add to a 50-mL tube up to a volume of 40 mL.
3. Centrifuge the tube of cell suspension at 400–800 × *g* for 6–8 min. Aspirate the supernatant, ensuring that the cell pellet is not disturbed. Resuspend the cell pellet in 40 mL of culture medium. Repeat the centrifugation, aspiration, and resuspension twice more. Remove a small aliquot (20 μL) of cell suspension and mix 1:1 with trypan blue solution for cell counting.
4. Centrifuge the suspension again. Concurrently, count viable cells using the trypan blue exclusion method with a hemacytometer and an inverted light microscope. Dead cells will take up the dye and appear blue–black in color (*see* **Note 13**).

3.2 Creation of Cell-Seeded Agarose Constructs

1. Aspirate the supernatant from the cell pellet and resuspend the pellet in culture medium to twice the desired cell concentration: 20,000,000 cells/mL.
2. Prepare a molten 4 % agarose solution. On a stirrer/hot plate in the flow hood, add 20 mL of sterile PBS solution to the beaker containing the autoclaved agarose powder and stirrer

bar. Cover tightly with aluminum foil and heat to 120 °C. Maintain at 120 °C until the agarose powder is fully dissolved and thoroughly mixed (the solution will change from cloudy to clear). Reduce the temperature to 50–60 °C.

3. Thoroughly mix equal volumes of cell suspension and agarose solution (*see* **Note 14**). Quickly pipet the mixture into cylindrical molds or a 100-mm Petri dish (*see* **Note 15**). Allow for gelation to occur by maintaining at a temperature below 30 °C. Obtain individual cell-seeded constructs and record the dimensions of representative samples. Discard any samples with a height variance greater than 10 %.
4. Carefully transfer the constructs to a fresh Petri dish or well-plate containing enough complete culture medium to completely submerge the constructs. Culture the constructs undisturbed at 37 °C, 5 % CO_2, and 95 % humidity for 24 h prior to mechanical stimulation.

3.3 Application of Mechanical Stimuli (*See* Note 16)

1. Measure representative cell-seeded agarose constructs for height and diameter. Calculate the required displacements for desired strains from the following relation: $\delta = \varepsilon\, l_o$; where δ is the required actuator displacement, ε is the applied strain, and l_o is the recorded sample height. Compressive and shear strains are determined in the exact same fashion. For example, constructs should be stimulated under a 5–15 % compressive strain amplitude or a 2–10 % shear strain amplitude.
2. In a 24-well plate, place retaining rings (compression loading) or affix sandpaper discs (shear loading) (Fig. 3).
3. Place loading platens in the lid. Ensure that the lid spacer is in place. Secure the platens with set screws (Fig. 2).
4. Carefully transfer the constructs to the wells. Center and orient properly.
5. Add 400 μL of complete culture medium (*see* **Note 17**).
6. Place the loading lid on the culture plate.
7. Establish construct-platen contact by loosening the set screws to drop the platens. Re-tighten the set screws to set the platens (*see* **Note 18**).

3.3.1 Compression Loading [4, 8–10]

1. In the Sequence window, add the following functions such that they occur in this order; all will occur in the vertical axis (*see* **Note 18**) (Fig. 1):
   (a) Zero load.
   (b) Zero position.
   (c) Move relative (first).
   (d) Sinusoid.
   (e) Move relative (second).

2. Enter the following parameters into the Mach-1 software (*see* **Note 19**):
   (a) Move relative (first): amplitude = ½ calculated displacement mm, velocity = 1 mm/s.
   (b) Sinusoid: amplitude = ½ calculated displacement mm, frequency = 1 Hz, cycles = time in seconds for stimulation (e.g., 1200 cycles = 20 min).
   (c) Move relative (second): amplitude = – ½ calculated displacement mm, velocity = 1 mm/s (*see* **Notes 20** and **21**).
3. To save data, enter a file path in the Data File Path box.
4. Place the culture plate in the device. Lower the actuator until it joins with the loading lid. Tighten the screw to attach the loading lid to the actuator (Fig. 2).
5. Remove the spacer. Begin the stimulation program.
6. Once finished, replace the spacer before detaching the loading lid from the actuator.
7. In the flow hood, exchange the loading lid with a well-plate lid and return the culture to the incubator.

3.3.2 Shear Loading [7, 10]

1. In the Sequence window, add the following functions such that they occur in this order (*see* **Note 18**) (Fig. 1):
   (a) Zero load.
   (b) Zero position (vertical).
   (c) Zero position (horizontal).
   (d) Move relative (vertical).
   (e) Move relative (horizontal).
   (f) Sinusoid (horizontal).
   (g) Move relative (horizontal).
   (h) Move relative (vertical).
2. Enter the following parameters into Mach-1 software (*see* **Note 22**):
   (a) Move relative (vertical): amplitude = calculated displacement mm, velocity = 1 mm/s.
   (b) Move relative (horizontal): amplitude = ½ calculated displacement mm, velocity = 1 mm/s.
   (c) Sinusoid (horizontal): amplitude = ½ calculated displacement mm, frequency = 1 Hz, cycles = time in seconds for stimulation (e.g., 1200 cycles = 20 min).
   (d) Move relative (horizontal): amplitude = – ½ calculated displacement mm, velocity = 1 mm/s (*see* **Note 21**).
   (e) Move relative (vertical): amplitude = – calculated displacement mm, velocity = 1 mm/s.

3. To save data, enter a file path in the Data File Path box.
4. Place the culture plate in the device. Lower the actuator until it joins with the loading lid. Tighten the screw to attach the loading lid to the actuator (Fig. 2).
5. Remove the spacer. Begin the stimulation program.
6. Once finished, replace the spacer before detaching the loading lid from the actuator.
7. In the flow hood, exchange the loading lid with a well-plate lid and return the culture to the incubator.

3.4 Radioisotope Incorporation to Determine Collagen and Proteoglycan Synthesis (See Note 23)

1. Immediately after stimulation, add 5 μCi of ^{3}H-proline (to determine collagen synthesis) and 5 μCi of ^{35}S-sulfate (to determine proteoglycan synthesis) to each construct. Mix thoroughly with the medium that is already in the well. Incubate for 24 h.
2. Remove excess medium with a disposable transfer pipet and dispose in liquid radioactive waste. Wash for 5 min in PBS solution three times. Discard the wash solution in liquid radioactive waste. Place each construct into a separate 1.5-mL locking-top tube.
3. Add 600 μL of papain digestion solution. Maintain at 65 °C in a heating block for 48–72 h. Remove from the heating block. Vortex each tube for 30 s.
4. Read the activity with a β-liquid scintillation counter with results in counts per minute (CPM) or disintegrations per minute (DPM), according to the instrument specifications for both ^{3}H and ^{35}S.
5. Scintillation results should be normalized to DNA content (i.e., ^{3}H-proline CPM/μg DNA) using the PicoGreen DNA assay.

3.5 PicoGreen DNA Assay [11]

1. Dilute aliquots of construct digest six times with Tris–EDTA buffer: for triplicate samples, add 50 μL of digest to 250 μL of Tris–EDTA buffer.
2. Prepare PicoGreen dye fresh according to the manufacturer's directions.
3. Add 100 μL of sample or DNA standard solution (0–2000 ng/mL) to each well of a black 96-well plate.
4. Add 100 μL of dye to each well quickly using a multichannel pipet. Carefully agitate the plate to mix. Incubate for 3–5 min at room temperature protected from light.
5. Read on spectrofluorometer: 480/520 nm (excitation/emission). Construct a standard curve based on the measured fluorescence of the DNA standard solutions and use this to determine sample DNA concentrations.

3.6 Preparing Stimulated Samples for Gene Expression Analyses (*See* Note 24)

Samples must be pretreated as follows in order to remove residual agarose which tends to coprecipitate with the RNA when using TRI Reagent extractions [12].

1. Snap freeze harvested constructs with liquid nitrogen. Store at −80 °C until analysis.
2. Crush the frozen sample with a mortar and pestle. Add 1 mL of TRI Reagent and homogenize with a homogenizer for 1–2 min. Let the homogenate stand at room temperature for 5 min.
3. Add 200 μL of chloroform to the homogenate and vortex. Let stand at room temperature for 10 min. Centrifuge at 12,000 × *g* for 10 min.
4. Collect the supernatant. Deposit the supernatant into a clean 1.5-mL tube with 250 μL of 100 % ethanol. Transfer the mixture to an EconoSpin silica mini spin column and centrifuge at 21,000 × *g* for 30 s. Discard the elutant.
5. Add 500 μL of guanidine–HCl solution to the spin column with 10 U of DNase I (amplification grade) in 100 μL of DNase buffer. Centrifuge at 8000 × *g* for 15 s. Discard the elutant.
6. Add 500 μL of column wash buffer. Centrifuge at 8000 × *g* for 15 s. Discard the elutant.
7. Add 300 μL of column wash buffer. Centrifuge at 21,000 × *g* for 2 min. Discard the elutant.
8. Attach a new 1.5-mL tube to the spin column. Elute RNA with 50 μL of nuclease-free water, centrifuging at 8000 × *g* for 1 min.
9. Assess the total RNA concentration and purity on a NanoDrop spectrophotometer.
10. Use an RNA to cDNA kit to synthesize cDNA. Follow the manufacturer's instructions.
11. Conduct PCR analyses (real-time quantitative PCR or conventional PCR) for the desired gene(s) of interest.

3.7 Determination of DNA, Proteoglycan, and Collagen Accumulation After Long-Term Stimulation

3.7.1 Construct Mass, Water Content, and Sample Digestion

1. Harvest the constructs, wash in PBS solution, and blot off excess liquid.
2. Measure the construct weight (wet mass).
3. Lyophilize overnight and weigh again (dry mass).
4. Determine the percentage water content from: $\%H_2O = \left[\left(m_w - m_d\right) / m_w\right] \times 100$, where $\%H_2O$ is the water content, m_w is the construct wet weight, and m_d is the construct dry weight.
5. Transfer the construct to a 1.5-mL tube and add 600 μL of papain digestion solution. Maintain at 65 °C in a heating block for 48–72 h. Remove from the heating block. Vortex each tube for 30 s (*see* **Note 25**).

3.7.2 DNA Content: Hoechst 33258 Assay [13]

1. Dilute aliquots of construct digest with PBS solution. Make up 150 μL of diluted sample for triplicate readings.
2. Add 50 μL of sample or DNA standard solution (0–6 μg/mL) per well to a black 96-well plate.
3. Prepare Hoechst 33258 dye according to the manufacturer's directions and protect from light.
4. While protected from light, add 200 μL of dye to each well using a multichannel pipet. Thoroughly mix.
5. Read immediately on a spectrofluorometer: 350/450 nm (excitation/emission). Construct a standard curve based on the measured fluorescence of the DNA standard solutions and use this to determine sample DNA concentrations.
6. Report DNA content as μg/construct or normalized to construct mass (wet or dry weight) as μg/mg.

3.7.3 Proteoglycan Content: Dimethylmethylene Blue Assay [14] (See Notes 24 *and* 26*)*

1. Prepare dimethylmethylene blue dye according to the manufacturer's directions.
2. Dilute construct digest with 1 % BSA solution if necessary. For triplicate readings, 30 μL is required.
3. Add 10 μL of sample or GAG standard solution per well to a transparent 96-well plate.
4. Quickly add 200 μL of DMMB dye solution to each well with a multichannel pipet. Gently mix (*see* **Note 27**).
5. Immediately read the absorbance on a spectrophotometer at 525 nm (*see* **Note 28**). Construct a standard curve based on the measured absorbance of the GAG standard solutions and use this to determine sample proteoglycan concentrations.
6. Report the proteoglycan content as μg/construct or normalized to construct mass (wet or dry weight) as μg/mg or to DNA content as μg/μg.

3.7.4 Collagen Content: Hydroxyproline Assay [15] (See Note 24*)*

1. Hydrolyze 100 μL aliquots of construct digest with 100 μL of 6 N HCl solution for 18 h at 110 °C in glass tubes with Teflon-lined caps. Neutralize with 100 μL of 5.7 N NaOH solution. Bring up the volume to 1 mL with distilled water (*see* **Note 29**).
2. While protected from the light, into each well of a transparent 96-well plate, add 100 μL of sample or hydroxyproline standard solution. Add 50 μL of 0.05 *N* chloramine-T solution, mix, and let stand for 20 min. Add 50 μL of 3.15 *N* perchloric acid solution, mix, and let stand for 5 min. Add 50 μL of freshly prepared Ehrlich's reagent solution, mix, and incubate for 20 min at 60 °C.
3. Remove from the heat and cool at 4 °C for 5 min. Let stand at room temperature until the color stabilizes (about 30 min).

4. Read the absorbance on a spectrophotometer at 560 nm. Construct a standard curve based on the measured absorbance of the hydroxyproline standard solutions and use this to determine the sample hydroxyproline concentrations.
5. Calculate the collagen content assuming that hydroxyproline accounts for 10 % of total collagen weight [16].
6. Report the collagen content as μg/construct or normalized to construct mass (wet or dry weight) as μg/mg or to DNA content as μg/μg.

3.7.5 Fixation for Subsequent Histological and/or Immunohistochemical Analyses (See Note 24*)*

1. Submerge the harvested construct in paraformaldehyde solution. Incubate at 4 °C for 24 h.
2. Remove the paraformaldehyde solution and wash with PBS solution. Transfer the sample to 70 % ethanol solution.
3. Embed in paraffin, cut 5-μm sections, deparaffinize, and stain appropriately (*see* **Notes 30** and **31**).

4 Notes

1. Excess HEPES stock solution may be kept at 4 °C.
2. The same procedure should be followed with other culture medium such as DMEM or CBM which have been extensively used in the culturing of chondrocytes.
3. To avoid bubbles, agitation may be performed on a rocker plate. If quick dissolution is desired, vortexing can be carried out, but wait for the bubbles to subside before filtering.
4. Other concentrations of FBS may be used. This would depend on the concentration of cells in the construct as well as the size of the construct.
5. Any small mechanical testing system should work, provided that the resolution of the system will allow for accurate control of loading and that the sample is able to be kept at physiologic temperatures (either through the use of a bath system or by locating the testing apparatus inside an incubator) and in sterile conditions.
6. These platens may be of single use or multi-use. Ensure that they are not so heavy that the weight of the platen will destroy the sample. Platens made of different materials can allow for applying different magnitudes of preload.
7. Depending on the sample size, retaining rings may be required to ensure a stable construct position during mechanical stimulation. These rings can be disposable or reusable depending on the culture format desired (e.g., disposable for radioisotope work).
8. For shear loading, extra friction is needed in the bottom of the wells to ensure that the constructs remain stationary. This can

be achieved by attaching precut discs of sterilized (i.e., autoclaved) waterproof sandpaper (coarse grit) to the bottom of the wells in the well plate.

9. Check the papain concentration, as it varies from batch to batch. Adjust accordingly to achieve a final 40 μg/mL concentration of papain in the digestion solution.
10. Ensure that nuclease-free products are used when isolating RNA. Use gloves to protect samples and tools from extraneous nucleases.
11. Filtering paraformaldehyde before use will remove small particles that may show up as artifacts in paraffin-embedded samples.
12. After overnight digestion, the cartilage slices should be almost completely dissolved. Small tissue pieces are commonly left behind and will be filtered out.
13. Ensure that the sample is mixed thoroughly to obtain the most accurate count. Cell viability should be 95 % or greater.
14. Mix carefully to avoid the creation of large bubbles that may affect cell viability and result in the premature failure of the construct during mechanical stimulation.
15. Constructs can be made with individual cylindrical Teflon molds (4 mm in height by 4 mm in diameter), or can be cut out of a larger molded gel (made in a 100-mm Petri dish) with a 4-mm biopsy punch after gelation has occurred. For stimulation purposes, constructs should be approximately 4 mm in height by 4 mm in diameter.
16. Before beginning the stimulations, it is important to decide on the type of stimulation(s) desired—amplitude (strain), frequency (Hz), duration (time), as well as the length of the experiment (i.e., single application, or repeated applications). For experiments containing repeated applications of stimulation, the rest period should also be determined prior to starting. It is also important to determine the type of analytical techniques needed before proceeding. Short-term techniques are suitable for single-application experiments as well as for determining the effect that the final stimulation has in a series of applications. Long-term techniques are more suited to long experimental times (weeks), since long-term experiments are more likely to have a measurable effect on accumulation of ECM molecules. Long-term techniques are also more suited to observe the cumulative effect of a series of applications of stimulation.
17. This should be enough to just barely reach the top of the construct. Excess medium may cause the construct to float making positioning for stimulation difficult. Constructs should always be fed prior to stimulation.

18. By using independent loading pins, initial contact with the specimen is easily established. Otherwise, prior to stimulation, a "Find Contact" function must be included. In this case, the find contact sequence should be run at low speed (to avoid overshoot) with a target load of ~0.5 g.
19. Ensure that the stage axis in use is vertical.
20. Ensure that the positive and negative directions of the system correspond to the actuator moving down and up, respectively. Otherwise, adjust accordingly.
21. If the haversine function is available, it may be used in place of the above three functions. Ensure that the correct (full) displacement for strain value is used instead of half.
22. Ensure that the stage axis in use corresponds with the desired displacement.
23. The purchase and disposal of radioactive materials (i.e., waste isotopes and items that may have come into contact radioactive materials) must follow the relevant institutional and/or governmental policies and procedures for the safe handling and disposal of radioactive substances.
24. Typically, mechanical stimulation conducted under the conditions outlined in this chapter will elicit an anabolic response from the cells. In terms of gene expression, this anabolic response can be determined from upregulated genes, including collagen type II (COL2), collagen type X (COLX), aggrecan (ACAN), proteoglycan 4 (PRG4), and the transcription factor SOX9. Similarly, mechanical stimulation will also result in increased synthesis and accumulation of collagen and proteoglycans, which can be quantified biochemically (hydroxyproline and DMMB assays, respectively) or through immunohistochemical staining of thin sections. For immunohistochemical staining, more intense staining for collagen II and aggrecan (without an associated change in collagen I staining) will be expected. The reader is also directed to more extensive reviews on this subject [17, 18].
25. Constructs can be digested immediately after harvesting.
26. The DMMB assay can be used as a surrogate measure of proteoglycan content in cartilaginous tissues or cultures [14]. This assay quantifies the concentration of chondroitin sulfate glycosaminoglycan which is the primary constituent of cartilage proteoglycans (e.g., aggrecan).
27. The presence of BSA will tend to promote bubble formation. Avoid the creation of bubbles as this will result in incorrect readings.
28. It is important to read the plate immediately as the complex is unstable and tends to precipitate quickly. Precipitates will result in incorrect readings.

29. This results in a 1/10 diluted sample.
30. Common histological stains for cartilage: hematoxylin and eosin (general connective tissue stain), safranin-O (proteoglycan stain), picrosirius red (collagen stain).
31. Common immunohistochemical stains for cartilage: collagen types I, II, and X, aggrecan.

References

1. Mankin H, Mow V, Buckwalter J et al (1994) Form and function of articular cartilage. In: Simon S (ed) Orthop. Basic Sci. chapter 1. American Academy of Orthopaedic Surgeons, Columbus, OH, pp 1–44
2. Hunziker E (1999) Articular cartilage repair: are the intrinsic biological constraints undermining this process insuperable? Osteoarthr Cartil 7:15–28
3. Buschmann MD, Gluzband YA, Grodzinsky AJ et al (1992) Chondrocytes in agarose culture synthesize a mechanically functional extracellular matrix. J Orthop Res 10:745–758. doi:10.1002/jor.1100100602
4. Waldman SD, Grynpas MD, Pilliar RM, Kandel RA (2002) Characterization of cartilagenous tissue formed on calcium polyphosphate substrates in vitro. J Biomed Mater Res 62:323–330. doi:10.1002/jbm.10235
5. Grodzinsky A, Levenston M (2000) Cartilage tissue remodeling in response to mechanical forces. Annu Rev Biomed Eng 2:691–713
6. Quinn T, Grodzinsky A, Buschmann M et al (1998) Mechanical compression alters proteoglycan deposition and matrix deformation around individual cells in cartilage explants. J Cell Sci 111:573–583
7. Waldman SD, Spiteri CG, Grynpas MD et al (2003) Long-term intermittent shear deformation improves the quality of cartilaginous tissue formed in vitro. J Orthop Res 21:590–596. doi:10.1016/S0736-0266(03)00009-3
8. Waldman SD, Spiteri CG, Grynpas MD et al (2004) Long-term intermittent compressive stimulation improves the composition and mechanical properties of tissue-engineered cartilage. Tissue Eng 10:1323–1331. doi:10.1089/ten.2004.10.1323
9. Kaupp JA, Weber JF, Waldman SD (2012) Mechanical stimulation of chondrocyte-agarose hydrogels. J Vis Exp 68, e4229. doi:10.3791/4229
10. Waldman SD, Spiteri CG, Grynpas MD et al (2003) Effect of Biomechanical Conditioning on Cartilaginous Tissue Formation in Vitro. J Bone Joint Surg 85:101–105
11. McGowan K (2002) Biochemical quantification of DNA in human articular and septal cartilage using PicoGreen® and Hoechst 33258. Osteoarthr Cart 10:580–587. doi:10.1053/joca.2002.0794
12. Mio K, Kirkham J, Bonass WA (2006) Tips for extracting total RNA from chondrocytes cultured in agarose gel using a silica-based membrane kit. Anal Biochem 351:314–316. doi:10.1016/j.ab.2005.12.031
13. Cesarone CF, Bolognesi C, Santi L (1979) Improved microfluorometric DNA determination in biological material using 33258 Hoechst. Anal Biochem 100:188–197, 10.1016/0003-2697(79)90131-3
14. Farndale RW, Buttle DJ, Barrett AJ (1986) Improved quantitation and discrimination of sulphated glycosaminoglycans by use of dimethylmethylene blue. Biochim Biophys Acta 883:173–177
15. Woessner JF (1961) The determination of hydroxyproline in tissue and protein samples containing small proportions of this imino acid. Arch Biochem Biophys 93:440–447
16. Brandt K, Doherty M, Lohmander L (1998) Composition and structure of articular cartilage. Osteoarthritis. Oxford University Press, New York, NY, pp 110–111
17. Grad S, Eglin D, Alini M, Stoddart MJ (2011) Physical stimulation of chondrogenic cells in vitro: a review. Clin Orthop Relat Res 469:2764–2772
18. Brady M, Waldman SD, Either CR (2014) The application of multiple biphysical cues to engineer functional neocartilage for treatment of osteoarthritis (Part I: cellular response). Tissue Eng Part B Rev. doi:10.1089/ten.teb.2013.0757

Chapter 16

Shear and Compression Bioreactor for Cartilage Synthesis

Kifah Shahin and Pauline M. Doran

Abstract

Mechanical forces, including hydrodynamic shear, hydrostatic pressure, compression, tension, and friction, can have stimulatory effects on cartilage synthesis in tissue engineering systems. Bioreactors capable of exerting forces on cells and tissue constructs within a controlled culture environment are needed to provide appropriate mechanical stimuli. In this chapter, we describe the construction, assembly, and operation of a mechanobioreactor providing simultaneous dynamic shear and compressive loading on developing cartilage tissues to mimic the rolling and squeezing action of articular joints. The device is suitable for studying the effects of mechanical treatment on stem cells and chondrocytes seeded into three-dimensional scaffolds.

Key words Cyclical shear force, Dynamic compression, Mechanical loading, Mechanobioreactor, Simultaneous shear and compression, Synovial joint

1 Introduction

The function of articular cartilage is to bear the mechanical loads and resist the cyclic compressive and tensile stresses generated in moving joints. The ability of cartilage to perform these tasks depends directly on the molecular and structural characteristics of cartilage extracellular matrix (ECM). The orientation and organization of networks of type II collagen fibrils play an important role in determining the strength and tensile stiffness of cartilage [1]. The bottlebrush shape of aggrecan, the dominant proteoglycan in articular cartilage, and its interactions with hyaluronic acid and link protein within the matrix mediate the ability of cartilage to counteract compressive loading. Water drawn into the tissue through the electrostatic activity of negatively charged anions on the glycosaminoglycan (GAG) chains of aggrecan causes the matrix to expand and swell and thus resist deformation [1, 2].

The formation and functional properties of articular cartilage are influenced significantly by mechanical stimuli. In vivo, the experience of mechanical forces affects cellular hyaluronan synthesis, cartilage matrix deposition, and joint development [3].

Pauline M. Doran (ed.), *Cartilage Tissue Engineering: Methods and Protocols*, Methods in Molecular Biology, vol. 1340, DOI 10.1007/978-1-4939-2938-2_16, © Springer Science+Business Media New York 2015

Depending on the type, magnitude, frequency, and duration of the forces applied, mechanical stimuli also modulate chondrogenic differentiation and production of cartilaginous tissues in three-dimensional in vitro culture systems. Many different forms of mechanical stimulus can be exerted on developing tissue constructs, such as hydrodynamic shear, hydrostatic pressure, compression, and tensile forces (reviewed in ref. 4). Of these, the most commonly applied mechanical treatment in cartilage tissue engineering is uniaxial compression. Dynamic compressive forces applied to chondrocyte cultures with cyclical frequency have been found to stimulate collagen and/or GAG production [5, 6]. Stimulatory effects on chondrogenesis have also been observed in scaffold-seeded mesenchymal stem cells subjected to dynamic compression [7–10].

To mimic more closely the mechanical forces exerted on cartilage during the rolling and squeezing action of articular joints, transient shear or sliding forces are required in addition to compressive loading. Compared with unstimulated controls, studies of the effects of these combined forces on chondrocyte and stem cell cultures have demonstrated beneficial outcomes for expression of chondrocytic marker genes, synthesis of cartilage matrix, distribution of matrix components, and construct mechanical properties [11–14]. Simultaneous compression and sliding or frictional forces have also been combined with fluid perfusion to simulate the hydrodynamic effects of synovial fluid entering the tissue during joint activity [15].

Several factors must be considered when interpreting the results from mechanostimulation experiments. Because dynamic loading of cartilage enhances the rate of mass transfer of vital components such as oxygen, nutrients, and growth factors within the tissue [16, 17], the direct stimulatory effects of mechanical treatment need to be distinguished from the consequences of improved transport of trophic factors to the cells in treated samples [14, 18]. Attention is also required to account for any increase in the amount of cartilage components escaping from the construct into the culture medium due to physical manipulation of the tissue [5, 14]. Because scaffold materials and accumulating ECM dissipate the energy exerted during mechanical loading, the degree to which the mechanical forces applied are modulated by the embedding matrix is also an important factor affecting the cellular response [8, 19].

In this chapter, we describe the construction and operation of a mechanobioreactor capable of exerting simultaneous cyclic shear and compressive forces within a liquid environment to mimic the conditions found in synovial joints [14]. The device is fabricated using readily available components and typical workshop machining and glass-making tools and equipment. The bioreactor is designed to apply intermittent shear loading on cartilage constructs at the same time as a dynamic compressive strain of tuneable amplitude and frequency superimposed on a static axial compressive strain also of tuneable magnitude. The device allows study of the long-term effects of mechanical stimulation, as the treated constructs are submerged in culture medium that is constantly stirred and can be replenished as required. The bioreactor has been used to study the effects of mechanical stimulation on chondrocytes seeded into polyglycolic acid (PGA) and PGA–alginate scaffolds [14]; however, other types of porous fibrous or hydrogel scaffold may be applied. In our work, stimulation of chondrocyte-seeded constructs using the mechanobioreactor for up to 2.5 weeks led to substantial improvements in collagen type II and GAG levels relative to shaking T-flask and mechanobioreactor control cultures without loading [14].

2 Materials

2.1 Bioreactor Components

Components for construction of the bioreactor are illustrated in Figs. 1 and 2.

1. Bioreactor vessel (Fig. 1a): 1.5-L cylindrical vessel made of autoclavable glass with a flat, 2.5-cm-wide flange at the top to support a head plate. The height and inner diameter of the vessel should be approximately 20 and 10 cm, respectively (*see* **Note 1**). A 3.5-mm-diameter, 5-cm-long glass pipe is attached through the wall at the bottom of the vessel using glass-blowing tools. This pipe can be used for medium sampling during bioreactor operation.
2. Head plate (Fig. 1b): Round 10-mm-thick stainless steel plate custom-cut to a diameter of 15 cm. Holes are cut in the head plate during assembly of the bioreactor (Subheading 3.1).
3. Head plate gasket (Fig. 1c): Custom-cut 2.5-cm-wide silicone rubber seal with a bore diameter of 10 cm (*see* **Note 2**).
4. Base plate (Fig. 1d): Custom-cut stainless steel plate, $9 \times 5 \times 1$ cm. Part of the plate surface (55×28 mm) is recessed by 2 mm to accommodate the tissue holding plate (described below). The regions of the recessed area directly under the holes in the tissue holding plate are each perforated with fifteen 2-mm-diameter holes (*see* **Note 3**). Other holes and grooves are made in the base plate as the bioreactor is assembled (Subheading 3.1).
5. Tissue holding plate (Fig. 1e): Custom-cut stainless steel plate, $55 \times 28 \times 2$ mm, with three aligned and equally spaced 15-mm-diameter holes (*see* **Note 4**). Two 5-mm-long curved prongs welded to the side of each hole prevent the compressed tissues from sliding out of the holes during shear loading.

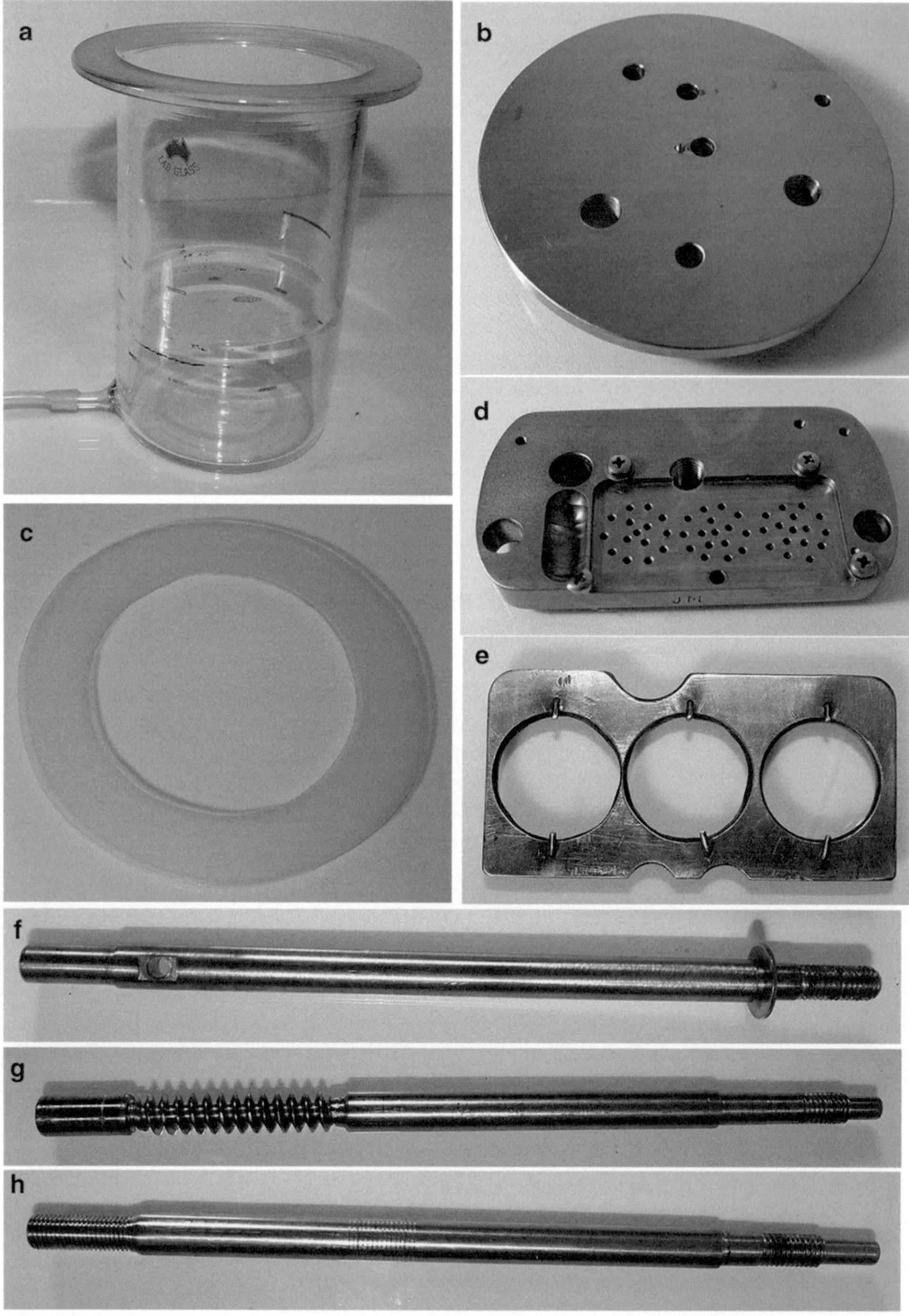

Fig. 1 Bioreactor components: (**a**) bioreactor vessel, (**b**) head plate, (**c**) head plate gasket, (**d**) base plate, (**e**) tissue holding plate, (**f**) axle support rod, (**g**) drive shaft, and (**h**) base plate control arm

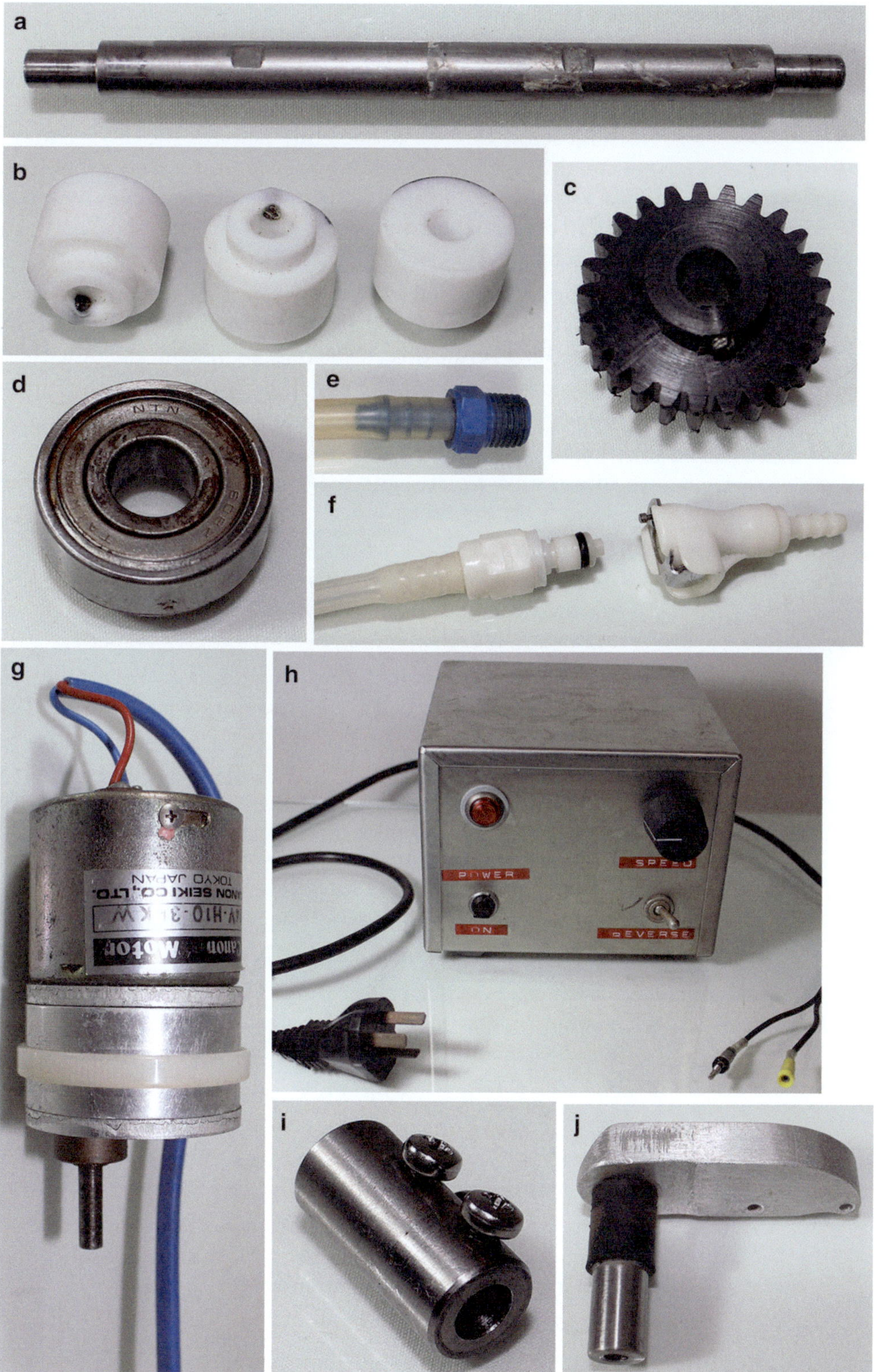

Fig. 2 Bioreactor components: (**a**) roller axle, (**b**) rollers with cams, (**c**) worm gear, (**d**) flat ball bearing, (**e**) plastic tubing adapter, (**f**) one-way snap-on tubing connector, (**g**) motor, (**h**) transformer and control unit, (**i**) shaft sleeve connector, and (**j**) motor support arm

6. Axle support rods (Fig. 1f): Two stainless steel rods, 18.5 cm long × 10 mm diameter. Each rod is threaded with 20-mm-long screw threads at one end. Towards the other end, a hole cut through the rod 3 cm away from the end is fitted with a 3.5-mm-bore bushing to form a mounting point for the roller axle (described below).
7. Drive shaft (Fig. 1g): Stainless steel rod, 19.5 cm long × 10 mm diameter, threaded with a 4.5-cm-long worm (three threads per cm) starting 2 cm from one end, and threaded with a 1-cm-long screw thread starting 1 cm from the other end.
8. Base plate control arm (Fig. 1h): Stainless steel rod, 19.5 cm long × 10 mm diameter, threaded with a 2-cm-long screw thread (ten threads per cm) at one end (the end that screws into the base plate) and a 1-cm-long screw thread starting at 1 cm from the other end.
9. Roller axle (Fig. 2a): Stainless steel rod, 9 cm long × 5 mm diameter. The diameter of the axle is narrowed to 3.5 mm towards the ends so that it fits through the bushings on the axle support rods.
10. Rollers with cams (Fig. 2b): Three Teflon® rollers with diameter 2 cm and length 1.5 cm, cut to give a smooth surface for tissue compression. Each roller surface is shaped to create a 0.1-mm cam spanning one-quarter of the surface area. Each roller has a 5-mm-diameter hole through its center for passage of the roller axle. The rollers have a small extension, 1 cm in diameter and 0.5 cm long, on one side to allow fixing to the roller axle using a small screw.
11. Worm gear (Fig. 2c): 3 cm diameter with 25 external spurs, each about 3 mm in length (*see* **Note 5**).
12. Bearings: Two flat ball bearings (Fig. 2d) and one thrust ball bearing, to facilitate rotational movement of the drive shaft and base plate control arm through the head plate.
13. Silicone tubing: 3.2 mm bore.
14. Plastic tubing adaptors (Fig. 2e): Autoclavable, with a screw thread at one end and a barbed tubing connector to fit 3.2-mm-bore tubing at the other end.
15. One-way snap-on tubing connectors (Fig. 2f): Autoclavable, to fit 3.2-mm-bore tubing, used for connecting venting filters and medium sampling (*see* **Note 6**).
16. Marked plastic control knob: Autoclavable.
17. Venting filters: Two 0.20-μm gas filters.
18. Motor (Fig. 2g): 24-V direct current (DC) motor, e.g., Canon® H10-3RKW.
19. Transformer and control unit (Fig. 2h): Custom-made transformer box that connects to AC power and generates a DC

output of 0–24 V, with controls to change the output voltage (speed) and current direction (reverse).

20. Shaft sleeve connector (Fig. 2i): 2-cm-long stainless steel tube used to connect the shaft from the motor to the drive shaft. Two screw holes are tapped through the tube for securing the shafts inside the tube.
21. Motor support arm (Fig. 2j): Horizontal stainless steel plate, 7×2.5× 1 cm (for attachment to the flat underside of the motor using small screws) connected perpendicularly to a 3-cm-long stainless steel post (for attachment to the head plate using a screw). The plate and post are connected into an L-shape using a 1-cm-long cylinder of thick rubber padding to absorb vibrations from the motor.
22. Small air pump: For example, Rena® model 301.

2.2 Cell Culture

1. Cartilage tissue samples: Up to three cylindrical cartilage constructs or samples, 15 mm in diameter. The sample height can be varied for different applications but should be within the range of 1–6 mm.
2. DMEM base medium: Dulbecco's modified Eagle's medium (DMEM) containing 4.5 g/L of glucose and 584 mg/L of L-glutamine with 3.7 g of sodium hydrogen carbonate, 2.39 g of *N*-2-(hydroxyethyl)piperazine-*N′*-2-ethane sulfonic acid (HEPES), 0.046 g of L-proline, and 10 mL of 100× nonessential amino acid solution (Sigma-Aldrich) made up to 895 mL in Milli-Q water. Sterilize by filtration using 0.2-μm pressure-filtration units.
3. Antibiotic solution: 10,000 U/mL penicillin, 10 mg/mL streptomycin, and 25 μg/mL amphotericin B.
4. Ascorbic acid solution: 50 mg/mL of ascorbic acid in Milli-Q water. Filter sterilize using a 0.2-μm single-use syringe filter.
5. Complete culture medium (CCM): 100 mL of fetal bovine serum (FBS), 5 mL of antibiotic solution, and 1 mL of freshly prepared ascorbic acid solution added just before use to 894 mL of DMEM base medium.

3 Methods

3.1 Bioreactor Assembly

Photographs of the bioreactor at different stages of assembly are shown in Figs. 3 and 4.

1. Tap a thread through the middle of the base plate and screw the base plate control arm into the tapped hole (Fig. 3a).
2. Insert the roller axle through the three rollers with cams, and then through the worm gear (Fig. 3b). Fix the rollers and the gear to the axle using screws (*see* **Note** 7).

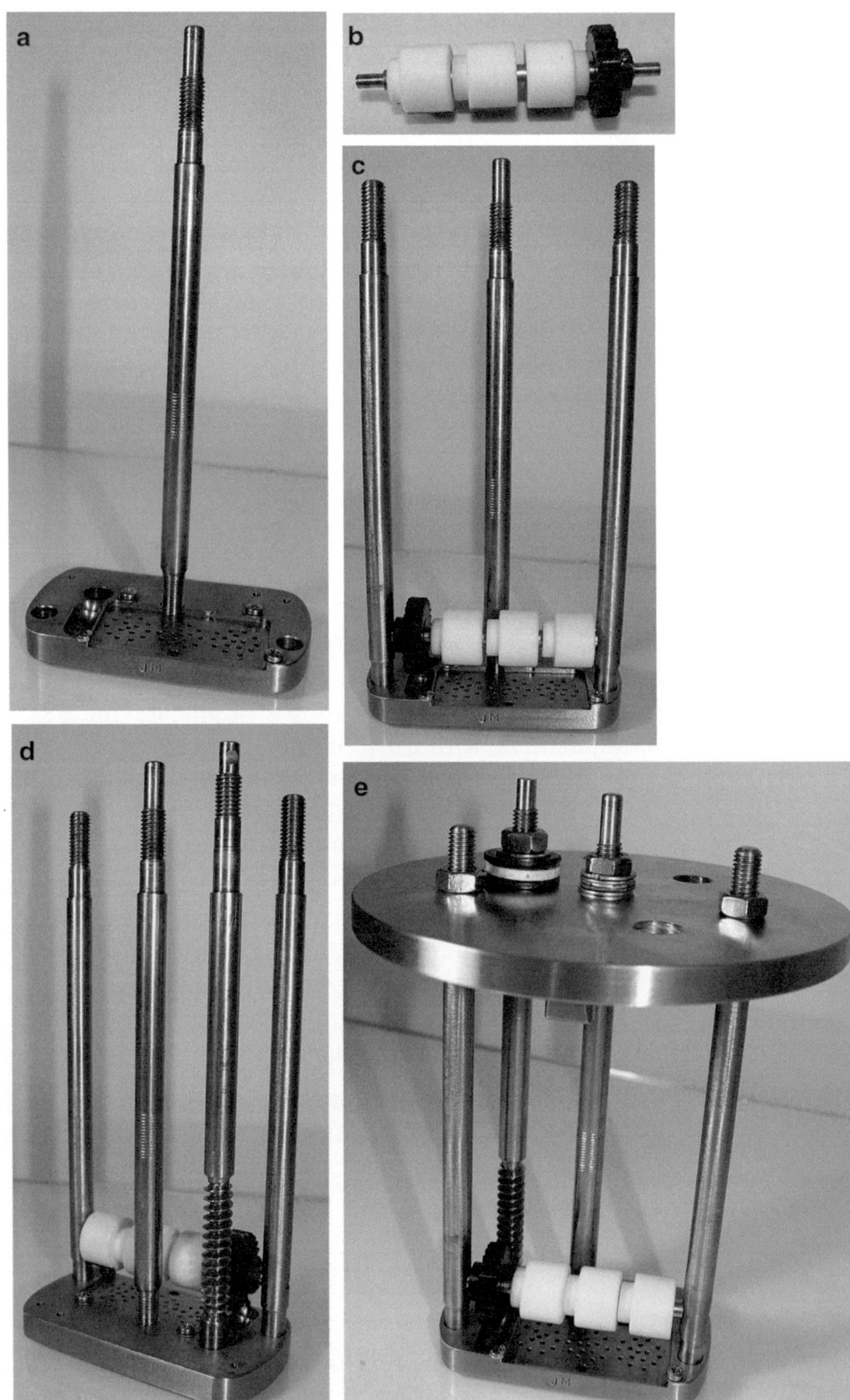

Fig. 3 Bioreactor assembly: (**a**) base plate control arm screwed into the middle of the base plate, (**b**) roller axle inserted through the three rollers with cams and worm gear, (**c**) roller axle mounted onto the axle support rods attached to the base plate, (**d**) drive shaft attached to the base plate to mesh with the worm gear, and (**e**) head plate fitted onto the base plate control arm, the two axle support rods, and the drive shaft using washers and screw nuts

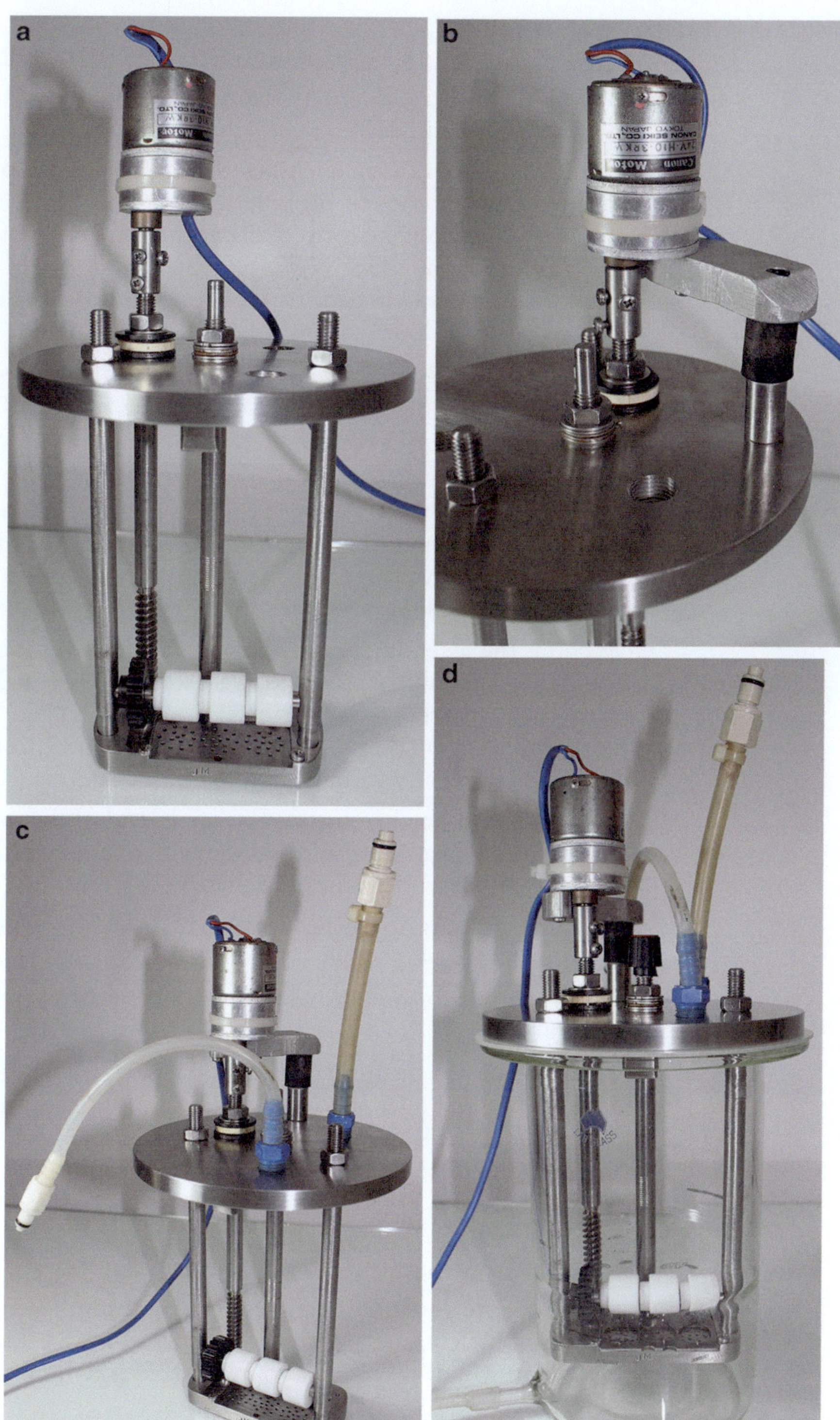

Fig. 4 Bioreactor assembly: (**a**) motor connected to the drive shaft using the shaft sleeve connector, (**b**) motor fixed to the head plate using the motor support arm, (**c**) silicone tubing for inlet and outlet gas streams attached to the head plate using plastic tubing adapters, and (**d**) marked plastic control knob attached to the top of the base plate control arm and the entire mechanical assembly inserted into the bioreactor vessel to complete the mechanobioreactor

3. Attach the roller axle to the mounting points on the axle support rods (Fig. 3c).
4. Cut two 10-mm-diameter holes through the base plate and insert the lower ends of the axle support rods (Fig. 3c). This will keep the rods at a fixed distance from each other and therefore keep the roller axle in place. Cut a groove in the base plate under the worm gear to allow rotation of the gear when the roller axle is lowered towards the tissue samples.
5. Cut a hole through the base plate to insert the drive shaft at a position that allows the worm and worm gear to mesh against each other (Fig. 3d).
6. Cut four holes in the head plate corresponding to the positions of the base plate control arm, the two axle support rods, and the drive shaft (Fig. 3e). Fit flat ball bearings into the holes for the drive shaft and base plate control arm to reduce friction and facilitate rotation.
7. Keeping the assembly upright, place washers over the tops of the base plate control arm, the two axle support rods, and the drive shaft, and then place the head plate on top (Fig. 3e). The washers will sit on ridges cut into each of the rods to hold the head plate level and parallel with the base plate.
8. Install a thrust ball bearing on the upper end of the drive shaft on top of the head plate and tighten with a screw nut. Tighten the axle support rods to the head plate with screw nuts (Fig. 3e).
9. Connect the motor to the drive shaft using the shaft sleeve connector (Fig. 4a).
10. Fix the motor to the head plate using the motor support arm (Fig. 4b).
11. Tap two holes in the head plate and screw in two plastic tubing adaptors for the inlet and outlet gas streams. Connect 15-cm-long pieces of silicone tubing to the adaptors and fit venting filters to the free ends using one-way snap-on tubing connectors (Fig. 4c).
12. Install the marked plastic control knob tight onto the top of the base plate control arm (Fig. 4d). Turning the knob moves the base plate vertically and changes the distance between the rollers with cams and the base plate (*see* **Note 8**). Each complete turn will displace the base plate vertically by 1 mm.
13. Place the head plate gasket onto the flange of the bioreactor vessel and then insert the mechanical parts of the device into the vessel (Fig. 4d).

3.2 Bioreactor Setup and Operation

1. Sterilize all components of the assembled mechanobioreactor, except the motor, by autoclaving. Place the sterilized bioreactor in a biosafety cabinet.

2. Remove the head plate with its attachments from the assembled bioreactor and place a sterile 7.5-cm-long magnetic stirrer bar into the bioreactor vessel.
3. Add 600 mL of CCM to the bioreactor vessel (*see* **Note 9**).
4. Using sterile forceps, place up to three cartilage tissue samples on the base plate and lock them into place using the tissue holding plate (*see* **Note 10**).
5. Using the marked plastic control knob on top of the base plate control arm, adjust the distance between the tissue samples and rollers with cams to a minimum.
6. Place the head plate with its attachments on top of the bioreactor vessel so that the cartilage tissue samples and mechanical parts of the bioreactor are inside the vessel. The head plate gasket forms a seal between the head plate and the vessel flange. The cartilage tissue samples should be completely submerged in the culture medium.
7. Reconnect the motor to the drive shaft. Transfer the assembled bioreactor to a CO_2 incubator: the incubator should have an internal power source and/or inlet–outlet ports suitable for electrical cables. Place the bioreactor on a magnetic stirrer inside the incubator.
8. Connect the small air pump (*see* **Note 11**) to one of the venting filters (inlet) and turn it on. Pressure will be released through the other (outlet) venting filter.
9. Connect the motor wires to the transformer and control unit located outside the incubator.
10. Incubate the cartilage tissue samples at 5 % CO_2 and 37 °C with the magnetic stirrer operated at 65 rpm. Replace 50 % of the medium with fresh medium every 3 days.
11. To apply static compression to the cartilage tissue samples, turn the marked plastic control knob to lift the base plate until the tissue pieces contact the rollers with cams. Lift the base plate further until the tissue samples are depressed by 0.2–0.4 mm.
12. To apply dynamic shear and compression to the cartilage tissue samples, switch the motor on to start rotation of the rollers with cams. Adjust the rotational speed to 3 rpm. The cams on the rollers will superimpose an additional 0.1 mm of compressive displacement on the tissues for 5 s in every 20-s cycle.
13. To stop the mechanical stimulation, switch the motor off and then turn the marked plastic control knob until the rollers come off the tissues.
14. Operate with mechanical loading for a total of 10 min daily over 2–3 weeks. Reverse the rotational direction of the rollers with cams after 5 min of operation (*see* **Note 12**).

4 Notes

1. A standard laboratory cell culture vessel can be used.
2. A seal with the exact dimensions required can be cut out of a 2–4-mm-thick silicone rubber sheet.
3. The holes will allow culture medium to circulate around the bottom surface of the cartilage tissue samples during mechanical treatment.
4. The holes in the tissue holding plate can be cut with different dimensions to accommodate cartilage constructs of other sizes and shapes.
5. A worm gear can be custom-made from stainless steel or purchased of exact dimension in hard plastic.
6. These connectors allow tubing joints to be broken and reconnected quickly to maintain aseptic conditions. When used for medium exchange, two connections will be needed: one for taking medium out through the glass pipe at the bottom of the bioreactor vessel and another for adding medium, e.g., through an additional port on the head plate. You will also need a medium container custom-modified for the purpose.
7. Space the rollers apart in such a way that each roller is positioned above one hole of the tissue holding plate.
8. When tissues are placed on the base plate, moving the base plate vertically allows adjustment of the static compression exerted by the rollers on the tissue.
9. If you are transferring tissues to the mechanobioreactor from another culture system, it is recommended that you also transfer up to 300 mL of the spent medium and mix it with fresh culture medium to a total of 600 mL. The benefit of including spent medium is that it contains growth factors secreted by the cells.
10. Chondrocytes seeded on scaffolds can be cultivated in the mechanobioreactor without applying mechanical loading for 2–3 weeks until there is enough extracellular matrix in the constructs to withstand mechanical treatment. Alternatively, the constructs may be cultivated in other culture systems before being transferred to the mechanobioreactor.
11. The small air pump operates inside the incubator. Its function is to pump gas from the incubator space into the bioreactor headspace through the inlet gas filter.
12. This operation regimen has been shown to enhance proteoglycan and collagen production and improve the quality of cartilage tissues [14]. Other operating conditions may also be beneficial depending on the system studied.

Acknowledgements

This work was funded by the Australian Research Council (ARC). We are grateful to Russell Cail for assistance with design of the mechanobioreactor and John Matiossa for device fabrication and workshop services.

References

1. Martel-Pelletier J, Boileau C, Pelletier J-P et al (2008) Cartilage in normal and osteoarthritis conditions. Best Pract Res Clin Rheumatol 22:351–384
2. Kiani C, Chen L, Wu YJ et al (2002) Structure and function of aggrecan. Cell Res 12: 19–32
3. Archer CW, Dowthwaite GP, Francis-West P (2003) Development of synovial joints. Birth Defects Res C Embryo Today 69:144–155
4. Schulz RM, Bader A (2007) Cartilage tissue engineering and bioreactor systems for the cultivation and stimulation of chondrocytes. Eur Biophys J 36:539–568
5. Kisiday JD, Jin M, DiMicco MA et al (2004) Effects of dynamic compressive loading on chondrocyte biosynthesis in self-assembling peptide scaffolds. J Biomech 37:595–604
6. Waldman SD, Spiteri CG, Grynpas MD et al (2004) Long-term intermittent compressive stimulation improves the composition and mechanical properties of tissue-engineered cartilage. Tissue Eng 10:1323–1331
7. Mouw JK, Connelly JT, Wilson CG et al (2007) Dynamic compression regulates the expression and synthesis of chondrocyte-specific matrix molecules in bone marrow stromal cells. Stem Cells 25:655–663
8. Terraciano V, Hwang N, Moroni L et al (2007) Differential response of adult and embryonic mesenchymal progenitor cells to mechanical compression in hydrogels. Stem Cells 25: 2730–2738
9. Pelaez D, Huang C-YC, Cheung HS (2009) Cyclic compression maintains viability and induces chondrogenesis of human mesenchymal stem cells in fibrin gel scaffolds. Stem Cells Dev 18:93–102
10. Haugh MG, Meyer EG, Thorpe SD et al (2011) Temporal and spatial changes in cartilage-matrix-specific gene expression in mesenchymal stem cells in response to dynamic compression. Tissue Eng A 17:3085–3093
11. Li Z, Yao S-J, Alini M et al (2010) Chondrogenesis of human bone marrow mesenchymal stem cells in fibrin–polyurethane composites is modulated by frequency and amplitude of dynamic compression and shear stress. Tissue Eng A 16:575–584
12. Grad S, Loparic M, Peter R et al (2012) Sliding motion modulates stiffness and friction coefficient at the surface of tissue engineered cartilage. Osteoarthritis Cartilage 20:288–295
13. Huang AH, Baker BM, Ateshian GA et al (2012) Sliding contact loading enhances the tensile properties of mesenchymal stem cell-seeded hydrogels. Eur Cell Mater 24:29–45
14. Shahin K, Doran PM (2012) Tissue engineering of cartilage using a mechanobioreactor exerting simultaneous mechanical shear and compression to simulate the rolling action of articular joints. Biotechnol Bioeng 109:1060–1073
15. Pourmohammadali H, Chandrashekar N, Medley JB (2013) Hydromechanical stimulator for chondrocyte-seeded constructs in articular cartilage tissue engineering applications. Proc Inst Mech Eng H 227:310–316
16. Bonassar LJ, Grodzinsky AJ, Frank EH et al (2001) The effect of dynamic compression on the response of articular cartilage to insulin-like growth factor-I. J Orthop Res 19:11–17
17. Mauck RL, Hung CT, Ateshian GA (2003) Modeling of neutral solute transport in a dynamically loaded porous permeable gel: implications for articular cartilage biosynthesis and tissue engineering. J Biomech Eng 125:602–614
18. Kelly T-AN, Ng KW, Wang CC-B et al (2006) Spatial and temporal development of chondrocyte-seeded agarose constructs in free-swelling and dynamically loaded cultures. J Biomech 39:1489–1497
19. Abdel-Sayed P, Darwiche SE, Kettenberger U et al (2014) The role of energy dissipation of polymeric scaffolds in the mechanobiological modulation of chondrogenic expression. Biomaterials 35:1890–1897

Chapter 17

Microbioreactors for Cartilage Tissue Engineering

Yu-Han Chang and Min-Hsien Wu

Abstract

In tissue engineering research, cell-based assays are widely utilized to fundamentally explore cellular responses to extracellular conditions. Nevertheless, the simplified cell culture models available at present have several inherent shortcomings and limitations. To tackle the issues, a wide variety of microbioreactors for cell culture have been actively proposed, especially during the past decade. Among these, micro-scale cell culture devices based on microfluidic biochip technology have particularly attracted considerable attention. In this chapter, we not only discuss the advantageous features of using micro-scale cell culture devices for cell-based assays, but also describe their fabrication, experimental setup, and application.

Key words Cell-based assays, Microbioreactors, Microfluidic biochip, Perfusion cell culture, Tissue engineering

1 Introduction

Ex vivo engineering of biological tissues is a rapidly growing area with the potential to impact significantly a wide range of biomedical applications. However, the major challenges to the generation of functional tissues and their clinical utilization are related to our limited understanding of the regulatory role of specific biochemical or biophysical conditions on tissue development. In a tissue engineering bioreactor, the optimum culture conditions are the key to successful cartilage tissue engineering since the synthesis of cartilage extracellular matrix is tightly dependent on extracellular environments [1]. To fundamentally investigate the quantitative links between extracellular environments and cellular responses, cell culture-based assays are normally carried out. However, the current simplified cell culture format (e.g., the use of petri dish or multi-well microplate as cell culture vessel) is associated with several inherent shortcomings and limitations mainly including the inability to provide a stable, well-defined, and biomimetic culture condition, and the consumption of more experimental samples.

Pauline M. Doran (ed.), *Cartilage Tissue Engineering: Methods and Protocols*, Methods in Molecular Biology, vol. 1340, DOI 10.1007/978-1-4939-2938-2_17, © Springer Science+Business Media New York 2015

To address the above issues, perfusion microbioreactor systems with various formats have been proposed [2–4]. The advantageous features of using perfusion and miniaturized cell culture models are discussed herein. Firstly, the miniaturization of a cell culture model has been proved previously to greatly minimize the chemical gradients existing in a 3D cultured construct or cell culture environment, leading to a more homogenous and quantifiable culture environment [5]. This could in turn contribute to more precise cell culture-based assays. Compared with conventional cell culture methods, moreover, micro-scale cell culture devices are particularly suitable for cell-based assays since their scale features are much closer to the native cellular microenvironments, in which the ratio of cell volume to extracellular fluid volume is usually greater than one [6]. This paves the route to create a more biomimetic cell culture condition in vitro. Due to the small dimensions in a miniaturized cell culture system, furthermore, a cell culture microbioreactor consumes relatively less research resources. This feature has been found particularly meaningful for tissue engineering research because biological samples (e.g., tissues or cells) or other experimental resources (e.g., reagents) are normally limited or costly. Secondly, the most widely used cell culture system nowadays is a static cell culture system (e.g., the utilization of a multiwell microplate as a cell culture vessel), where the culture medium is supplied in a batch-wise and manual manner. Although such conventional cell culture systems are normally simple to use, the cell culture environments within the culture system may fluctuate due to periodical medium replacement processes [7]. Under such unstable conditions, the cellular response to the culture conditions investigated may become more complex. Conversely, a perfusion cell culture system not only can reduce the need for labor-intensive manual operations but, more importantly, can also continuously provide a system for nutrient supply and waste removal and hence keep the culture environment more stable [7]. This contributes to more quantifiable and well-defined culture conditions, which is particularly meaningful for cell/tissue culture-based assays.

Microfluidics refers to the science and technology that allows scientists to manipulate tiny (10^{-9} to 10^{-18} l) amounts of fluids using microstructures with characteristic dimensions on the order of tens to hundreds of micrometers [8]. With the tremendous advances in microfluidic biochip technology, microfluidic systems have been progressively used as emerging tools for high-throughput perfusion micro-scale cell cultures. The application of microfluidic perfusion cell culture systems has found several niches as promising alternatives to conventional cell culture methods. In this chapter, the fabrication and experimental setup of a typical microfluidic perfusion cell culture system (mainly including a microfluidic perfusion cell culture chip: Fig. 1, and the associated devices: Fig. 2) as previously published [9] will be used as an example for description purposes.

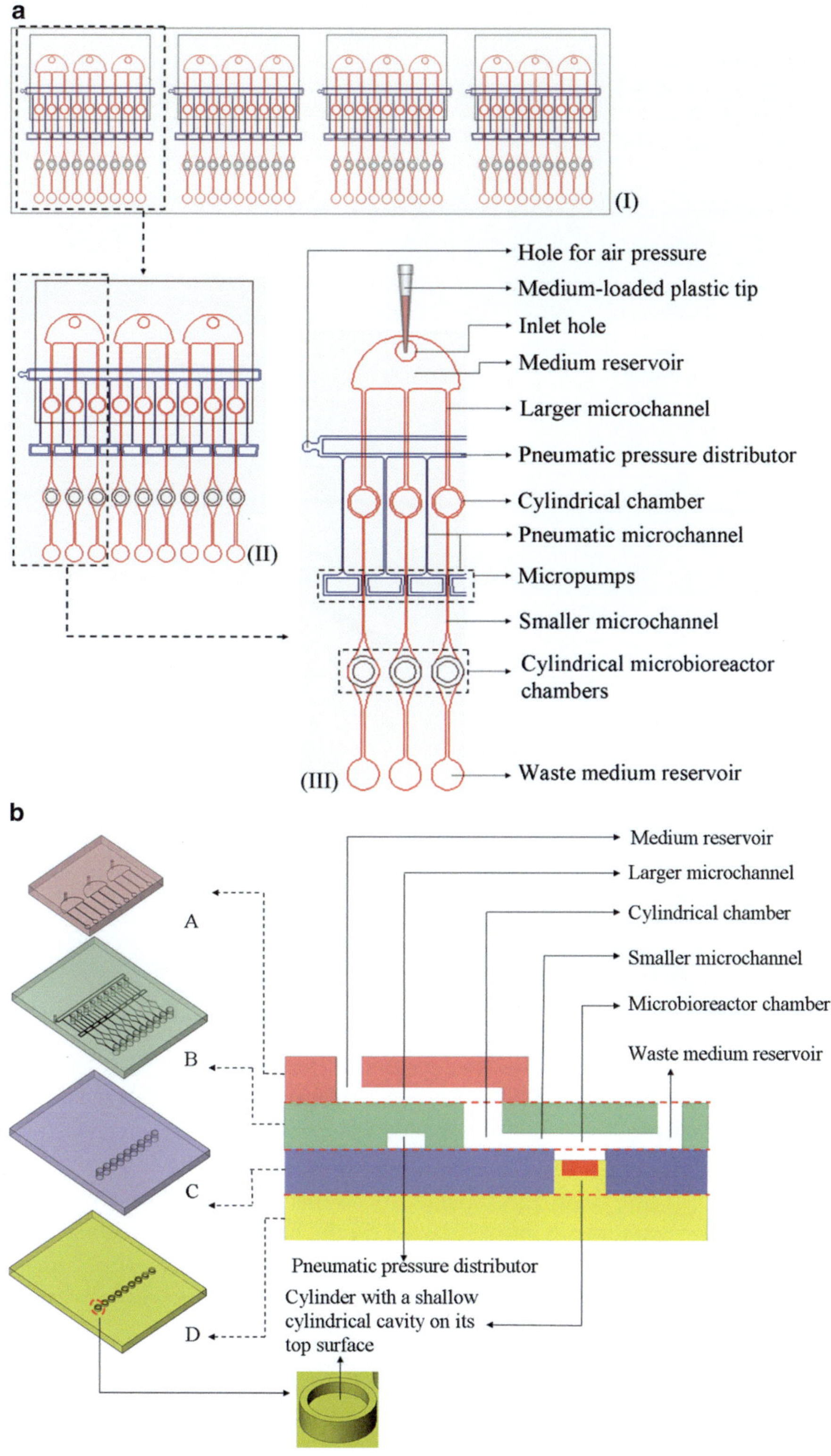

Fig. 1 The (**a**) design and (**b**) structure of a micro-scale perfusion cell culture chip based on microfluidic technology (reprinted with permission from ref. [9]. Copyright 2011 Elsevier)

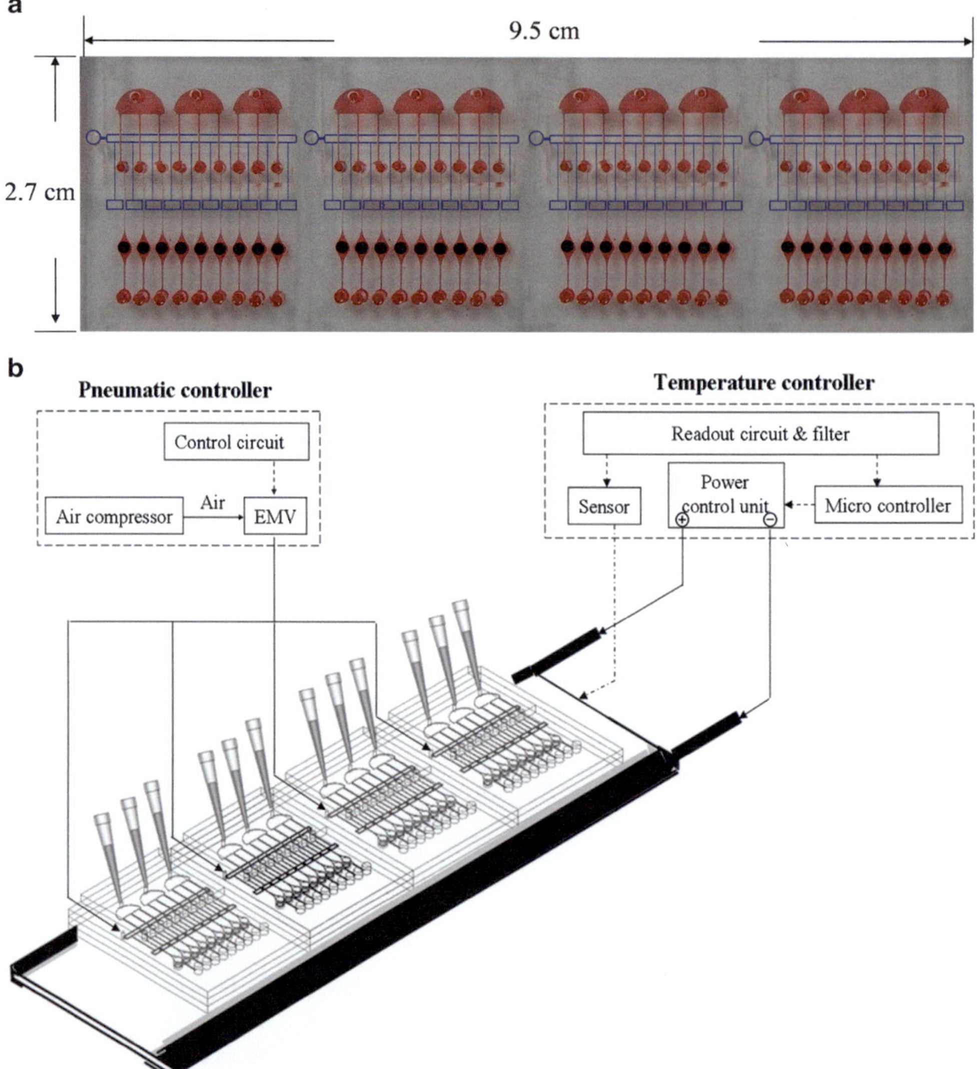

Fig. 2 (**a**) The photograph (*red*: medium reservoirs, medium microchannels, and waste medium reservoirs; *blue*: pneumatic microchannels; *black*: cell culture area) of microfluidic cell culture chip and (**b**) overall experimental setup (reprinted with permission from ref. [9]. Copyright 2011 Elsevier)

The design and specifications of a microfluidic perfusion cell culture system are entirely based on the user's need. Apart from the fabrication, and experimental setup, the practical application of such micro-scale cell culture system for a drug chemosensitivity assay (*see* **Note 1**) will also be described.

In the case demonstrated, the cell culture chip integrates the functions of continuous culture medium delivery, and cell/scaffold loading mechanisms in a microfluidic system. Its layout is schematically illustrated as Fig. 1a-(I). It comprises four sections, in which each section contains nine microbioreactors (Fig. 1a-(II)).

In the design (Fig. 1a-(III)), one medium reservoir supplies three individual microbioreactors, allowing replication of each experimental condition in triplicate. In this study, 36 parallel microbioreactors are in the chip, and thus, 12 different culture conditions can be assayed simultaneously. In each cell culture unit (Fig. 1a-(III)), the medium reservoir is loaded with cell culture medium directly through the insertion of a medium-loaded plastic tip to the inlet hole. The loaded medium is then delivered first through the larger microchannel (L: 4.4 mm, W: 150 μm, H: 100 μm), cylindrical chamber (D: 1.5 mm), another smaller microchannel (L: 5.5 mm, W: 50 μm, H: 100 μm), the cylindrical microbioreactor chamber (D: 1.5 mm, H: 1 mm), and finally to a waste medium reservoir (D: 1.5 mm, H: 1 mm). In the design, the medium flow is driven by the incorporated pneumatically driven membrane-based micropump (*see* **Note 2**), which is located at the both sides of the smaller microchannel (Fig. 1a-(III)). Moreover, the structure of the microfluidic cell culture chip illustrated in Fig. 1b comprises three layers of microfabricated polydimethylsiloxan (PDMS) plates (A–C) that are bonded and then assembled with the polymethylmethacrylate (PMMA) plate D to form a microfluidic perfusion cell culture system. Structurally, the medium reservoir and the larger microchannel for medium flow are in layer A. The pneumatic pressure distributor (L: 21 mm, W: 150 μm, H: 100 μm), pneumatic microchannel (the parts located at the both sides of medium microchannel, L: 1 mm, W: 150 μm, H: 100 μm) (also refer to Fig. 1a-(III)), the cylindrical chamber, the smaller microchannel for medium flow, the microbioreactor chamber, and the waste medium reservoir are in layer B. Layer C consists of multiple through holes (D: 1.5 mm), allowing the insertion of their corresponding cylinders (D: 1.5 mm, H: 1.0 mm) on layer D. In this work, there is a shallow cylindrical cavity with specific dimensions (e.g., D: 1 mm, H: 250 μm) on the top surface of each cylinder (Fig. 1b), which is not only used to accommodate a cell/scaffold for 3D cell culture but also to quantitatively define the volume of such sample loading [9].

2 Materials

2.1 Microfluidic Perfusion Cell Culture Chip

1. Draft the layout of the micropatterns (Fig. 1b: Layers A and B) using AutoCAD (Autodesk Inc.) (*see* **Note 3**) and then convert into a transparency-based photomask.
2. Purchase the SU-8 50 photoresist (MicroChem, USA), and a silicon wafer for the fabrication of the positive molds for the Layers A and B (Fig. 1b).
3. Prepare a polydimethylsiloxan (PDMS) polymer (Sylgard® 184, Dow Corning, USA) (*see* **Note 4**) for the replica molding

of the Layers A, B, and C. PDMS polymer is prepared by thoroughly mixing the PDMS pre-polymer with a curing agent in a ratio of 10:1 by weight according to the manufacturer's instructions.

4. Prepare a polymethylmethacrylate (PMMA) plate for fabrication of the mold for the Layer C, and for the direct fabrication of the Layer D.

2.2 The Associated Devices

1. Assemble a custom-made pneumatic controller integrating an air compressor (MDR2-1A/11, Jun-Air Inc., Japan), four electromagnetic valves (EMV) (S070M-5BG-32, SMC Inc., Taiwan), and a programmable control circuit system to activate and control the pneumatically driven micro-pumps in the microfluidic perfusion cell culture chip.
2. Prepare a thermal control system to provide a stable thermal condition of 37 °C for cell culture outside a cell incubator. The thermal control system mainly consists of a transparent indium tin oxide (ITO)-based micro-heater chip and a thermal controller (*see* **Note 5**).

2.3 Drug Chemosensitivity Assay (Application Demonstration)

1. Cells and scaffold: SW 620 cancer cell line in a scaffold of 1.5 % (w/v) agarose gel (low-gelling temperature agarose). Chondrocytes or chondrogenic stem cells may be used instead. Suspend the cells in the gel at a density of 5×10^6 cells ml^{-1}.
2. Cell culture medium: L-15 medium supplemented with various concentrations (0, 5×10^{-5}, 10^{-4}, 5×10^{-4}, 10^{-3} M) of Epirubicin (trade name: Ellence®, Pfizer), an anticancer drug. For cartilage applications, a suitable medium supplemented with growth factors or other additives may be used. Prepare 50 μl of the cell culture medium in plastic tips.
3. 70 % ethanol solution: 70 % (w/v) ethanol in water.
4. 3 % Pluronic F68 surfactant solution: 3 % (w/v) Pluronic® F68 solution in water.
5. LIVE/DEAD® Viability/Cytotoxicity Kit (Molecular Probes).

3 Methods

3.1 Fabrication of Microfluidic Perfusion Cell Culture Chip

1. Fabricate the PDMS Layers A and B (Fig. 1b) by a standard soft lithography process involving photolithography (*see* **Note 6**) for mold fabrication, and subsequent replica molding using PDMS polymer as material.
2. For the photolithography process, use a spin coater to prepare a thin SU-8 50 photoresist with the desirable thickness on a clean silicon wafer. In preparation for this, first establish a standard curve correlating the operating conditions (e.g., rotational speed and time) with the resulting photoresist thickness.

3. For the preparation of the positive mold for the PDMS Layers A and B (Fig. 1b), spin SU-8 50 photoresist onto a silicon wafer at a rotational speed of 1000 rpm for 1 min to achieve the required photoresist thickness of 100 μm.
4. Soft-bake the SU-8 coated wafers initially at 65 °C for 10 min and then at 95 °C for 30 min.
5. Using a mask aligner, convert the micropatterns on the photomask onto the SU-8 50 photoresist through ultraviolet (UV) exposure with a total exposure dose of 550 mJ/cm^2.
6. Bake the exposed wafers initially at 65 °C for 1 min and then at 95 °C for 15 min.
7. Develop the baked wafers using SU-8 developer (MicroChem, USA) for about 10 min. Rinse the developed wafers with isopropyl alcohol and blow-dry.
8. For the PDMS Layer C (Fig. 1b), fabricate a PMMA mold using a CNC (Computer Numerical Control) miller (EGX-400, Roland Inc., Japan; 0.5 mm drill bit; rotational speed: 26,000 rpm) instead of using the above photolithography process. This is mainly because the desired microstructures are relatively large.
9. For subsequent replica molding, pour the prepared PDMS polymer onto the above-mentioned fabricated molds and cure at 70 °C using a hotplate for 1 h. The cured PDMS plates (PDMS Layers A, B, and C; Fig. 1b) are then obtained after careful de-molding.
10. Fabricate the PMMA plate D directly using the aforementioned CNC miller.
11. After the microfabricated substrates (Layers A–D; Fig. 1b) are prepared, bond the three PDMS plates (A–C) by plasma oxidation treatment using a general oxygen plasma machine.
12. Assemble the bonded PDMS plates A–C with the PMMA plate D simply through a plug-in process to form a microfluidic cell culture system (Fig. 2a).

3.2 Experimental Setup

1. Connect four air tubes from the custom-made pneumatic controller to the microfluidic cell culture chip via four holes for inserting air tubes (Fig. 2b). Through the controller, the actions of micro-pumps and the resulting medium pumping performance can be manipulated.
2. Place the microfluidic cell culture chip on the surface of the transparent ITO-based microheater chip to provide a thermal condition of 37 °C for cell culture (Fig. 2b).
3. In operations, load the medium reservoir with cell culture medium directly by inserting a medium-loaded plastic tip into the inlet hole of the medium reservoir (Fig. 2b).

3.3 Cell-Based Chemosensitivity Assay (Application Demonstration)

A cell culture-based chemosensitivity assay is described for demonstration purposes. In tissue engineering research, the microfluidic cell culture system can be used to fundamentally study the effect of various biochemical factors on cellular functions. In such applications, the biochemical substances tested can be added in the culture medium and the same cell culture process followed as described below.

1. Before loading the prepared cell/agarose suspension to the microfluidic system, sterilize the surfaces of the reservoirs for medium loading, microchannels for medium flow, microbioreactor chambers, and the PMMA plate D using 70 % ethanol solution for 30 min. Following this, treat the surfaces with 3 % Pluronic F68 surfactant solution for about 24 h (*see* **Note** 7).
2. Load the prepared cell/agarose suspension into the designed cylindrical cavities (Fig. 1b) for accommodating 3D cell culture constructs.
3. Load the prepared cell culture medium into the fresh medium reservoirs through inserting each medium-loaded tip to the hole (Fig. 2b).
4. Pump the culture medium from the medium reservoir to the mircobioreactors using the incorporated pneumatic micropumps at the frequency and pneumatic pressure of 4 Hz and 20 psi, respectively, to achieve a desired medium pumping rate of 2.2 μl hr^{-1} for a 36-h perfusion cell culture.
5. Evaluate the viability of the cells in the agarose gel under various concentrations of anticancer drug using a fluorescent LIVE/DEAD® Viability/Cytotoxicity Kit and subsequent microscopic observation.
6. Capture the images of live (green) and dead (red) cells using a fluorescence microscope and a digital camera. Quantify the cell viability by counting the live (green) and dead (red) cells in the images.

4 Notes

1. As an example, a cell-based anticancer drug chemosensitivity assay is described to evaluate the response of cancer cells to an anticancer drug. This can assist doctors to tailor a chemotherapy regimen for individual patients. The same cell culture/scaffold/microbioreactor approach may be applied to test the response of cartilage cells to selected culture conditions.
2. A pneumatically driven membrane-based micropump is a type of mechanical micropump that is commonly integrated in microfluidic biochips for pumping liquid. It is fabricated from

an elastic polymer (e.g., PDMS) and a standard soft-lithography process, first reported by Unger and coworkers [10]. The pumping mechanism is based on the pulsations of elastic membranes actuated by the pneumatic chambers above to generate a continuous peristaltic-like activation effect for driving the fluid forward.

3. AutoCAD is a commercial software for 2D and 3D computer-aided design.
4. Polydimethylsiloxan (PDMS) polymer is commonly referred to as silicone. It is a biocompatible, nontoxic, optically transparent, and highly gas permeable material. PDMS is widely used as the material for fabricating microfluidic biochips.
5. ITO is a transparent conductive material which can generate heat due to its electric resistance when an electric current passes through. Through the manipulation of the electric current applied, the temperature on the ITO glass can be controlled in a manageable manner. In order to command the proposed ITO microheater thermally, a micro-controller (MPC82G516, Megawin Tech. Co., Taiwan) with a 10-bit analog-to-digital converter (ADC) and an 8-bit pulse-width-modulator (PWM) is used in this study [11].
6. Photolithography is the basic technique in microfabrication technology. Its purpose is to transfer the geometric shapes on a photomask to the surface of a silicon wafer via light (e.g., UV light). A specific microstructure on the silicon wafer can then be fabricated through the subsequent etching process. The basic steps involved in the process are wafer cleaning; photoresist application; soft baking; mask alignment; exposure and development; and hard-baking.
7. PDMS, the material used for constructing the microfluidic perfusion cell culture system, is hydrophobic in nature and tends to nonspecifically adsorb proteins. To avoid the adsorption of proteins in medium on PDMS surfaces, PDMS surfaces can be treated with surfactant solution, e.g., Pluronic F68®, which has been proved to largely reduce nonspecific serum protein adsorption [12, 13].

Acknowledgement

This work is sponsored by the National Science Council (NSC) in Taiwan (NSC 102-2911-I-182-502), and Chang Gung Memorial Hospital (CMRPD 1C0121).

References

1. Wu MH, Urban JPG, Cui ZF, Cui Z, Xu X (2007) Effect of extracellular pH on matrix synthesis by chondrocytes in 3D agarose gel. Biotechnol Prog 23:430–434
2. Wu MH, Huang SB, Cui ZF, Cui Z, Lee GB (2008) A high throughput perfusion-based microbioreactor platform integrated with pneumatic micropumps for three-dimensional cell culture. Biomed Microdevices 10:309–319
3. Huang SB, Wu MH, Wang SS, Lee GB (2011) Microfluidic cell culture chip with multiplexed medium delivery and efficient cell/scaffold loading mechanisms for high-throughput perfusion 3-dimensional cell culture-based assays. Biomed Microdevices 13:415–430
4. Huang SB, Wang SS, Hsieh CH, Lin YC, Lai CS, Wu MH (2013) An integrated microfluidic cell culture system for high-throughput perfusion three-dimensional cell culture-based assays: Effect of cell culture model on the results of chemosensitivity assays. Lab Chip 13:1133–1143
5. Wu MH, Urban JPG, Cui Z, Cui ZF (2006) Development of PDMS microbioreactor with well-defined and homogenous culture environment for chondrocyte 3-D culture. Biomed Microdevices 8:331–340
6. Walker GM, Zeringue HC, Beebe DJ (2004) Microenvironment design considerations for cellular scale studies. Lab Chip 4:91–97
7. Wu MH, Kuo CY (2011) Application of high throughput perfusion micro 3-D cell culture platform for the precise study of cellular responses to extracellular conditions-effect of serum concentrations on the physiology of articular chondrocytes. Biomed Microdevices 13:131–141
8. Whitesides GM (2006) The origins and the future of microfluidics. Nature 442:368–373
9. Wu MH, Chang YH, Liu YT, Chen YM, Wang SS, Wang HY, Lai CS, Pan TM (2011) Development of high throughput microfluidic cell culture chip for perfusion 3-dimensional cell culture-based chemosensitivity assay. Sens Actuators B Chem 155:397–407
10. Unger MA, Chou HP, Thorsen T, Scherer A, Quake SR (2000) Monolithic microfabricated valves and pumps by multilayer soft lithography. Science 288:113–116
11. Lin JL, Wu MH, Kuo CY, Lee KD, Shen YL (2010) Application of indium tin oxide (ITO)-based microheater chip with uniform thermal distribution for perfusion cell culture outside a cell incubator. Biomed Microdevices 12: 389–398
12. Boxshall K, Wu MH, Cui Z, Cui ZF, Watts JF, Baker MA (2006) Simple surface treatments to modify protein adsorption and cell attachment properties within a poly(dimethylsiloxane) microbioreactor. Surf Interface Anal 38: 198–201
13. Wu MH (2009) Simple poly(dimethylsiloxane) surface modification to control cell adhesion. Surf Interface Anal 41:11–16

Part V

In Vivo Application

Chapter 18

Transplantation of Tissue-Engineered Cartilage in an Animal Model (Xenograft and Autograft): Construct Validation

Hitoshi Nemoto, Deborah Watson, and Koichi Masuda

Abstract

Tissue engineering holds great promise for cartilage repair with minimal donor-site morbidity. The in vivo maturation of a tissue-engineered construct can be tested in the subcutaneous tissues of the same species for autografts or of immunocompromised animals for allografts or xenografts. This section describes detailed protocols for the surgical transplantation of a tissue-engineered construct into an animal model to assess construct validity.

Key words Tissue engineering, Cartilage repair, Alginate-recovered-chondrocyte, Animal model, Transplantation, Xenograft, Autograft, Mouse, Rabbit

1 Introduction

Cartilage tissue has a poor capacity for self-repair [1]. Although the use of autografts for repair of cartilage in congenital disease, trauma, tumor resection, or articular cartilage defect has great advantages in terms of rejection reactions and ethics, there are limitations of tissue amounts used and donor-site morbidity.

To overcome these disadvantages, many studies have reported the use of tissue engineering techniques for cartilage repair. Tissue engineering techniques can minimize donor-site morbidity because cell expansion in vitro reduces the amount of autograft tissue required. Importantly, rejection reactions from the transplantation of expanded autograft cells are negligible. Several procedures using tissue engineering techniques to repair defects of articular cartilage have been developed, including autologous mesenchymal cell transplantation [2, 3], matrix-assisted chondrocyte implantation [3, 4], and transplantation of allogenic chondrocytes in alginate beads [5].

Pauline M. Doran (ed.), *Cartilage Tissue Engineering: Methods and Protocols*, Methods in Molecular Biology, vol. 1340, DOI 10.1007/978-1-4939-2938-2_18, © Springer Science+Business Media New York 2015

In the maxillofacial surgical field, although autologous cartilage transplantation remains the clinical gold standard for cartilage or bone defects in cranio-maxillo-facial reconstruction [6], several studies of reconstruction using tissue engineering methods have also been performed [7, 8].

1.1 In Vivo Evaluation Using Subcutaneous Transplantation

The in vivo evaluation of cartilage or osteochondral constructs using subcutaneous (SQ) transplantation can be used to answer questions that are difficult to address by in vitro culture experimentation. The main questions are as follows:

1. *Mechanical resistance of the construct*: Whether the construct can maintain its three-dimensional (3D) structure under internal pressure during SQ transplantation and movement of the animal [9].
2. *Maturation of the construct*: Whether there are any differences in the degree of maturation between constructs cultured in vitro and in vivo. Deoxyribonucleic acid (DNA) content, a surrogate measure of cell number, was higher in in vivo than in in vitro neocartilage constructs, and glycosaminoglycans (GAGs) accumulation per DNA, a proxy for the amount of proteoglycan contained in cartilage tissue, decreased significantly after in vivo implantation [9, 10]. The mechanical strength of a construct grafted in vivo was higher than that of a construct cultured in vitro [9]. Calcification of the construct can be observed in vivo [9].
3. *Reaction to the construct by the host*: Whether there are any influences of wound healing and immuno-reaction. Inflammation-associated wound healing and immuno-reaction are etiologies for grafted cartilage resorption [11, 12]. The autograft resorption rate has been reported to range from 12 to 50 % [13] for human nasal septum cartilage (NSC) and from 15 to 50 % for rabbit NSC [14].
4. *Safety and absorption of the scaffold*: Whether a scaffold can be absorbed in vivo while keeping its 3D structure in parallel with grafted cartilage maturity [15]. For clinical use when using a scaffold, ideally the scaffold should be absorbed and not cause a foreign body reaction.

For example, several translational research studies [10, 16–20] have been conducted to investigate the in vivo and in vitro maturation and integration into host tissues of scaffold-free cartilaginous constructs engineered using the alginate-recovered-chondrocyte (ARC) method (Fig. 1) [21]. The ARC method consists of the following steps. Primary chondrocytes, or chondrocytes expanded in monolayer culture, are encapsulated in alginate beads in a 3D environment and cultured for a period of time to establish a cell-associated matrix. The cells with their cell-associated matrix, released

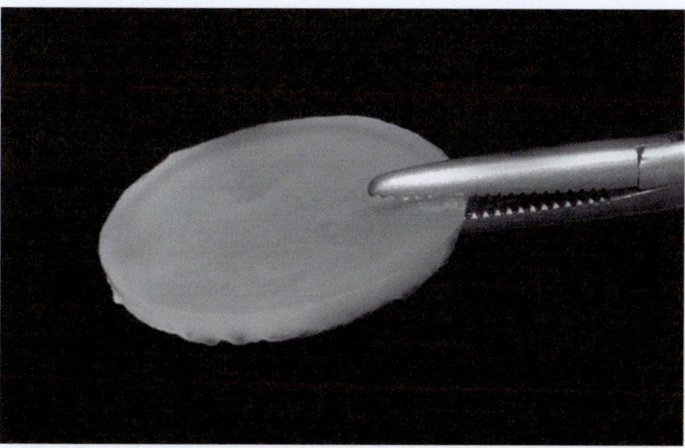

Fig. 1 Tissue-engineered cartilage using the alginate-recovered-chondrocyte (ARC) method. This ARC tissue was made from human nasal septum chondrocytes. The isolated chondrocytes were expanded in monolayer culture and then encapsulated in alginate beads and cultured adding insulin-like growth factor 1 (IGF-1) and growth differentiation factor 5 (GDF-5) for 2 weeks. Following release from alginate, the chondrocytes were cultured for 6 weeks at very high density on cell culture insert semi-permeable membranes

from the alginate by adding chelating agents, are cultured on a semi-permeable membrane in cell culture inserts to form a cartilaginous tissue. This scaffold-free tissue-engineered construct has a 3D structure resembling that of cartilage and mechanical strength [16, 17]. In addition, the method can be applied to create an osteochondral tissue construct, made as a composite structure of a synthetic bone substitute on hydroxyapatite (HA) and rabbit ARC tissue. These tissues transplanted into subcutaneous pockets of immunodeficient nude mice for 8 weeks (Fig. 2) exhibited in vivo cartilage maturation [19]. For an autograft experiment, constructs from rabbit nasal septum chondrocytes that were transplanted SQ into the nasal dorsum for 60 days were well tolerated [9].

The following sections describe detailed protocols to perform in vivo studies using nude mice for xenografts and rabbits for autografts. However, before performing an in vivo construct transplantation study, careful planning and consideration for animal ethics are required.

1.2 Before You Start an Animal Experiment: Approval from the Institutional Animal Care and Use Committee (IACUC)

All procedures using animals should be performed using the "International Guiding Principles for Biomedical Research Involving Animals (1985)" [22]. Practical information can be found in the "Guide for the Care and Use of Laboratory Animals" [23].

For researchers in the United States, please follow the "Public Health Service Policy on Humane Care and Use of Laboratory Animals" [24]. Approval from the IACUC of the host institution should be obtained before initiation of animal

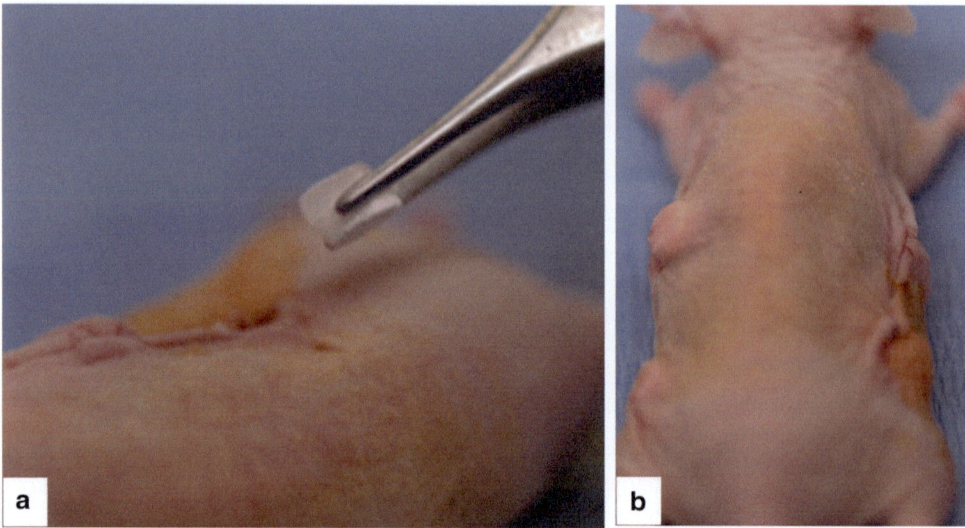

Fig. 2 Xenograft transplantation of an osteochondral construct [alginate-recovered-chondrocyte (ARC) cartilaginous tissue on hydroxyapatite (HA)] into a nude mouse. The ARC tissue was made from rabbit articular chondrocytes. Subcutaneous pockets were bluntly created from four small 1-cm skin incisions on bilateral dorsa. (**a**) One construct was placed into each subcutaneous pocket. (**b**) A subcutaneous absorbable suture prevented movement of the implant

studies. **Please note that, depending on the individual institution, the animal experimental protocol may need to be submitted to the IACUC at least 3 months prior to the experiment. Please refer the local IACUC policy for the detailed animal handling procedure.**

Additional information can be obtained at the National Institutes of Health (NIH) website: "REGULATIONS AND STANDARDS" (http://oacu.od.nih.gov/regs/index.htm).

2 Materials

2.1 Nude Mouse Implantation Model (Xenograft to Dorsa)

1. Athymic Nude Mice [Foxn1^{nu}]. This immunodeficient nude mouse, which originated from the NIH, lacks a thymus and is unable to produce T cells. Other types of immunodeficient mice can be used (see Ref. [25]).
2. A povidone iodine-based surgical scrub (Scrub care®; Care Fusion, Leawood, KS, USA) for scrubbing the surgical site.
3. 70 % isopropyl alcohol (Rubbing Alcohol®; Vi-Jon, Smyrna, TN, USA) for scrubbing the surgical site.
4. Artificial tear ointment (Akwa Tears®; Akorn, Inc., Lake Forest, IL, USA) to cover and protect eyes during surgery.
5. Isoflurane (Isoflurane®; Piramal Healthcare, Maharashtra, India) for induction and maintenance of anesthesia.

6. Ketamine (Ketaset®; Fort Dodge Animal Health, Fort Dodge, IA, USA) for induction of anesthesia.
7. Xylazine (AnaSed®; LLOYD, Inc., Shenandoah, IA, USA) for induction of anesthesia.
8. 4-0 absorbable suture (Dexon®; Mansfield, MA, USA or VICRYL®; Johnson and Johnson, Somerville, NJ, USA) for closing incision.
9. 0.9 % NaCl (0.9 % Sodium Chloride®; Hospira, Inc., Lake Forest, IL, USA) or lactated Ringer's solution (LRS) (Lactated Ringer's Solution®; Hospira, Inc.), for prevention of postoperative (post-OP) dehydration.
10. Diluted Marcaine: 1:2 dilution of 0.25 % Marcaine (0.25 % Marcaine®; Hospira, Inc.) in sterile water (Sterile Water®; Hospira, Inc.), for post-OP analgesia.
11. Bacitracin, Neomycin, and Polymyxin B ointment (Triple Antibiotic Ointment®; Perrigo, Allegan, MI, USA), as needed for skin ulcerations or wounds in post-OP care.
12. Buprenorphine (Buprenex®; Reckitt Benckiser Pharmaceuticals, Inc., Richmond, VA, USA) for analgesia if post-OP pain is observed.

2.2 Rabbit Implantation Model (Autograft to Nasal Dorsum, Ear, and/or Dorsa)

1. New Zealand white rabbits or Japanese white rabbits. To perform experiments using United States Department of Agriculture (USDA) regulated animals, additional requirements must be considered. Detailed information for experiments using USDA regulated animals (the rabbit is categorized as a large animal by the USDA) can be found in the "United States Department of Agriculture (2013) Animal Welfare Act and Animal Welfare Regulations" [26].
2. Ketamine for induction of anesthesia and preparation of euthanasia.
3. Xylazine for induction of anesthesia.
4. Isoflurane for induction and maintenance of anesthesia.
5. A povidone iodine-based surgical scrub for scrubbing the surgical site.
6. 70 % isopropyl alcohol for scrubbing the surgical site.
7. Artificial tear ointment to cover and protect eyes during surgery.
8. 4-0 absorbable suture for closing incision.
9. 0.9 % NaCl or LRS for prevention of postoperative (post-OP) dehydration.

10. Chlorhexidine (0.12 % W/W Chlorhexidine Rinse®; Virbac AH, Inc., Fort Worth, TX, USA) to clean the wound and ulcer in post-OP care.
11. Bacitracin, Neomycin, and Polymyxin B ointment, as needed for skin ulcerations or wounds in post-OP care.
12. Carprofen (RIMADYL®; Pfizer, New York, NY, USA) for analgesia for 3 days post-OP.
13. Cefazolin (Cefazolin®; Hospira, Inc., Lake Forest, IL, USA) as antibiotic immediately post-OP and for 3 days post-OP.
14. Hay, pineapple sticks, Fruity Bites (Bio-Serv, Flemington, NJ, USA), or Critical Care (Oxbow Enterprises, Inc., Murdock, NB, USA) for appetite loss in post-OP.
15. A mixture of pentobarbital sodium and phenytoin sodium (Beuthanasia-D Special®; Schering-Plough Animal Health Corp., Union, NY, USA) for euthanasia.
16. Acepromazine (Acepromazine Maleate Injection®; Clipper Distributing Company, LLC, St Joseph, MO, USA) for sedation for euthanasia.

3 Methods

3.1 Nude Mouse Implantation Model (Xenograft to Dorsa)

3.1.1 Preparation of Instruments

1. Clean all surgical instruments and sterilize by steam autoclave or ethylene oxide prior to use. To confirm proper sterility, mark surgical packs with sterilization indicators and the date of sterilization.
2. If instruments are to be used for multiple surgeries on a single day, clean with 70 % isopropyl alcohol and sterilize by a glass bead sterilizer between mice.

3.1.2 Acclimation and Preparation of Animals

1. Allow all mice to acclimate in their cage for at least 48 h prior to surgery. (Check local IACUC policy.) Monitor each mouse and carefully assess prior to surgery for any abnormalities in eating habits, urine output, and stool output.
2. 24–48 h before surgery, weigh all mice to determine anesthesia and therapeutic drug dosages and as the baseline for post-OP monitoring.

3.1.3 Preparation of Surgeon

1. Surgeons wear clean scrubs, face mask and head covering, and close-toed shoes and shoe covers.
2. Surgeons wash both hands with soap and put on sterile gloves using sterile technique.

3.1.4 Anesthesia and Surgical Procedure

The surgical procedure is performed under general anesthesia (gas anesthesia by Isoflurane or Ketamine/xylazine mixture, 100–150/

10–15 mg/kg). Monitoring of the mouse is required and must be documented during the procedure. Therapeutic drugs administered during the procedure must be recorded. (Check local IACUC policy.) The length of this procedure is 5–10 min per mouse.

1. Induce anesthesia using isoflurane (4–5 %) in oxygen administered in an anesthesia chamber.
2. Once sedated, remove each mouse from the chamber and administer isoflurane by a nosecone to maintain anesthesia (1.0–2.5 %) (*see* **Note 1**).
3. Place artificial tear ointment into each eye of the mouse and clip the surgical area according to the procedure.
4. Prep the surgical site with a povidone iodine solution and rinse with 70 % isopropyl alcohol.
5. Place the mouse in a prone position onto a hot water blanket/pad or a heated surgical table to ensure an adequate body temperature throughout the procedure.
6. Drape the mouse using a sterile sheet. Using sterile technique and equipment, make one 1.5–2.0 cm midline incision over the dorsa or four small 1-cm skin incisions on bilateral dorsa

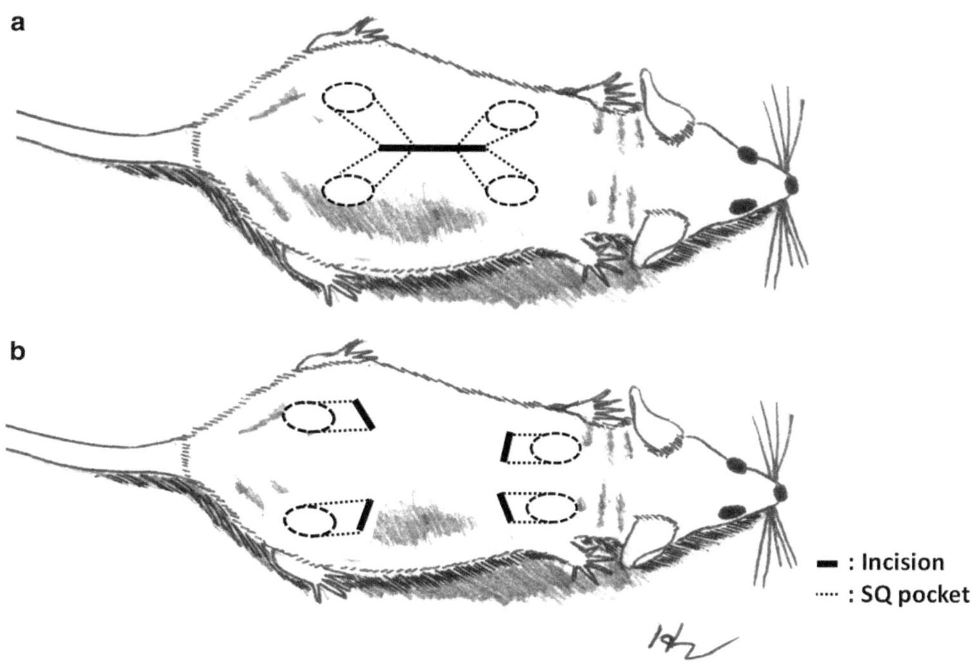

Fig. 3 Schema of incisions and subcutaneous (SQ) pockets (xenograft model). There are two ways to design graft implantation: (**a**) One midline incision to create four SQ pockets and (**b**) Multiple incisions, one for each pocket

and bluntly create subcutaneous pockets with a spatula or hemostat (Fig. 3) (*see* **Note 2**).

7. Take pictures of the implants as day 0. Place three or four constructs into separate pockets over the mouse dorsa, with special attention taken to not tear/disrupt the implants.
8. To prevent the implant from moving, which may happen if a single incision is used to create four SQ pockets, first close the entrance of the pockets with absorbable sutures before skin closure. Close the incision with a surgical stapler.
9. Allow the mice to recover from anesthesia on a heated pad located in the procedure room where they can be continuously observed. Give a SQ injection of 1 ml of 0.9 % NaCl or LRS and immediately inject 0.1 ml of diluted Marcaine intradermally (ID).

3.1.5 Postsurgical Monitoring

1. Monitor all mice daily for the first 5 days following surgery. Monitor carefully for any wounds or surgical dehiscence, pain, decrease in appetite or water intake, gait disturbances, neurological problems, and/or abnormal behavior (*see* **Note 3**).
2. Remove staples within 10 days following surgery.

3.1.6 Necropsy

Necropsy is performed using the "American Veterinary Medical Association Guidelines for the Euthanasia of Animals" [27].

1. At the end of the study, euthanize the mice with CO_2 and cervical dislocation.
2. Remove the implants from the SQ pockets of the mice.
3. Take pictures of the implants.
4. Fix the implants in an appropriate solution (e.g., 10 % formalin) for histological analysis and/or prepare for other specific analyses.

3.1.7 Record Maintenance

Post-OP observation records are written following the individual institution's policies. Observations on experiment animals must be documented daily for 5 days (see your institutional regulation). If problems occur at a later date in the study, the animal must be observed daily until resolved.

3.2 Rabbit Implantation Model (Autograft to Nasal Dorsum, Ear, and/or Dorsa)

3.2.1 Preparation of Instruments

1. Clean all surgical instruments and sterilize by steam autoclave or ethylene oxide prior to use. To confirm proper sterility, mark surgical packs with sterilization indicators and the date of sterilization.
2. Prepare and sterilize one set of instruments for each rabbit.

3.2.2 Acclimation and Preparation of Animals

1. Allow all rabbits to acclimate in their cage for at least 72 h prior to surgery. (Check local animal care policy.) Monitor each rabbit and carefully assess prior to surgery for any abnormalities in eating habits, urine output, and stool output.
2. 24–48 h before surgery, weigh all rabbits to determine anesthesia and therapeutic drug dosages and as the baseline for post-OP monitoring.

3.2.3 Preparation of Surgeon

1. Surgeons wear clean scrubs, face mask and head covering, and close-toed shoes and shoe covers.
2. Surgeons perform a surgical scrub in the surgeon's scrub area. This includes scrubbing both hands, in between fingers, and both forearms with a designated surgical scrub brush for approximately 5 min.
3. Surgeons wear sterile gown and put on sterile gloves using sterile technique.

3.2.4 Anesthesia and Surgical Procedure

The anesthesia is performed by trained personnel and monitored by a separate anesthetist. The surgical procedure is performed under general anesthesia (gas anesthesia by Isoflurane). During the surgical procedure, monitoring [including heart rate (HR), respiration rate (RR), and temperature] of each rabbit is required and must be documented, with appropriate frequency (usually every 5–10 min, see your institutional regulations). Therapeutic drugs administered during the procedure must be recorded. The length of the procedure is 10–15 min per rabbit.

1. Sedate the rabbit initially with a SQ mixture of ketamine (35 mg/kg) and xylazine (5 mg/kg) (*see* **Note 4**).
2. Once sedated, intubate the animal and place under general anesthesia of 1–3 % isoflurane (*see* **Note 5**).
3. Place artificial tear ointment into the rabbit's eyes and clip the nasal dorsum.
4. Prep the nasal dorsum with a povidone iodide solution and rinse with 70 % isopropyl alcohol. Perform this procedure three times. Apply a final prep of a povidone iodide solution to the area and do not remove with alcohol.
5. Place the rabbit in a prone position onto a hot water blanket/pad or a heated surgical table to ensure an adequate body temperature throughout the procedure.
6. Drape the rabbit with a sterile sheet using sterile technique. Depending on the purpose, the following incisions can be made to a SQ depth: a 1.5-cm midline incision over the nasal dorsum, a 2-cm incision over the ventral ear region, or 1.5-cm multiple incisions over the dorsa. Blunt dissection using a ster-

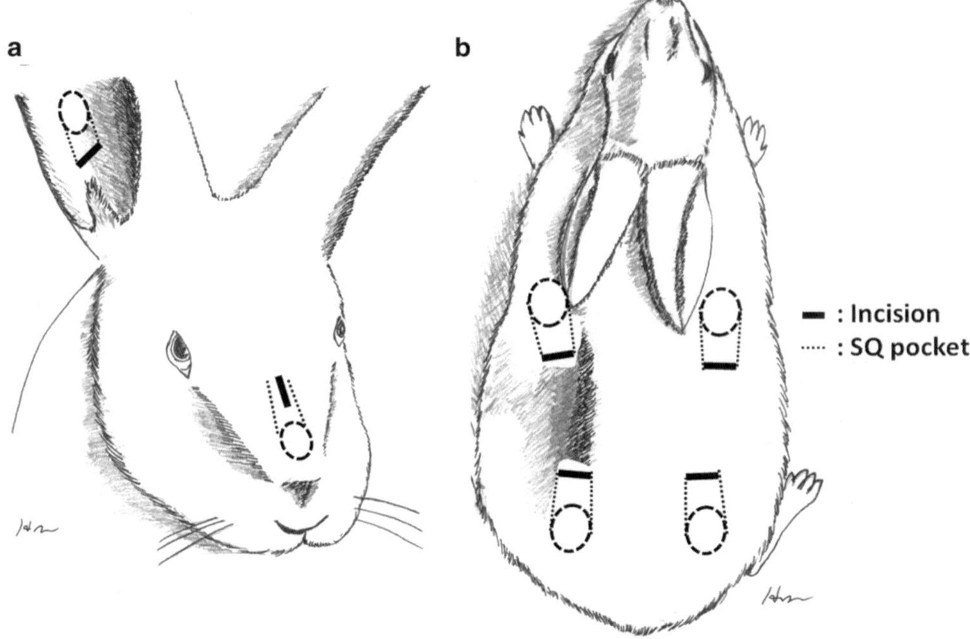

Fig. 4 Schema of incisions and subcutaneous (SQ) pockets (autograft model). Multiple incisions sites are chosen depending on the purpose: (**a**) Graft to the nasal dorsum or the ventral ear region and (**b**) Graft to the dorsa

ile spatula or hemostat generates a precisely sized SQ pocket for implantation (Fig. 4).

7. Take pictures of the implants as day 0. Depending on the purpose, place one construct in each SQ pocket created.
8. Use absorbable sutures to close the incision and approximate the skin and subcutaneous tissue.
9. Extubate the rabbit unless the animal's respiration is unstable. Allow the rabbits to recover from anesthesia on a heated water blanket and/or underneath a Bair Hugger where they can be continuously observed. Record the rabbit's HR, RR, and temperature every 15 min until it is sternal. In addition, give a SQ injection of Cefazolin (22 mg/kg) and 50 ml of 0.9 % NaCl or LRS. An Elizabethan-collar (E-collar) can be used to protect the surgical wound. Keep each rabbit in recovery until it is sternal and holding up its head. Once the rabbit is considered alert and responsive, return it to its cage. Monitor the RR and responses every 20–30 min until every animal has been returned to its cage and is Bright Alert Responsive (BAR).

3.2.5 Postsurgical Monitoring

1. Give the rabbits Cefazolin (22 mg/kg) and Carprofen (4 mg/kg) SQ for 3 days post-OP and monitor carefully for any wounds or surgical dehiscence, pain, decrease in appetite or water intake, gait disturbances, neurological problems, and/or

abnormal behavior. Monitor all rabbits daily for the first 2 weeks following surgery, then three times per week for the duration of the study (*see* **Note 6**).

2. 7–10 days following surgery, remove the sutures. Monitor the rabbits closely for any wound dehiscence or self-mutilation (*see* **Note 7**).

3.2.6 Necropsy

Necropsy is performed using the "American Veterinary Medical Association Guidelines for the Euthanasia of Animals" [27]. The rabbits are euthanized at the study end point.

1. Following deep sedation with a SQ mixture of ketamine (25 mg/kg) and acepromazine (1 mg/kg) and inhalation of isoflurane, clip the ventral ear to expose the marginal ear vein.
2. Euthanize the rabbit with Beuthanasia-D Special (0.222 ml/kg) applied intravenously into the marginal ear vein. Remove the implants.
3. Take pictures of the implants. Fix the implants in an appropriate solution (e.g., 10 % formalin) for histological analysis and/or prepare for other specific analyses.

3.2.7 Record Maintenance

Observation records of post-OP care are written following the individual institution's policies. Observations of experimental animals must be documented daily for 5 days. If problems occur at a later date in the study, the animal must be observed daily until resolved.

4 Notes

1. A test for the depth of anesthesia is made by examining the mouse for response to external stimuli, such as pedal reflexes and tail pinch response. In addition, the animal's heart rate (HR) and respiration rate (RR) are continuously monitored for any changes. Under optimal anesthesia: The HR is steady and between 300 beats per minutes (bpm) and 450 bpm and the RR is 55–65 breaths per minute [28].
2. If the implant contains cells, each construct should be grafted through separated SQ tunnels because of the possibility of cell cross-contamination. In this case, it is better to choose the multiple incision design (Fig. 3b).
3. If pain or discomfort is observed, proper analgesics (Buprenorphine 0.05–0.1 mg/kg SQ) should be administered immediately, and repeated as indicated. If the mouse should develop a skin ulceration, the wound is monitored and treated with Triple antibiotic ointment daily; should it become painful, the animal is euthanized.

4. It takes 10–15 min to get appropriate sedation with ketamine and xylazine.
5. The rabbit's depth of anesthesia is continuously monitored using an SPO2 monitor, as well as manually watching the rabbit's respiration and reflexes. Reflexes and response to external stimuli are monitored and include palpebral and pedal responses. Any response to external stimuli by way of animal movement, eye movement, or a sharp increase in HR or RR stops the surgery from beginning, or if in progress, surgery is paused until an adequate level of anesthesia is reached. Only when the animal is deep enough under anesthesia, should the surgery begin and/or resume. Under optimal anesthesia: The HR is steady and between 130 and 325 bpm and the RR is 30–60 breaths per minute.
6. It is common for rabbits to occasionally experience a decrease in their appetite post-OP. Therefore, their appetite is monitored daily and each rabbit is weighed pre-OP, as well as post-OP on days 4 and 7. After the initial post-OP period, the rabbits are weighed weekly. If their appetite and weight decrease, the proper analgesics are administered and the animal is given supplemental feed in the form of treats, hay, pineapple sticks, Fruity Bites, or Critical Care.
7. If dehiscence of the incision occurs, minor surgical repairs (small resuturing, etc.) or major surgical repairs (large area of tissue debridement, etc.) are allowed. Also possible are self-inflicted wounds near the incision site. Unless the injured area is too large, the rabbits are surgically repaired if the wound/dehiscence exposes the underlying fascia. If there is not enough intact skin around the wound to close the area, or there is an underlying seroma, the area may be left to heal by second intention. If the wound is superficial, the wound is allowed to heal by second intention. If this should occur, the fur is clipped from the surrounding area, the wound is cleaned with dilute chlorhexidine and saline, and Triple antibiotic ointment is applied until a scab forms over the wound. An E-collar is placed on the rabbit.

References

1. Mankin HJ (1982) The response of articular cartilage to mechanical injury. J Bone Joint Surg Am 64(3):460–466
2. Wakitani S, Imoto K, Yamamoto T et al (2002) Human autologous culture expanded bone marrow mesenchymal cell transplantation for repair of cartilage defects in osteoarthritic knees. Osteoarthritis Cartilage 10(3):199–206
3. Kuroda R, Ishida K, Matsumoto T et al (2007) Treatment of a full-thickness articular cartilage defect in the femoral condyle of an athlete with autologous bone-marrow stromal cells. Osteoarthritis Cartilage 15(2):226–231
4. Kon E, Verdonk P, Condello V et al (2009) Matrix-assisted autologous chondrocyte transplantation for the repair of cartilage defects of the knee systematic clinical data review and study quality analysis. Am J Sports Med 37(1 suppl):156S–166S

5. Almqvist KF, Dhollander AA, Verdonk PC et al (2009) Treatment of cartilage defects in the knee using alginate beads containing human mature allogenic chondrocytes. Am J Sports Med 37(10):1920–1929
6. Sajjadian A, Rubinstein R, Naghshineh N (2010) Current status of grafts and implants in rhinoplasty: part I. Autologous grafts. Plast Reconstr Surg 125(2):40e–49e
7. Chang SC, Tobias G, Roy AK et al (2003) Tissue engineering of autologous cartilage for craniofacial reconstruction by injection molding. Plast Reconstr Surg 112(3): 793–799
8. Yanaga H, Imai K, Koga M et al (2012) Cell-engineered human elastic chondrocytes regenerate natural scaffold in vitro and neocartilage with neoperichondrium in the human body post-transplantation. Tissue Eng Part A 18(19–20):2020–2029
9. Kushnaryov A, Yamaguchi T, Briggs K et al (2014) Evaluation of autogenous engineered septal cartilage grafts in rabbits: a minimally invasive preclinical model. Adv Otolaryngol 2014:415821
10. Chang AA, Reuther MS, Briggs KK et al (2012) In vivo implantation of tissue-engineered human nasal septal neocartilage constructs a pilot study. Otolaryngol Head Neck Surg 146(1):46–52
11. Haisch A, Gröger A, Radke C et al (2000) Macroencapsulation of human cartilage implants: pilot study with polyelectrolyte complex membrane encapsulation. Biomaterials 21(15):1561–1566
12. Hom D (1994) The wound healing response to grafted tissues. Otolaryngol Clin North Am 27(1):13–24
13. Brent B (1979) The versatile cartilage autograft: current trends in clinical transplantation. Clin Plast Surg 6(2):163–180
14. de Souza MMA, Gregório LC, Sesso R et al (2008) Study of rabbit septal cartilage grafts placed on the nasal dorsum. Arch Facial Plast Surg 10(4):250–254
15. Puelacher W, Mooney D, Langer R et al (1994) Design of nasoseptal cartilage replacements synthesized from biodegradable polymers and chondrocytes. Biomaterials 15(10):774–778
16. Alexander TH, Sage AB, Chen AC et al (2010) Insulin-like growth factor-i and growth differentiation factor-5 promote the formation of tissue-engineered human nasal septal cartilage. Tissue Eng C 16(5):1213–1221
17. Caffrey JP, Kushnaryov AM, Reuther MS et al (2013) Flexural properties of native and tissue-engineered human septal cartilage. Otolaryngol Head Neck Surg 148(4):576–581
18. Chia SH, Schumacher BL, Klein TJ et al (2004) Tissue-engineered human nasal septal cartilage using the alginate-recovered-chondrocyte method. Laryngoscope 114(1): 38–45
19. Kitahara S, Nakagawa K, Sah RL et al (2008) In vivo maturation of scaffold-free engineered articular cartilage on hydroxyapatite. Tissue Eng Part A 14(11):1905–1913
20. Masuda K, Pfister B, Sah R et al (2006) Osteogenic protein-1 promotes the formation of tissue-engineered cartilage using the alginate-recovered-chondrocyte method. Osteoarthritis Cartilage 14(4):384–391
21. Masuda K, Sah RL, Hejna MJ et al (2003) A novel two-step method for the formation of tissue-engineered cartilage by mature bovine chondrocytes: the alginate-recovered-chondrocyte (arc) method. J Orthop Res 21(1):139–148
22. Bankowski Z (1985) International guiding principles for biomedical research involving animals. http://www.cioms.ch/publications/guidelines/1985_texts_of_guidelines.htm
23. Committee for the Update of the Guide for the Care and Use of Laboratory Animals (2011) Guide for the care and use of laboratory animals. http://oacu.od.nih.gov/regs/guide/guide_2011.pdf
24. National Institutes of Health (2002) Public health service policy on humane care and use of laboratory animals. http://grants.nih.gov/grants/olaw/references/phspol.htm#USGovPrinciples. Accessed 12 May 2015
25. Belizário JE (2009) Immunodeficient mouse models: an overview. Open Immunol J 2:79–85
26. United States Department of Agriculture (2013) Animal welfare act and animal welfare regulations http://www.aphis.usda.gov/animal_welfare/downloads/Animal%20Care%20Blue%20Book%20-%202013%20-%20FINAL.pdf
27. American Veterinary Medical Association (2013) AVMA guidelines for the euthanasia of animals. http://oacu.od.nih.gov/regs/AVMA_Euthanasia.pdf
28. Ewald AJ, Werb Z, Egeblad M (2011) Monitoring of vital signs for long-term survival of mice under anesthesia. Cold Spring Harb Protoc 2011(2):pdb. prot5563

Part VI

Evaluation

Chapter 19

Proteomic Analysis of Engineered Cartilage

Xinzhu Pu and Julia Thom Oxford

Abstract

Tissue engineering holds promise for the treatment of damaged and diseased tissues, especially for those tissues that do not undergo repair and regeneration readily in situ. Many techniques are available for cell and tissue culturing and differentiation of chondrocytes using a variety of cell types, differentiation methods, and scaffolds. In each case, it is critical to demonstrate the cellular phenotype and tissue composition, with particular attention to the extracellular matrix molecules that play a structural role and that contribute to the mechanical properties of the resulting tissue construct. Mass spectrometry provides an ideal analytical method with which to characterize the full spectrum of proteins produced by tissue-engineered cartilage. Using normal cartilage tissue as a standard, tissue-engineered cartilage can be optimized according to the entire proteome. Proteomic analysis is a complementary approach to biochemical, immunohistochemical, and mechanical testing of cartilage constructs. Proteomics is applicable as an analysis approach to most cartilage constructs generated from a variety of cellular sources including primary chondrocytes, mesenchymal stem cells from bone marrow, adipose tissue, induced pluripotent stem cells, and embryonic stem cells. Additionally, proteomics can be used to optimize novel scaffolds and bioreactor applications, yielding cartilage tissue with the proteomic profile of natural cartilage.

Key words Proteomics, Mass spectrometry, Cartilage, SDS-electrophoresis, Extracellular matrix, Chondrocyte

1 Introduction

Proteins of the extracellular matrix play an essential role in determining the mechanical properties of tissues. Cartilage is an avascular tissue, rich in extracellular matrix molecules. Collagens and proteoglycans work together to create a structure that supports the function of cartilage as a resilient tissue that is able to withstand forces. Injury to the cartilage can result in irreversible damage and tissue engineering holds promise for the treatment of damaged and diseased cartilage. It is critical to recapitulate the normal composition and organization of the extracellular molecules that play a structural role in tissue-engineered cartilage.

Mass spectrometry (MS) and proteomics provide an economical and efficient method to monitor the complete profile of extracellular

Pauline M. Doran (ed.), *Cartilage Tissue Engineering: Methods and Protocols*, Methods in Molecular Biology, vol. 1340, DOI 10.1007/978-1-4939-2938-2_19, © Springer Science+Business Media New York 2015

matrix molecules from a sample of cartilage tissue. By comparing to native cartilage, it is possible to assess how closely the tissue-engineered cartilage resembles the native tissue.

Several laboratories have developed methods of extraction and separation of cartilage proteins and subsequent analysis by mass spectrometry to evaluate the cartilage proteome. While efforts to generate laboratory grown cartilage have been carried out for more than 20 years, the application of mass spectrometry for proteomic analysis has seen a steady increase over the past 10 years. In 2005 and 2006, methods to characterize the proteome of human normal articular chondrocytes were developed and applied to osteoarthritis [1–5]. Osteoarthritis continues to be one of the primary foci for proteomic analysis [6], along with questions regarding the proteome of growth plate cartilage and changes in cartilage proteome during development [7, 8]. Recently, matrix-assisted laser desorption/ionization (MALDI) was used to characterize the proteome of osteoarthritic cartilage [9].

Quantitative proteomics has been applied to cartilage studies utilizing two-dimensional gel electrophoresis for the separation of extracted proteins prior to mass spectrometry [10, 11], isobaric tag for relative and absolute quantitation (iTRAQ) labeling in conjunction with two-dimensional gel techniques [12, 13], and stable isotope labeling by amino acids in cell culture (SILAC) [14]. Label-free techniques for quantitative proteomics of cartilage have also been developed [15]. Extraction methods have been optimized for proteomic analysis of cartilage. Wilson and Bateman applied sequential extraction and fractionation with two-dimensional electrophoresis of cartilage proteins to improve the data obtained [16].

In cartilage tissue engineering, proteomics has been applied to the study of cartilage tissue explants as well as chondrocytes maintained in serum-containing and serum-free culture [14, 17]. To better understand the mechanobiology of cartilage tissue, proteomics has been used to characterize the effect of mechanical compression injury as well as the effect of stretching on cartilage [18, 19]. Various sources of cells for tissue engineering of cartilage have been characterized using proteomics including primary chondrocytes, bone marrow-derived chondrogenic precursor cells [20, 21], adipose-derived chondrogenic stem cells [22], and human mesenchymal stem cells used for chondrogenic differentiation and cartilage tissue engineering [23]. Application of proteomics to the field of cartilage biology and tissue engineering includes the analysis of posttranslational modifications [18], degradation of cartilage matrix molecules [24], explant models of inflammation [25, 26], and for questions of molecular interactions within the extracellular matrix of cartilage, or the “interactome” [27, 28].

Here we outline a versatile method of extraction, separation, mass spectrometry, and proteomic analysis for the evaluation of the extracellular matrix generated by primary chondrocytes maintained in three-dimensional culture using native cartilage as a standard.

2 Materials

Materials for cell isolation, expansion, and differentiation seeding on scaffolds and bioreactor use are provided in accompanying chapters and will not be covered here. Materials described are for analysis of proteins from chondrocytes maintained in three-dimensional culture compared to proteins extracted from native cartilage. Nanopure water (18.2 MΩ) is used to prepare all reagent solutions. All reagents are of analytical grade unless otherwise specified.

2.1 Protein Extraction from Native Cartilage and Chondrocytes

1. Cartilage tissue, e.g., we obtained cartilage from the femoral head and condyles of an early third trimester fetal calf.
2. Tissue homogenizer (Polytron).
3. Low salt cartilage extraction buffer: 0.1 M NaCl containing 0.05 M Tris–HCl, 0.01 M EDTA, and protease inhibitors 4-(2-aminoethyl) benzenesulfonyl fluoride (AEBSF, 1 mM), Pepstatin A (10 μM), E-64 (150 μM), Bestatin (50 μM), Leupeptin (20 μM), and aprotinin (0.8 μM), pH 7.0.
4. High salt cartilage extraction buffer: 1 M NaCl containing 0.05 M Tris–HCl, 0.01 M EDTA, and protease inhibitors AEBSF (1 mM), Pepstatin A (10 μM), E-64 (150 μM), Bestatin (50 μM), Leupeptin (20 μM), and aprotinin (0.8 μM), pH 7.0.
5. PBS: 0.1 M phosphate buffered saline, pH 7.4.
6. Cell lysis buffer: 25 mM Tris–HCl pH 7.6, 150 mM NaCl, 1 % NP-40, 1 % sodium deoxycholate, 0.1 % sodium dodecyl sulfate (SDS) and protease inhibitors AEBSF (1 mM), Pepstatin A (10 μM), E-64 (150 μM), Bestatin (50 μM), Leupeptin (20 μM), and aprotinin (0.8 μM).

2.2 Extract Treatment and Protein Quantification

1. Cibacron Blue agarose resin: Cibacron Blue agarose beads in a 50 % aqueous slurry.
2. Albumin binding/wash buffer: 25 mM Tris, 25 mM NaCl, pH 7.5.
3. Bicinchoninic acid (BCA) protein assay kit.
4. Acetone solution: 80 % acetone in Nanopure water.

2.3 SDS Polyacrylamide Gel Electrophoresis

1. 4–12 % Bis-Tris mini gel.
2. SDS sample loading buffer: 10 % glycerol, 0.14 M Tris Base, 0.1 M Tris–HCl, 2 % lithium dodecyl sulfate (LDS), 0.5 mM EDTA, 0.02 % Serva Blue G250; 0.006 % phenol red, 1.25 % 2-mercaptoethanol, pH 8.5.
3. Running buffer: 50 mM 2-(*N*-morpholino)ethanesulfonic acid (MES), 50 mM Tris Base, 0.1 % SDS, 1 mM EDTA, pH 7.2.
4. Coomassie Blue stain (Bio-Safe™, Bio-Rad).

2.4 In-Gel Tryptic Protein Digestion

1. Gel washing solution: 50 mM NH_4HCO_3, prepared freshly.
2. Coomassie Blue destaining solution: 40 % MS-grade acetonitrile, 50 mM NH_4HCO_3, prepared freshly.
3. Trypsin solution: 20 μg/mL of proteomic-grade trypsin prepared just before use and kept on ice.
4. Reducing solution: 10 mM dithiothreitol (DTT), prepared freshly.
5. Alkylating solution: 55 mM iodoacetamide (IAA), prepared freshly.
6. Acetonitrile/formic acid solution: 60 % MS-grade acetonitrile in water containing 5 % formic acid.
7. Peptide reconstitution solution: 5 % MS-grade acetonitrile, 0.1 % MS-grade formic acid.

2.5 Nanoscale Liquid Chromatography (LC) and Nano Electrospray—Tandem Mass Spectrometry

1. Nanoscale LC system.
2. C18 desalting column: 2 cm, ID 100 μm, 5 μm.
3. C18 analytical column: 10 cm, ID 75 μm, 3 μm.
4. Mobile phase A: 99.9 % water, 0.1 % MS-grade formic acid.
5. Mobile phase B: 99.9 % MS-grade acetonitrile, 0.1 % MS-grade formic acid.
6. Nano electrospray ionization (ESI) source, e.g., Flex II nano-ESI source (Thermo Scientific).
7. Mass spectrometer, e.g., Velos Pro Dual-Pressure Linear Ion Trap mass spectrometer (Thermo Scientific).

2.6 Peptide and Protein Identification by Database Searching

1. Database search software, e.g., Proteome Discoverer 1.4 (Thermo Scientific).
2. Database search engine, e.g., Sequest HT (Thermo Scientific) and Mascot 2.4 (Matrix Science).
3. Protein sequence database, e.g., National Center for Biotechnology Information (NCBI) non-redundant protein database (in this case, we used the non-redundant protein database for *Bos taurus* downloaded on June 16, 2014).

3 Methods

Methods for cell isolation, expansion, and differentiation seeding on scaffolds and bioreactor use are provided in accompanying chapters and will not be covered here. In the method described here, proteins from chondrocytes maintained in three-dimensional culture are compared to proteins extracted from native cartilage. A commonly used gel-LC-MS/MS proteomic profiling workflow is described in detail (Fig. 1).

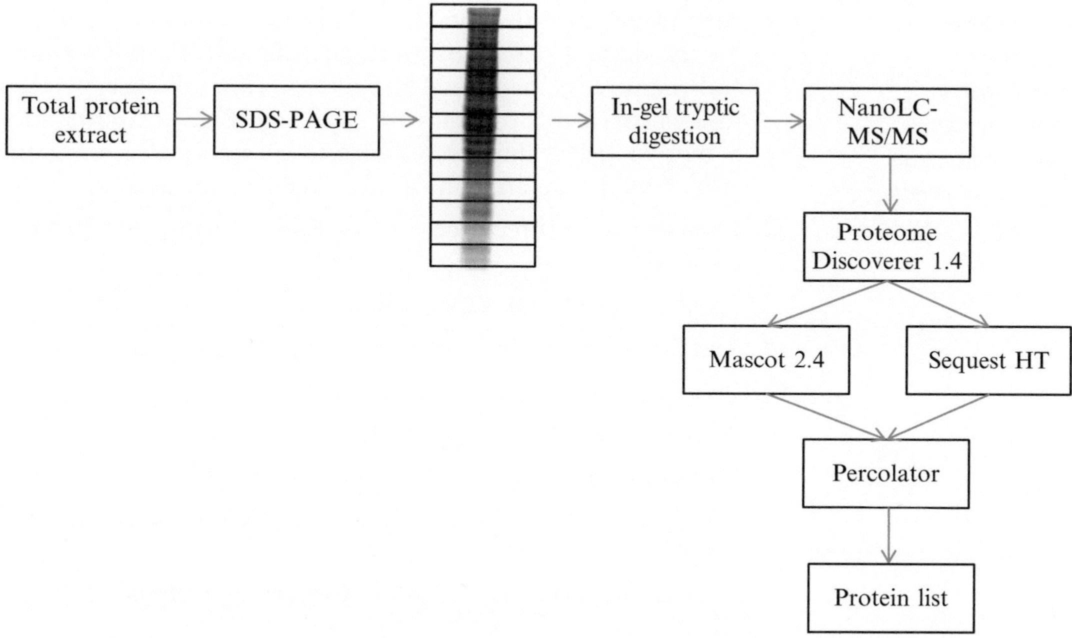

Fig. 1 A Gel-LC-MS/MS workflow for proteomic profiling

Keratin contamination is a significant challenge in MS-based bottom-up proteomics [29]. Large amounts of keratin in a sample can mask proteins of interest that are less abundant. Thus, efforts need to be taken to minimize keratin contamination in all steps prior to tryptic digestion (*see* **Note 1**).

3.1 Cartilage Tissue Preparation and Extraction

1. Remove perichondrium and adhering tissues from primary cartilage material.
2. Mince and homogenize the cartilage using a tissue homogenizer.
3. Extract cartilage proteins in low salt cartilage extraction buffer at 4 °C with stirring for 4 h.
4. Centrifuge the sample at 100,000 × *g* at 4 °C for 1 h to pellet insoluble material.
5. Carefully transfer the supernatant to a new tube.
6. Extract cartilage proteins from the resulting pellet in high salt cartilage extraction buffer at 4 °C with stirring for 4 h.
7. Centrifuge the sample at 100,000 × *g* at 4 °C for 1 h to pellet insoluble material.
8. Carefully transfer the supernatant to a new tube.
9. Take an aliquot of each supernatant to determine the protein concentration using the bicinchoninic acid (BCA) assay.
10. Store the rest of the supernatant at −80 °C for further analysis.

3.2 Protein Extraction from Primary Chondrocytes and Chondrocytes in Three-Dimensional Cultures

1. After maintaining the chondrocytes in three-dimensional culture for the desired period of time, carefully transfer the medium and the three-dimensional scaffold carrying the chondrocytes to a clean 15-mL conical tube and centrifuge at 2500 × *g* for 5 min to pellet the three-dimensional scaffold carrying the cells. Save the supernatant for secreted protein analysis.
2. Wash the pellet three times in cold PBS. Centrifuge at 2500 × *g* for 5 min.
3. Carefully decant the supernatant.
4. Add 0.5 ml of cell lysis buffer to the pellet and gently mix.
5. Incubate on ice for 30 min. Gently mix every 10 min.
6. Centrifuge at 14,000 × *g* for 15 min at 4 °C.
7. Carefully transfer the supernatant to a new tube.
8. Take an aliquot of the supernatant to determine the protein concentration using the BCA assay.
9. Store the supernatant at −80 °C for further analysis.

3.3 Bovine Serum Albumin Depletion and Protein Quantification

Serum albumin may be removed from culture medium samples to improve the detection of lower abundance proteins (*see* **Note 2**).

1. Centrifuge the medium collected in Subheading 3.2, **step 1**, at 3000 × *g* to eliminate cell debris.
2. Transfer 5 mL of the medium into dialysis tubing with a molecular cutoff of 3000 Da. Dialyze the medium against 1 L of albumin binding/wash buffer for 24 h at 4 °C.
3. Replace the buffer and continue the dialysis for 24 h at 4 °C.
4. Place a spin column (column volume 1000 μl) into a 2.0-mL collection tube. Shake the Cibacron Blue agarose resin bottle to resuspend the resin. Using a wide-bore micropipette tip, transfer 400 μL of the slurry (corresponding to 200 μL of settled resin volume) into the spin column and loosely cap the column.
5. Centrifuge at 12,000 × *g* for 1 min to remove excess liquid. Discard flow-through and place the spin column back into the same collection tube.
6. Add 200 μL of albumin binding/wash buffer to the spin column.
7. Centrifuge at 12,000 × *g* for 1 min. Discard flow-through and place the spin column into a 1.5-ml collection tube.
8. Apply 50 μL of dialyzed medium sample to the resin and incubate for 1–2 min at room temperature.
9. Centrifuge at 12,000 × *g* for 1 min. Reapply the flow-through to the spin column and incubate for 1–2 min at room temperature to ensure maximal albumin binding.

10. Centrifuge at 12,000 ×*g* for 1 min. Retain the flow-through. Place the spin column in a new collection tube.
11. Wash the resin to release unbound proteins by adding 50 μL of albumin binding/wash buffer.
12. Centrifuge at 12,000 ×*g* for 1 min. Retain the flow-through. Place the spin column in a new collection tube.
13. Repeat **steps 11** and **12** four additional times.
14. Combine all the fractions. Concentrate the depleted medium by a factor of 20 using a protein ultrafiltration filter with a cut-off of 3 kDa.
15. Take an aliquot of the concentrated medium to determine the protein concentration using the BCA assay.
16. Store the rest of the medium at −80 °C for further analysis.

3.4 Acetone Protein Precipitation

Precipitate protein using acetone to concentrate proteins and eliminate compounds that may interfere with SDS-PAGE and LC-MS analysis.

1. Chill acetone and 80 % acetone solution to −20 °C.
2. Place chondrocyte lysate solution containing 100 μg of total protein into a clean 1.5-ml microcentrifuge tube.
3. Add four times the sample volume of cold (−20 °C) acetone to the tube.
4. Mix the sample and incubate overnight at −20 °C.
5. Centrifuge for 15 min at 14,000 ×*g* at 4 °C.
6. Carefully decant the supernatant without dislodging the protein pellet.
7. Add 1 mL of cold (−20 °C) 80 % acetone solution and resuspend the protein pellet.
8. Centrifuge for 15 min at 14,000 ×*g* at 4 °C
9. Carefully decant the supernatant without dislodging the protein pellet. Pipette off the residual washing solution using a gel-loading tip.
10. Air-dry the protein pellet for 5 min at room temperature (*see* **Note 3**).

3.5 SDS Polyacrylamide Gel Electrophoresis

1. Prepare samples (i.e., air-dried protein pellet obtained from chondrocytes, cartilage extract, and culture medium) in SDS sample loading buffer. Use ~100 μg of protein for each sample.
2. Heat the sample at 70 °C for 10 min.
3. Load the entire volume of sample onto a 4–12 % Bis-Tris mini gel.
4. Run the gel at 200 V for 35 min using running buffer.
5. At the end of the electrophoresis, wash the gel in deionized water three times.

6. Stain the gel with Coomassie Blue stain for 1 h.
7. Wash the gel with deionized water extensively until the water is clear.
8. Record the gel image before proceeding with in-gel tryptic digestion.

3.6 In-Gel Tryptic Protein Digestion

3.6.1 Gel Band Excision

1. Cut the entire gel lane into a number of slices of equal length or excise protein bands of interest from the stained gel using a clean razor blade.
2. Cut each gel slice into small pieces (~1 mm^3) and transfer the pieces to a 1.5-mL low protein binding microcentrifuge tube prerinsed with HPLC-grade acetonitrile.

3.6.2 Destaining of Gel Pieces

1. Wash the gel pieces with 0.5 mL of gel washing solution. Shake for 15 min then discard the washing solution.
2. Add 200 μL of Coomassie Blue destaining solution and shake for 15 min (*see* **Note 4**). Discard the destaining solution.
3. Repeat the destaining step until the gel pieces are completely destained.
4. Add 200 μL of MS-grade acetonitrile and shake for 15 min to dehydrate the gel pieces.
5. Spin the tubes briefly and discard the supernatant.
6. Completely dry the gel pieces in a vacuum evaporator.

3.6.3 Reduction and Alkylation

1. Add 100 μL of reducing solution to cover the gel pieces, and incubate for 1 h at 56 °C with shaking.
2. Cool the gel pieces to room temperature. Discard the reducing solution.
3. Add 100 μL of alkylating solution to the gel pieces.
4. Incubate for 45 min at room temperature in the dark. Mix occasionally.
5. Centrifuge the tubes briefly and discard the alkylating solution.
6. Wash the gel pieces with 200 μL of gel washing solution on a shaker for 15 min. Discard the washing solution.
7. Repeat the washing step. Centrifuge briefly, discarding the liquid phase.
8. Add 200 μL of MS-grade acetonitrile and shake for 15 min to dehydrate the gel pieces.
9. Centrifuge briefly and discard the liquid phase.
10. Dry the gel pieces completely in a vacuum evaporator.

3.6.4 Tryptic Digestion

1. Precool the gel pieces and trypsin solution on ice.
2. Add enough trypsin solution (30–50 μL) to the gel pieces to cover.

3. Incubate on ice for 60 min.
4. Centrifuge the tube briefly. Pipette off any remaining trypsin solution (*see* **Note 5**).
5. Add 20 μL of gel washing solution to cover the gel pieces.
6. Incubate in a shaker at 30 °C overnight.

3.6.5 Peptide Extraction

1. At the end of tryptic digestion, add 50 μL of gel washing solution to the gel pieces.
2. Shake vigorously for 10 min.
3. Centrifuge the tubes briefly. Sonicate in a water bath for 5 min.
4. Centrifuge briefly and transfer the extract to a new clean microcentrifuge tube using a long narrow gel-loading tip.
5. Add 50 μL of acetonitrile/formic acid solution to the gel pieces. Repeat **steps 2** and **3**.
6. Centrifuge briefly. Transfer the extract to the tube in **step 4** using a gel-loading tip.
7. Dry the extracted digests in a vacuum evaporator.
8. Add 30 μL of peptide reconstitution solution to each tube. Mix for 10 min and sonicate for 5 min.
9. Centrifuge the samples for 10 min at 15,000 × *g*.
10. Transfer the reconstituted peptide solution into a standard 2-mL HPLC sample vial with a 100-μL glass low-volume insert. Samples are now ready for LC-MS analysis (*see* **Note 6**).

3.7 Reverse Phase Nano LC Separation of Peptides

The resulting peptide mixtures are separated by reverse phase nano LC using a two-column setup. All LC mobile phases are degassed with continuous helium sparging (*see* **Note** 7). Figure 2 shows a typical base peak chromatogram of peptide mixtures.

1. House samples in a refrigerated autosampler during the entire of period of analysis. Inject 8-μL samples onto a C18 desalting column, desalting with 20 μL of mobile phase A.
2. Elute desalted peptide mixtures from the desalting column onto a C18 analytical column and separate using a linear gradient with two mobile phases (A and B) at a flow rate of 250 nL/min. Begin the gradient at 0 % B, increase linearly to 40 % B over 60 min and then to 80 % B over 16 min. Maintain the mobile phases at this percentage for a further 14 min as a washing step.

3.8 Nano Electrospray—Tandem Mass Spectrometry

1. Use a fused silica emitter (10 μL ID) directly attached to the analytical column through a zero dead volume union to elute peptides sprayed at a voltage of 2.2 kV.

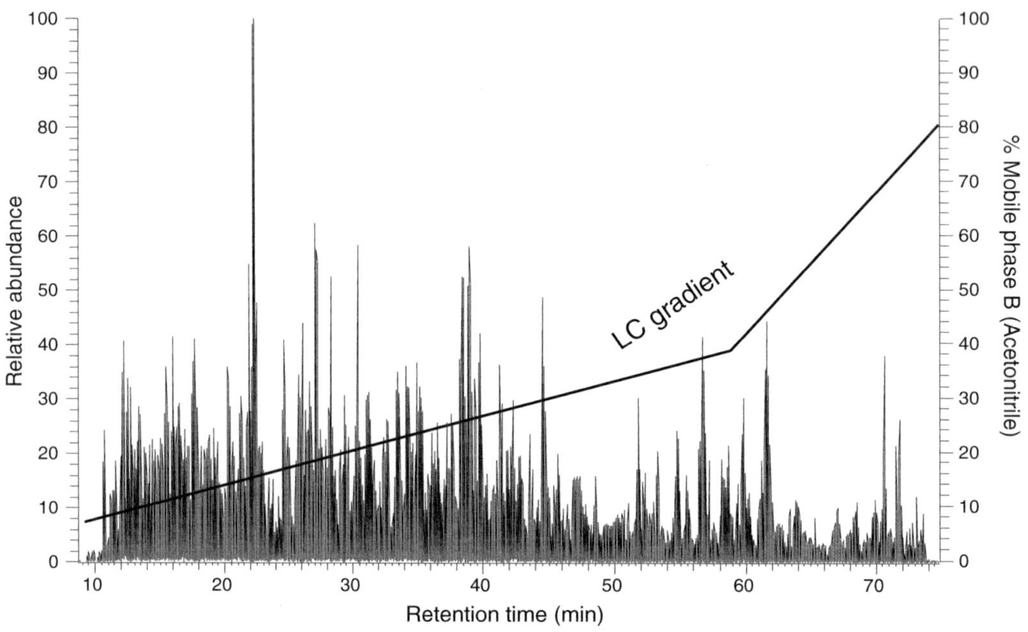

Fig. 2 A typical base peak chromatogram of peptide mixture resulting from in-gel tryptic digestion following gradient elution

2. Collect MS/MS data in data-dependent acquisition mode. Use Collision-Induced Dissociation (CID) with a normalized collision energy of 35 % to fragment the precursor ions. Collect MS/MS data for the 10 most abundant precursor ions selected from the proceeding full MS scan over the m/z range of 300–2000.

3.9 Peptide and Protein Identification by Database Searching

Peptide spectral matching and protein identification are achieved by database searching using specific algorithms, e.g., Mascot 2.4.0 (Matrix Science) and Sequest HT algorithms (Thermo Scientific) (*see* **Note 8**). MS/MS spectra of peptides may be used to search against the NCBI non-redundant protein database (*Bos taurus* in this example), which can be obtained from the NCBI website. The main search parameters for Mascot 2.4 and Sequest HT include semi trypsin, maximum missed cleavage site of two, precursor mass tolerance of 1.5 Da, fragment mass tolerance of 0.8 Da, fixed modification of carbamidomethyl cysteine (+57.021 Da), and variable modification of oxidized methionine (+15.995 Da). A decoy database search is performed to calculate the false discovery rate (FDR). Proteins containing at least two high confidence peptides (FDR$\leq$0.01) are considered positively identified and are reported. Figure 3 shows a typical peptide ion fragment spectrum. Prevalent detectable proteins are presented in Table 1.

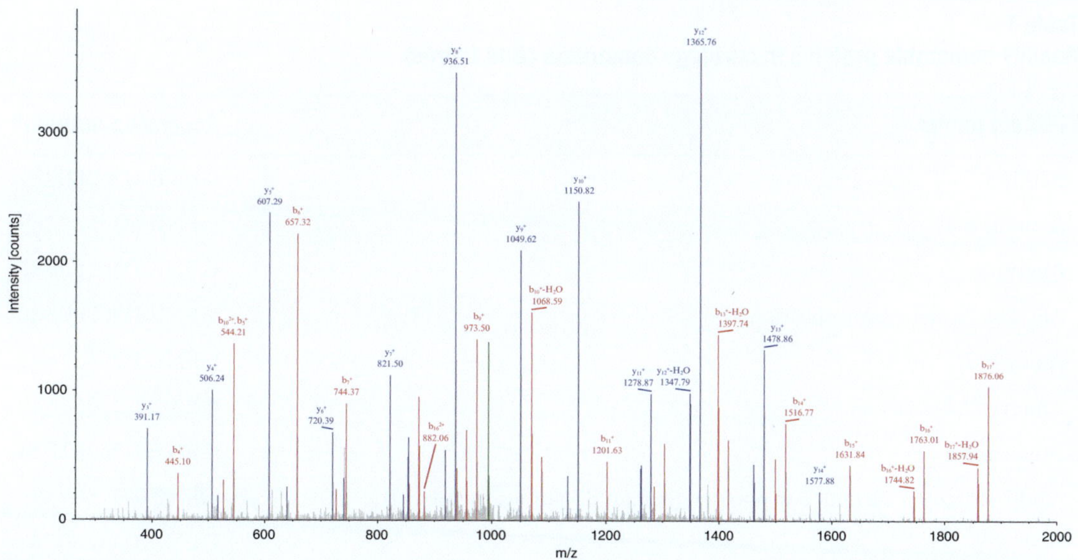

Fig. 3 Ion fragment spectrum of peptide DAEEVISQTIDTITDMIK from collagen alpha-1(VI) (*Bos taurus*)

4 Notes

1. Keratin contamination is a significant challenge in MS-based shotgun proteomics [29]. The main source of keratin contamination is dust that contains dead skin cells from humans. Although keratin contamination cannot be completely avoided, it is important to take every effort to minimize the contamination. A key factor in avoiding keratin is to avoid the contamination of samples, supplies, and reagents with dust particles. Thus, efforts should be taken to keep the workplace free of dust. Wear a clean lab coat and nitrile gloves at all times during sample handling. Perform sample preparation in a HEPA-filtered laminar-airflow hood when possible. Always use analytical grade or higher reagents and store all the reagents and samples properly to minimize their exposure to dust.
2. The high concentration of albumin in the fetal bovine serum commonly used in cell culture may obscure the detection of many secreted proteins in culture media. Thus, the depletion of albumin is necessary. There are different methods available to deplete albumin in cell culture media. We found that agarose resin of affinity ligand Cibacron Blue dye gives satisfactory results as evidenced by both SDS-PAGE and mass spectrometry analysis (Fig. 4).
3. Care should be taken to prevent overdrying of the protein pellet as this will make resolubilization difficult.

Table 1
Readily detectable proteins in cartilage constructs (*Bos taurus*)

Protein name[a]	Accession number[b]
Actin, β	GENE ID: 280979
Actinin, α 4	GENE ID: 522269
Aggrecan	GENE ID: 280985
Aldolase A, fructose-bisphosphate	GENE ID: 509566
Annexin A1	GENE ID: 327662
Annexin A2	GENE ID: 282689
Annexin A5	GENE ID: 281626
Acidic (leucine-rich) nuclear phosphoprotein 32 family, member B	GENE ID: 509685
Biglycan	GENE ID: 280733
Calreticulin	GENE ID: 281036
Chondroadherin	GENE ID: 281069
Collagen, type I, α 1	GENE ID: 282187
Chondrocalcin-Carboxy propeptide of collagen type II α1	GENE ID: 407142
Collagen, type II, α 1	GENE ID: 407142
Collagen, type VI, α 1	GENE ID: 511422
Collagen, type IX, α 1	GENE ID: 282195
Collagen, type XI, α 1	GENE ID: 287013
Collagen, type XI, α 2 (PARP)	GENE ID: 515435
Collagen, type XII, α 1	GENE ID: 359712
Collagen, type XIV, α 1	GENE ID: 7373*
Cartilage oligomeric matrix protein	GENE ID: 281088
Decorin	GENE ID: 280760
Epiphycan	GENE ID: 281747
Fibromodulin	GENE ID: 281168
Hyaluronan and proteoglycan link protein 1	GENE ID: 281717
Heat shock 70 kDa protein 5 (glucose-regulated protein, 78 kDa)	GENEID: 415113
Heparan sulfate proteoglycan 2	GENE ID: 444872
Lactate dehydrogenase A	GENE ID: 281274
Matrilin 1, cartilage matrix protein	GENE ID: 512059
Matrilin 3	GENE ID: 540041

(continued)

Table 1
(continued)

Protein name[a]	Accession number[b]
Nidogen 2 (osteonidogen)	GENE ID: 521854
Nucleolin	GENE ID: 497013
Phosphoglycerate kinase 1	GENE ID: 507476
Procollagen-lysine 1, 2-oxoglutarate 5-dioxygenase 1	GENE ID: 281409
Procollagen-lysine, 2-oxoglutarate 5-dioxygenase 2	GENE ID: 533642
Protein disulfide isomerase family A, member 3	GENE ID: 281803
Protein disulfide isomerase family A, member 4	GENEID: 415110
Protein disulfide isomerase family A, member 5	GENE ID: 511603
Prolyl 4-hydroxylase	GENE ID: 281373
Pyruvate kinase, muscle	GENE ID: 512571
Semenogelin I (α-Inhibin)	GENE ID: 281254
Serpin peptidase inhibitor, clade H (collagen binding protein 1)	GENE ID: 510850
Tenascin	GENE ID: 540664
Serotransferrin	GENE ID: 280705
Thrombospondin 1	GENE ID: 281530
Thrombospondin 1, amino terminal domain	GENE ID: 281530
Tumor rejection antigen (gp96 or HSP90B1)	GENE ID: 282646
Vitrin	GENE ID: 280957

National Center for Biotechnology Information (NCBI)
[a]Entrez Gene Full Name Protein Accession number
[b]GENE ID Accession Number. Organism: *Bos taurus*
*By homology to human sequence; *Homo sapiens*

4. 50 % acetonitrile ammonium bicarbonate buffer is commonly used to destain Coomassie Blue stained gel pieces [30]. We found that 40 % acetonitrile ammonium bicarbonate is a more efficient destaining buffer.
5. It is important to aspirate off excessive trypsin as the abundant peptides from trypsin autolysis may interfere with LC-MS analysis.
6. The resulting peptide samples should be analyzed immediately if possible. Storing peptides at room temperature or 4 °C may result in loss of some peptides. If samples cannot be analyzed immediately, they should be stored at −70 °C to minimize the loss of peptides.

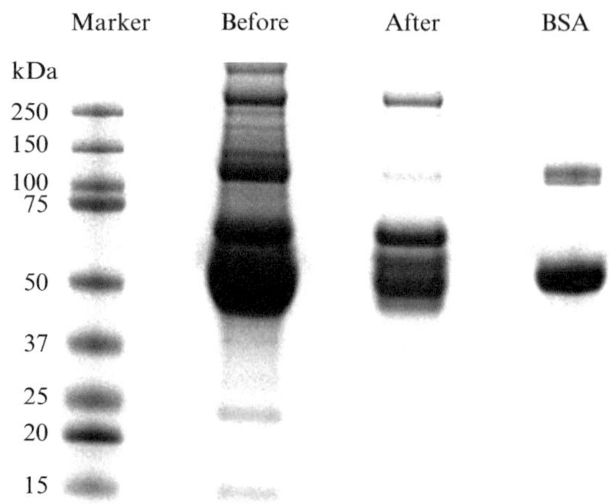

Fig. 4 SDS-PAGE of cell culture medium containing 10 % fetal bovine serum before and after bovine serum albumin depletion

7. It is well known that gases dissolved in the LC mobile phase can cause a variety of problems in liquid chromatography. Air bubbles can also severely interrupt the nanospray. Thus, mobile phase degassing is very important in LC-Nanospray. For instruments that are not equipped with an in-line degasser such as the Thermo Scientific Easy nLC II, it is critically important to properly degas the mobile phase. We found that sonication is not sufficient since the mobile phase will be resaturated with air during the extended period of analysis. We recommend using continuous helium sparging to degas the mobile phases. We found this method provided optimal results.
8. A crucial component of the analysis of shotgun proteomics data is the search engine. There are many different search engines available. Each search engine has its unique searching algorithm. It has been shown that searching the same datasets using multiple search engines and then combining the search results usually improves the analysis and gives better protein coverage [31]. We routinely use both Sequest HT and Mascot 2.0 algorithms in database search. We find that this approach gives improved protein and peptide coverage (Fig. 5). Thus we recommend using multiple search engines and combining the search results.

Acknowledgements

Authors wish to thank Barbara Jibben for technical support. This work was supported by Institutional Development Awards (IDeA) from the National Institute of General Medical Sciences of the

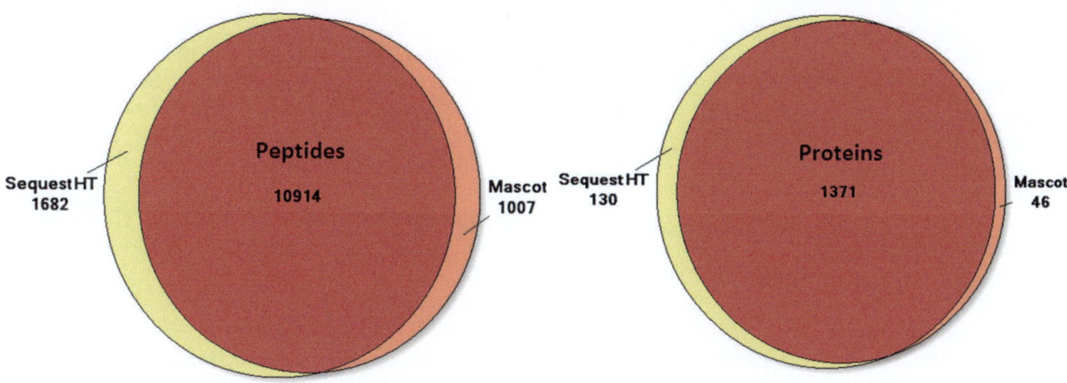

Fig. 5 Number of peptides and proteins identified by Sequest HT and Mascot 2.0

National Institutes of Health under grant numbers P20GM103408 and P20GM109095, NNX10AN29A from NASA, and the Boise State University Biomolecular Research Center.

References

1. Ruiz-Romero C, Lopez-Armada MJ, Blanco FJ (2005) Proteomic characterization of human normal articular chondrocytes: a novel tool for the study of osteoarthritis and other rheumatic diseases. Proteomics 5:3048–3059
2. De Ceuninck F, Marcheteau E, Berger S, Caliez A, Dumont V, Raes M, Anract P, Leclerc G, Boutin JA, Ferry G (2005) Assessment of some tools for the characterization of the human osteoarthritic cartilage proteome. J Biomol Tech 16:256–265
3. Vincourt JB, Lionneton F, Kratassiouk G, Guillemin F, Netter P, Mainard D, Magdalou J (2006) Establishment of a reliable method for direct proteome characterization of human articular cartilage. Mol Cell Proteomics 5:1984–1995
4. Lammi MJ, Hayrinen J, Mahonen A (2006) Proteomic analysis of cartilage- and bone-associated samples. Electrophoresis 27: 2687–2701
5. Garcia BA, Platt MD, Born TL, Shabanowitz J, Marcus NA, Hunt DF (2006) Protein profile of osteoarthritic human articular cartilage using tandem mass spectrometry. Rapid Commun Mass Spectrom 20:2999–3006
6. Yu CJ, Ko CJ, Hsieh CH, Chien CT, Huang LH, Lee CW, Jiang CC (2014) Proteomic analysis of osteoarthritic chondrocyte reveals the hyaluronic acid-regulated proteins involved in chondroprotective effect under oxidative stress. J Proteomics 99:40–53
7. Belluoccio D, Wilson R, Thornton DJ, Wallis TP, Gorman JJ, Bateman JF (2006) Proteomic analysis of mouse growth plate cartilage. Proteomics 6:6549–6553
8. Wilson R, Norris EL, Brachvogel B, Angelucci C, Zivkovic S, Gordon L, Bernardo BC, Stermann J, Sekiguchi K, Gorman JJ, Bateman JF (2012) Changes in the chondrocyte and extracellular matrix proteome during post-natal mouse cartilage development. Mol Cell Proteomics 11(M111):014159
9. Cillero-Pastor B, Eijkel GB, Kiss A, Blanco FJ, Heeren RM (2013) Matrix-assisted laser desorption ionization-imaging mass spectrometry: a new methodology to study human osteoarthritic cartilage. Arthritis Rheum 65:710–720
10. Pecora F, Forlino A, Gualeni B, Lupi A, Giorgetti S, Marchese L, Stoppini M, Tenni R, Cetta G, Rossi A (2007) A quantitative and qualitative method for direct 2-DE analysis of murine cartilage. Proteomics 7:4003–4007
11. Ruiz-Romero C, Calamia V, Carreira V, Mateos J, Fernandez P, Blanco FJ (2010) Strategies to optimize two-dimensional gel electrophoresis analysis of the human joint proteome. Talanta 80:1552–1560
12. Ikeda D, Ageta H, Tsuchida K, Yamada H (2013) iTRAQ-based proteomics reveals novel

biomarkers of osteoarthritis. Biomarkers 18: 565–572

13. Ji YH, Ji JL, Sun FY, Zeng YY, He XH, Zhao JX, Yu Y, Yu SH, Wu W (2010) Quantitative proteomics analysis of chondrogenic differentiation of C3H10T1/2 mesenchymal stem cells by iTRAQ labeling coupled with on-line two-dimensional LC/MS/MS. Mol Cell Proteomics 9:550–564
14. Polacek M, Bruun JA, Johansen O, Martinez I (2010) Differences in the secretome of cartilage explants and cultured chondrocytes unveiled by SILAC technology. J Orthop Res 28:1040–1049
15. Wilson R, Diseberg AF, Gordon L, Zivkovic S, Tatarczuch L, Mackie EJ, Gorman JJ, Bateman JF (2010) Comprehensive profiling of cartilage extracellular matrix formation and maturation using sequential extraction and label-free quantitative proteomics. Mol Cell Proteomics 9:1296–1313
16. Wilson R, Bateman JF (2008) A robust method for proteomic characterization of mouse cartilage using solubility-based sequential fractionation and two-dimensional gel electrophoresis. Matrix Biol 27:709–712
17. Koo J, Kim KI, Min BH, Lee GM (2010) Differential protein expression in human articular chondrocytes expanded in serum-free media of different medium osmolalities by DIGE. J Proteome Res 9:2480–2487
18. Piltti J, Hayrinen J, Karjalainen HM, Lammi MJ (2008) Proteomics of chondrocytes with special reference to phosphorylation changes of proteins in stretched human chondrosarcoma cells. Biorheology 45:323–335
19. Stevens AL, Wishnok JS, Chai DH, Grodzinsky AJ, Tannenbaum SR (2008) A sodium dodecyl sulfate-polyacrylamide gel electrophoresis-liquid chromatography tandem mass spectrometry analysis of bovine cartilage tissue response to mechanical compression injury and the inflammatory cytokines tumor necrosis factor alpha and interleukin-1beta. Arthritis Rheum 58:489–500
20. Chiang H, Hsieh CH, Lin YH, Lin S, Tsai-Wu JJ, Jiang CC (2011) Differences between chondrocytes and bone marrow-derived chondrogenic cells. Tissue Eng Part A 17:2919–2929
21. Rocha B, Calamia V, Mateos J, Fernandez-Puente P, Blanco FJ, Ruiz-Romero C (2012) Metabolic labeling of human bone marrow mesenchymal stem cells for the quantitative analysis of their chondrogenic differentiation. J Proteome Res 11:5350–5361
22. Gong L, Zhou X, Wu Y, Zhang Y, Wang C, Zhou H, Guo F, Cui L (2014) Proteomic analysis profile of engineered articular cartilage with chondrogenic differentiated adipose tissue-derived stem cells loaded polyglycolic acid mesh for weight-bearing area defect repair. Tissue Eng Part A 20:575–587
23. Rocha B, Calamia V, Casas V, Carrascal M, Blanco FJ, Ruiz-Romero C (2014) Secretome analysis of human mesenchymal stem cells undergoing chondrogenic differentiation. J Proteome Res 13:1045–1054
24. Wilson R, Belluoccio D, Little CB, Fosang AJ, Bateman JF (2008) Proteomic characterization of mouse cartilage degradation in vitro. Arthritis Rheum 58:3120–3131
25. Clutterbuck AL, Smith JR, Allaway D, Harris P, Liddell S, Mobasheri A (2011) High throughput proteomic analysis of the secretome in an explant model of articular cartilage inflammation. J Proteomics 74:704–715
26. Swan AL, Hillier KL, Smith JR, Allaway D, Liddell S, Bacardit J, Mobasheri A (2013) Analysis of mass spectrometry data from the secretome of an explant model of articular cartilage exposed to pro-inflammatory and anti-inflammatory stimuli using machine learning. BMC Musculoskelet Disord 14:349
27. Brachvogel B, Zaucke F, Dave K, Norris EL, Stermann J, Dayakli M, Koch M, Gorman JJ, Bateman JF, Wilson R (2013) Comparative proteomic analysis of normal and collagen IX null mouse cartilage reveals altered extracellular matrix composition and novel components of the collagen IX interactome. J Biol Chem 288:13481–13492
28. Brown RJ, Mallory C, McDougal OM, Oxford JT (2011) Proteomic analysis of Col11a1-associated protein complexes. Proteomics 11: 4660–4676
29. Glenn G (2014) Preparation of protein samples for mass spectrometry and N-terminal sequencing. Methods Enzymol 536:27–44
30. Gundry RL, White MY, Murray CI, Kane LA, Fu Q, Stanley BA, Van Eyk JE (2009) Preparation of proteins and peptides for mass spectrometry analysis in a bottom-up proteomics workflow. Curr Protoc Mol Biol. doi:10.1002/0471142727.mb1025s88, Chapter 10, Unit10.25
31. Shteynberg D, Nesvizhskii AI, Moritz RL, Deutsch EW (2013) Combining results of multiple search engines in proteomics. Mol Cell Proteomics 12:2383–2393

Chapter 20

Mechanical Testing of Cartilage Constructs

Dinorath Olvera, Andrew Daly, and Daniel John Kelly

Abstract

A key goal of functional cartilage tissue engineering is to develop constructs with mechanical properties approaching those of the native tissue. Herein we describe a number of tests to characterize the mechanical properties of tissue engineered cartilage. Specifically, methods to determine the equilibrium confined compressive (or aggregate) modulus, the equilibrium unconfined compressive (or Young's) modulus, and the dynamic modulus of tissue engineered cartilaginous constructs are described. As these measurements are commonly used in both the articular cartilage mechanics literature and the cartilage tissue engineering literature to describe the mechanical functionality of cartilaginous constructs, they facilitate comparisons to be made between the properties of native and engineered tissues.

Key words Equilibrium modulus, Aggregate modulus, Young's modulus, Dynamic modulus

1 Introduction

Collagen fibrils and proteoglycans are the structural components of articular cartilage that, together with water, support the loads that are applied to the tissue. The collagen network of articular cartilage provides tensile stiffness and strength, but also functions to restrain the swelling pressure of the embedded proteoglycans, which provide compressive stiffness to the tissue [1]. These trapped proteoglycans carry negative electrical charges in the physiological environment. The density of these fixed charges is known as the fixed charge density (FCD). The swelling pressure exerted by this FCD, known as the Donnan osmotic fluid pressure, plays a key role in maintaining cartilage hydration and in determining the ability of the tissue to support compressive loads. The equilibrium compressive properties of articular cartilage (defined below) therefore depend strongly on the proteoglycan content of the tissue, whereas the dynamic properties (where fluid pressurization plays a significant role in load support) have been shown to correlate with the tissues interstitial water and collagen content [2–5].

Pauline M. Doran (ed.), *Cartilage Tissue Engineering: Methods and Protocols*, Methods in Molecular Biology, vol. 1340, DOI 10.1007/978-1-4939-2938-2_20, © Springer Science+Business Media New York 2015

Here we first describe how to determine the equilibrium compressive modulus (in either confined or unconfined compression) and dynamic modulus of tissue engineered cartilaginous constructs. The equilibrium unconfined compressive (Young's) modulus is determined using a uniaxial unconfined compression test. The equilibrium confined compression (aggregate) modulus is determined using a uniaxial confined compression test. For both the confined and unconfined configurations, a stress relaxation test will be described. During such a test, a ramp displacement is applied to an engineered tissue until a particular strain (change in sample height divided by the original height) is reached, at which point the applied deformation is held constant until the measured force reaches an equilibrium value. During the ramp phase of the test, (relatively) high forces are recorded as fluid is forced out of the tissue. At equilibrium, no fluid flows and the load applied to the engineered construct is entirely borne by the solid phase of the tissue. Hence the equilibrium compressive modulus provides a measure of the inherent compressive stiffness of the tissue solid matrix.

A test to determine the dynamic modulus of articular cartilage in unconfined compression is also described. During a dynamic test, a cyclic displacement is applied to the tissue and the amplitude of the resulting force is measured. For normal articular cartilage, fluid pressurization supports a significant percentage of the load applied during such dynamic tests [6]. Such high levels of interstitial fluid pressurization are a result of the low permeability of the tissue, which means that the fluid cannot easily escape during loading. Hence the dynamic modulus provides insight into not only the compressive stiffness of engineered cartilage but also its permeability. Both the equilibrium compressive modulus and dynamic modulus are commonly used in the tissue engineering literature to describe the mechanical functionality of tissue engineered cartilage [7–11].

2 Materials

2.1 Sample Preparation

1. Biopsy punch of known diameter: minimum 3 mm in diameter, typically 4–10 mm in diameter. For confined compression tests, the diameter of the biopsy punch should match the inner diameter of the confining chamber of the material testing equipment.
2. Cartilage tissue: the tests are ideally performed using cylindrical samples of engineered cartilage.
   (a) If the samples are not cylindrical, first use the biopsy punch to remove a cylindrical sample from the engineered construct.

(b) Carefully remove the cartilage core from the biopsy punch. The top and bottom surfaces of the cartilage plug should be reasonably flat; if not, a microtome blade (or similar) can be used to remove superficial tissue from the engineered construct in order to create flat surfaces (*see* **Note 1**).

(c) Prior to mechanical testing, measure the diameter of the sample (*see* **Note 2**).

2.2 Equipment Setup

1. Material testing equipment: A material testing machine capable of applying a prescribed deformation to a sample while simultaneously recording the applied force is required.

   (a) In the setup for unconfined uniaxial compression, two concentric stainless steel platens (*see* Fig. 1a, b) are immersed in a bath of phosphate buffered saline (PBS) solution (*see* Fig. 1c). The bottom platen is fixed and the top platen's displacement can be controlled in the vertical direction by the material testing machine. Either platen is connected to a load cell to measure the force required to deform the sample. The sample (*see* Fig. 1d) is placed between the platens and compressed by lowering the top platen.

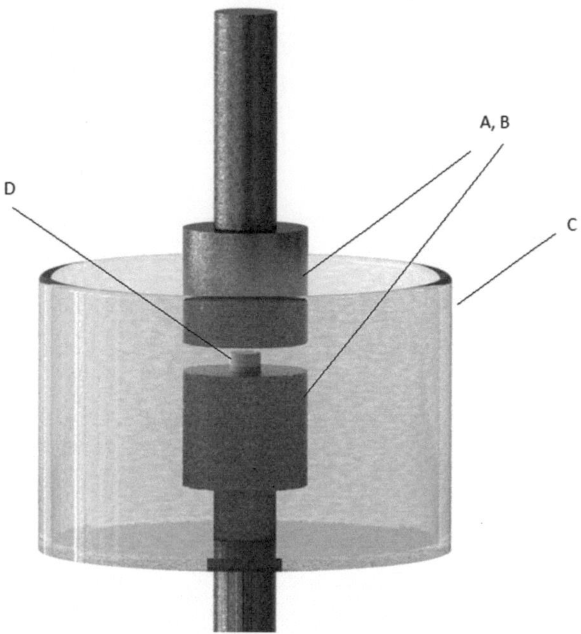

Fig. 1 Unconfined compression setup. In the unconfined test, two concentric stainless steel platens (**a**, **b**) are immersed in a bath of PBS solution (**c**). The sample (**d**) is placed between the platens and compressed by lowering the top platen

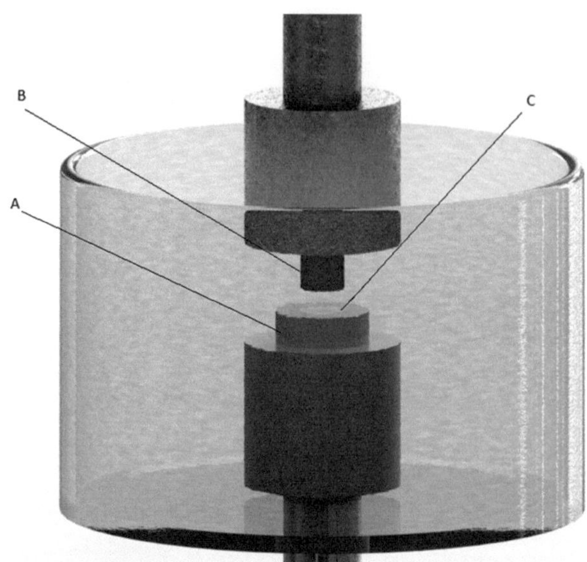

Fig. 2 Confined compression setup. During confined compression testing, the cartilage sample is placed inside a confining chamber (**a**). A rigid porous indenter (**b**) attached to the top platen (**c**) compresses the sample

(b) During the confined compression test, the cartilage sample is placed inside a confining chamber (*see* Fig. 2a). This prevents fluid from exuding out of the sample in the radial direction. In addition, the sample cannot bulge in the radial direction. A rigid porous indenter (e.g., a steel or bronze filter with a diameter slightly smaller than that of the confining chamber) (*see* Fig. 2b) attached to the top platen compresses the sample (*see* Fig. 2c) (*see* **Note 3**).

3 Methods

3.1 Unconfined Compression Testing

For both confined and unconfined compression tests, the cartilage sample is subjected to a stress relaxation test to obtain either the equilibrium confined compression (aggregate) modulus or the equilibrium unconfined compressive (Young's) modulus. The test program for both tests is composed of a ramp and hold phase. During the ramp phase the sample is compressed until a predefined strain is reached. At this point the displacement is held constant allowing the tissue to relax, this is the hold phase. During the hold phase the proportion of load borne by fluid pressurization gradually reduces and the force drops to an equilibrium value.

1. Connect an appropriate load cell (e.g., a 10–50 N load cell) to the mechanical testing machine. Record the applied force measured by the load cell on the testing machine as a function of time during the entire test.
2. Ensure all platens and the water bath are fixed tightly.

3. Fill the bath with PBS solution and mark the level of the liquid (*see* **Note 4**).
4. Measure the diameter of the cartilage specimen and place it flat in the middle of the bottom platen.
5. Slowly lower the top platen towards the top surface of the specimen and zero the force on the load cell. Ensure that the top platen is not touching the specimen and that the level of PBS solution in the water bath covers a small portion of the top platen.
6. Apply the required preload by slowly lowering the top platen towards the sample at a rate of 1 mm/min, until the force recorded by the load cell reaches this predefined value (*see* **Note 5**). This is the point where the platen is assumed to come into contact with the sample. The distance between the two platens at this point is assumed to equate to the sample height (*see* **Note 6**).
7. Hold the position of the upper platen constant for at least 120 s to allow the force to equilibrate (*see* **Note** 7).
8. Apply a ramp displacement to compress the sample to a certain percentage of its original height at a predefined rate (*see* **step 6** on calculating the height of the sample). We suggest compressing to reduce the height of the sample by 10 % (a 10 % strain) at a displacement rate of 0.06 mm/min (*see* **Note 8**).
9. Hold this displacement for 30 min to allow the sample to relax, i.e., until the measured force reaches an equilibrium value and no longer changes with time (*see* **Note** 7).

3.2 Dynamic Test

The dynamic modulus test cycles the sample between two strain levels at a defined magnitude and frequency: typically an amplitude of 1 % strain at a frequency between 0.1 and 1 Hz is used. There is no relaxation phase between each strain cycle. It is suggested that the dynamic test be run directly after the confined or unconfined stress relaxation test (e.g., after a strain of 10 % has been applied to the sample, hence deforming the sample between 10 % strain and 11 % strain during the dynamic test). For cartilage specimens, apply a displacement of 1 % strain at a frequency of 1 Hz for ten cycles. Record the applied force as a function of time during the test.

3.3 Confined Compression Testing

The methodology for the confined compression test is the same as for unconfined testing except that the sample must be carefully placed in the confining chamber before the test commences. It must also be ensured that the porous indenter and the confining chamber (*see* Fig. 2a, b) are concentric to ensure that no frictional forces are generated as the porous indenter moves into the confining chamber. The equilibrium and dynamic moduli are measured following the same procedures as described in Subheadings 3.1 and 3.2.

3.4 Data Analysis

All of the moduli calculated are ratios of stress (which has units of pressure) to strain (which is dimensionless); hence the Young's modulus, aggregate modulus, and dynamic modulus have units of pressure. Its SI unit is the pascal (Pa or N/m2). The moduli of tissue engineered cartilage are usually expressed in kilopascals (kPa) or megapascals (MPa). Typically, the equilibrium aggregate modulus and dynamic moduli of native articular cartilage range from ~0.1 to 0.8 MPa and 5 to 40 MPa, respectively [12], although the reported properties of tissue engineered cartilage are often many times lower than these native values.

3.4.1 Determining the Equilibrium Unconfined Compression (Young's) Modulus

1. Plot the force versus time data obtained during the unconfined compression test (*see* Fig. 3). The measured force will increase until the end of the ramp phase. During the hold phase the measured force will gradually decrease, eventually reaching an equilibrium value.
2. Find the average equilibrium force by taking an average of the last 10 force readings from the load cell (e.g., the average force during the last 10 s of the test, assuming a force reading is recorded every 1 s).

Calculate the equilibrium stress ($\sigma_{\text{equilibrium}}$) by dividing the average force (F) by the cross-sectional area (A) of the specimen (Eq. 1) (*see* **Note 9**):

$$\sigma_{\text{equilibrium}} = \frac{F}{A} \tag{1}$$

The equilibrium unconfined compression (Young's) modulus ($E_{\text{equilibrium}}$) can be found by dividing the equilibrium stress ($\sigma_{\text{equilibrium}}$) by the applied strain ($\epsilon_{\text{equilibrium}}$) (Eq. 2):

$$E_{\text{equilibrium}} = \frac{\sigma_{\text{equilibrium}}}{\epsilon_{\text{equilibrium}}} \tag{2}$$

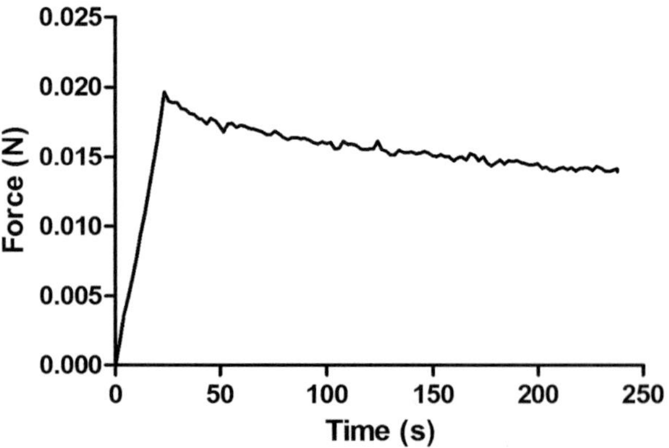

Fig. 3 Typical force-time data recorded during an equilibrium unconfined compression test

The methodology described in Subheading 3.1 suggests an applied strain of 0.1 (or 10 % strain).

3.4.2 Determining the Equilibrium Confined Compression (Aggregate) Modulus

1. Plot the force versus time data obtained during the confined compression test as described in Subheading 3.4.1 for unconfined compression.

Calculate the equilibrium stress ($\sigma_{\text{equilibrium}}$) by dividing the force at equilibrium by the cross-sectional area (A) of the specimen (Eq. 1).

Determine the equilibrium confined compression (aggregate) modulus by dividing the equilibrium stress by the applied strain (similar to Eq. 2 above).

3.4.3 Determining the Dynamic Modulus

1. Plot the force versus time data obtained during the dynamic test (*see* Fig. 4). In Fig. 4, the applied forces are shown as the sample is compressed between 10 and 11 % strain at 1 Hz for ten cycles.
2. Find the average change in force (ΔF) between a peak and trough over the ten cycles.

Compute the amplitude of the dynamic stress (σ_{dynamic}) applied to the sample by dividing ΔF by the cross-sectional area (A) of the sample (Eq. 3).

$$\sigma_{\text{dynamic}} = \frac{\Delta F}{A} \tag{3}$$

Determine the dynamic modulus (E_{dynamic}) by dividing the dynamic stress (σ_{dynamic}) by the applied dynamic strain ($\epsilon_{\text{dynamic}}$) (Eq. 4). The methodology described in Subheading 3.2 suggests an applied dynamic strain amplitude of 0.01 (or 1 % strain).

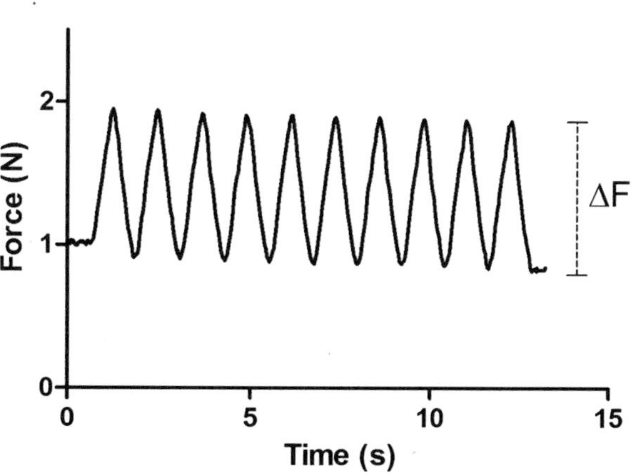

Fig. 4 Typical force-time data recorded during a dynamic unconfined compression test

$$E_{\text{dynamic}} = \frac{\sigma_{\text{dynamic}}}{\epsilon_{\text{dynamic}}} \quad (4)$$

4 Notes

1. It is critical that the top and bottom surfaces of the construct are parallel and the circular area is not compromised during slicing.
2. Some swelling of the construct may occur so it is necessary to accurately measure the cross-sectional diameter using Vernier calipers or a microscope.
3. Upon compression, fluids can exude out of the cartilage; in confined compression, the fluid cannot escape the solid side boundaries of the confining chamber; therefore the indenter needs to be porous so that the fluid exudes through the top.
4. It is important to keep the liquid level constant throughout testing.
5. The magnitude of the preload should be the smallest possible load that can be applied to the construct that ensures full contact between the top surface of the sample and the top loading platen. In our experience, a preload of ~10 % of the equilibrium force is sufficient. One suggestion is to first test a sample with a preload between 0.01 and 0.02 N. Observe the equilibrium force for this test and use 10 % of this value for the preload during subsequent testing (a higher value should be considered if this preload does not provide full contact between the loading platen and the sample).
6. The height of the sample determines how much the sample is compressed, i.e., if a 10 % strain is to be applied to the sample, the upper platen is displaced downward by an amount equal to 10 % of the construct's original height.
7. It may be necessary to wait longer if the sample is not fully relaxed, i.e., if the force value has not reached an equilibrium value.
8. It should be noted that the load cell itself may deform by a non-negligible amount during testing (depending on the design of the load cell). If using platen to platen displacement as measured by the materials testing machine to determine the level of strain applied to the construct, a correction factor (usually supplied by the manufacturer) may need to be applied to determine the actual strain applied to the sample if significant deformation occurs within the load cell itself.

9. Assume that the sample is cylindrical. The cross-sectional area of the sample (A) is therefore calculated from the diameter D as:

$$A = \pi \frac{D^2}{4}$$

References

1. Mow VC, Huiskes R (2005) Basic orthopaedic biomechanics and mechano-biology, 3rd edn. Lippincott Williams & Wilkins, Philadelphia
2. Julkunen P, Harjula T, Iivarinen J, Marjanen J, Seppanen K, Narhi T et al (2009) Biomechanical, biochemical and structural correlations in immature and mature rabbit articular cartilage. Osteoarthritis Cartilage 17(12):1628–1638
3. Kiviranta P, Rieppo J, Korhonen RK, Julkunen P, Töyräs J, Jurvelin JS (2006) Collagen network primarily controls poisson's ratio of bovine articular cartilage in compression. J Orthop Res 24:690–699
4. Williamson AK, Chen AC, Sah RL (2001) Compressive properties and function-composition relationships of developing bovine articular cartilage. J Orthop Res 19: 1113–1121
5. Schinagl RM, Gurskis D, Chen AC, Sah RL (1997) Depth-dependent confined compression modulus of full-thickness bovine articular cartilage. J Orthop Res 15:499–506
6. Soltz MA, Ateshian GA (2000) Interstitial fluid pressurization during confined compression cyclical loading of articular cartilage. Ann Biomed Eng 28:150–159
7. Farrell MJ, Fisher MB, Huang AH, Shin JI, Farrell KM, Mauck RL (2014) Functional properties of bone marrow-derived MSC-based engineered cartilage are unstable with very long-term in vitro culture. J Biomech 47:2173–2182
8. Thorpe SD, Nagel T, Carroll SF, Kelly DJ (2013) Modulating gradients in regulatory signals within mesenchymal stem cell seeded hydrogels: a novel strategy to engineer zonal articular cartilage. PLoS One 8:e60764
9. Buckley CT, Vinardell T, Thorpe SD, Haugh MG, Jones E, McGonagle D et al (2010) Functional properties of cartilaginous tissues engineered from infrapatellar fat pad-derived mesenchymal stem cells. J Biomech 43: 920–926
10. Carroll SF, Buckley CT, Kelly DJ (2014) Cyclic hydrostatic pressure promotes a stable cartilage phenotype and enhances the functional development of cartilaginous grafts engineered using multipotent stromal cells isolated from bone marrow and infrapatellar fat pad. J Biomech 47:2115–2121
11. Mauck RL, Soltz MA, Wang CC, Wong DD, Chao PH, Valhmu WB et al (2000) Functional tissue engineering of articular cartilage through dynamic loading of chondrocyte-seeded agarose gels. J Biomech Eng 122:252–260
12. Gannon AR, Nagel T, Kelly DJ (2012) The role of the superficial region in determining the dynamic properties of articular cartilage. Osteoarthritis Cartilage 20:1417–1425

Index

Pauline M. Doran (ed.), *Cartilage Tissue Engineering: Methods and Protocols*, Methods in Molecular Biology, vol. 1340, DOI 10.1007/978-1-4939-2938-2, © Springer Science+Business Media New York 2015

MIX
Papier aus verantwortungsvollen Quellen
Paper from responsible sources
FSC® C105338

If you have any concerns about our products, you can contact us on
ProductSafety@springernature.com

In case Publisher is established outside the EU, the EU authorized representative is:
Springer Nature Customer Service Center GmbH
Europaplatz 3, 69115 Heidelberg, Germany

Printed by Libri Plureos GmbH
in Hamburg, Germany